Orthogonal Polynomials in MATLAB

Exercises and Solutions

SOFTWARE • ENVIRONMENTS • TOOLS

The SIAM series on Software, Environments, and Tools focuses on the practical implementation of computational methods and the high performance aspects of scientific computation by emphasizing in-demand software, computing environments, and tools for computing. Software technology development issues such as current status, applications and algorithms, mathematical software, software tools, languages and compilers, computing environments, and visualization are presented.

Software, Environments, and Tools

Walter Gautschi, *Orthogonal Polynomials in MATLAB: Exercises and Solutions*

Daniel J. Bates, Jonathan D. Hauenstein, Andrew J. Sommese, Charles W. Wampler, *Numerically Solving Polynomial Systems with Bertini*

Uwe Naumann, *The Art of Differentiating Computer Programs: An Introduction to Algorithmic Differentiation*

C. T. Kelley, *Implicit Filtering*

Jeremy Kepner and John Gilbert, editors, *Graph Algorithms in the Language of Linear Algebra*

Jeremy Kepner, *Parallel MATLAB for Multicore and Multinode Computers*

Michael A. Heroux, Padma Raghavan, and Horst D. Simon, editors, *Parallel Processing for Scientific Computing*

Gérard Meurant, *The Lanczos and Conjugate Gradient Algorithms: From Theory to Finite Precision Computations*

Bo Einarsson, editor, *Accuracy and Reliability in Scientific Computing*

Michael W. Berry and Murray Browne, *Understanding Search Engines: Mathematical Modeling and Text Retrieval, Second Edition*

Craig C. Douglas, Gundolf Haase, and Ulrich Langer, *A Tutorial on Elliptic PDE Solvers and Their Parallelization*

Louis Komzsik, *The Lanczos Method: Evolution and Application*

Bard Ermentrout, *Simulating, Analyzing, and Animating Dynamical Systems: A Guide to XPPAUT for Researchers and Students*

V. A. Barker, L. S. Blackford, J. Dongarra, J. Du Croz, S. Hammarling, M. Marinova, J. Wasniewski, and P. Yalamov, *LAPACK95 Users' Guide*

Stefan Goedecker and Adolfy Hoisie, *Performance Optimization of Numerically Intensive Codes*

Zhaojun Bai, James Demmel, Jack Dongarra, Axel Ruhe, and Henk van der Vorst, *Templates for the Solution of Algebraic Eigenvalue Problems: A Practical Guide*

Lloyd N. Trefethen, *Spectral Methods in MATLAB*

E. Anderson, Z. Bai, C. Bischof, S. Blackford, J. Demmel, J. Dongarra, J. Du Croz, A. Greenbaum, S. Hammarling, A. McKenney, and D. Sorensen, *LAPACK Users' Guide, Third Edition*

Michael W. Berry and Murray Browne, *Understanding Search Engines: Mathematical Modeling and Text Retrieval*

Jack J. Dongarra, Iain S. Duff, Danny C. Sorensen, and Henk A. van der Vorst, *Numerical Linear Algebra for High-Performance Computers*

R. B. Lehoucq, D. C. Sorensen, and C. Yang, *ARPACK Users' Guide: Solution of Large-Scale Eigenvalue Problems with Implicitly Restarted Arnoldi Methods*

Randolph E. Bank, *PLTMG: A Software Package for Solving Elliptic Partial Differential Equations, Users' Guide 8.0*

L. S. Blackford, J. Choi, A. Cleary, E. D'Azevedo, J. Demmel, I. Dhillon, J. Dongarra, S. Hammarling, G. Henry, A. Petitet, K. Stanley, D. Walker, and R. C. Whaley, *ScaLAPACK Users' Guide*

Greg Astfalk, editor, *Applications on Advanced Architecture Computers*

Roger W. Hockney, *The Science of Computer Benchmarking*

Françoise Chaitin-Chatelin and Valérie Frayssé, *Lectures on Finite Precision Computations*

Orthogonal Polynomials in MATLAB

Exercises and Solutions

Walter Gautschi
Purdue University
West Lafayette, Indiana

Society for Industrial and Applied Mathematics
Philadelphia

10 9 8 7 6 5 4 3 2 1

Publisher	David Marshall
Acquisitions Editor	Elizabeth Greenspan
Developmental Editor	Gina Rinelli
Managing Editor	Kelly Thomas
Production Editor	Louis R. Primus
Copy Editor	Matthew Bernard
Production Manager	Donna Witzleben
Production Coordinator	Cally Shrader
Compositor	Techsetters, Inc.
Graphic Designer	Lois Sellers

Library of Congress Cataloging-in-Publication Data
Names: Gautschi, Walter.
Title: Orthogonal polynomials in MATLAB : exercises and solutions / Walter Gautschi, Purdue University, West Lafayette, Indiana.
Description: Philadelphia : Society for Industrial and Applied Mathematics, [2016] | Series: Software, environments, and tools ; 26 | Includes bibliographical references and index.
Identifiers: LCCN 2016000510 (print) | LCCN 2016005423 (ebook) | ISBN 9781611974294 | ISBN 9781611974300
Subjects: LCSH: Orthogonal polynomials–Problems, exercises, etc.–Data processing. | MATLAB.
Classification: LCC QA404.5 .G3564 2016 (print) | LCC QA404.5 (ebook) | DDC 515/.55028553–dc23
LC record available at http://lccn.loc.gov/2016000510

Contents

Preface

There are a number of texts covering the theory of orthogonal polynomials on the real line and some of their applications (for example, see [4, 11, 37, 43, 46, 52, 60, 61]). A comprehensive account of computational methods for generating orthogonal polynomials numerically has appeared only relatively recently in [24]. All the techniques discussed in [24], including tests and the generation of figures and tables, are supported by MATLAB programs which are collected in the package `OPQ` (*Orthogonal Polynomials and Quadrature*). Some of the more important `OPQ` routines have since been transcribed into symbolic MATLAB programs, collected in the package `SOPQ` (*Symbolic Orthogonal Polynomials and Quadrature*), which can be run in variable-precision arithmetic. For the content of the two packages, see Appendices A and B, respectively, and for their internet location the beginning of Chapter 1.

The use of the `OPQ` routines is illustrated in the article [27] in four sections entitled Orthogonal polynomials; Sobolev orthogonal polynomials; Quadrature; and Approximation. Appended to each of these sections is a collection of exercises, some fairly routine, others more advanced and not without practical interest. The present text takes up these exercises, slightly edited to make them self-contained, and supplements them with many new exercises. Detailed solutions are provided to all of them, complete with relevant MATLAB software. All pieces of software used in this book can be downloaded from

`www.siam.org/books/se26.`

High-precision computation is applied when appropriate, using the package `SOPQ`.

Most of the exercises involve polynomials orthogonal relative to an absolutely continuous measure,

$$d\lambda(t) = w(t)dt,$$

involving a weight function w supported on a finite or infinite interval $[a,b]$, $-\infty \le a < b \le \infty$, or on a finite number of such intervals. We generally assume that $w \ge 0$ on $[a,b]$ and that all moments of w,

$$\mu_r = \int_{\mathbb{R}} t^r w(t)dt, \quad r = 0,1,2,\ldots, \quad \mu_0 > 0, \tag{0.1}$$

exist, or at least those that are relevant. For simplicity of notation, we integrate over the real line $\mathbb{R}$, it being understood that outside the support of w the integrand is identically zero. Most of the time we use *monic* orthogonal polynomials (having leading coefficient 1) and denote them by $\pi_k(\cdot\,; d\lambda)$, that is,

$$\begin{aligned} &\pi_k(t;d\lambda) = t^k + \text{terms of lower degree}, \\ &\int_{\mathbb{R}} \pi_k(t;d\lambda)\pi_\ell(t;d\lambda)d\lambda(t) = 0, \quad k \ne \ell. \end{aligned} \tag{0.2}$$

Occasionally, we also consider *discrete* orthogonal polynomials, which are orthogonal relative to a discrete measure,

$$d\lambda_N(x) = \sum_{k=1}^{N} w_k \delta(x - x_k)dx, \quad w_k > 0, \tag{0.3}$$

where δ is the Dirac delta function and x_k mutually distinct numbers (usually ordered increasingly). There exist exactly N orthogonal polynomials in this case,

$$\sum_{k=1}^{N} w_k \pi_m(x_k; d\lambda_N)\pi_n(x_k; d\lambda_N) = 0, \quad m \neq n, \tag{0.4}$$

where $m, n = 0, 1, 2, \ldots, N-1$.

Every system $\{\pi_k(\cdot; d\lambda)\}$ of monic orthogonal polynomials is known to satisfy a *three-term recurrence relation*

$$\begin{aligned} &\pi_{k+1}(t) = (t - \alpha_k)\pi_k(t) - \beta_k \pi_{k-1}(t), \quad k = 0, 1, 2, \ldots, \\ &\pi_0(t) = 1, \quad \pi_{-1}(t) = 0, \end{aligned} \tag{0.5}$$

where $\alpha_k = \alpha_k(d\lambda)$ are real, and $\beta_k = \beta_k(d\lambda)$ positive coefficients, both depending on $d\lambda$. These are of crucial importance for the constructive theory of orthogonal polynomials. Once known, they provide access to Gaussian quadrature and to other important approximation-theoretic processes involving, for example, continued fractions. Unless they are known explicitly, the central theme in constructive orthogonal polynomials, therefore, is methods for computing these coefficients numerically, given the measure $d\lambda$. There are two such methods generally applicable. The first is the classical *Chebyshev algorithm*, or the more recent *modified Chebyshev algorithm*, that takes the first $2n$ moments, resp., modified moments, of the measure and produces from them the first n recurrence coefficient α_k, β_k, $k = 0, 1, 2, \ldots, n-1$. Since moments are often explicitly known, the classical Chebyshev algorithm is a convenient way to generate the desired coefficients. Because of severe ill-conditioning, however, the use of high-precision arithmetic is imperative. The second method is based on an appropriate *discretization* of the measure and takes the corresponding discrete orthogonal polynomials to approximate the desired ones.

Another circle of problems, also of practical importance, has to do with modifications of measures. If one knows the recurrence coefficients of the measure $d\lambda$, the problem is to find the recurrence coefficients of a modified measure, namely the measure $d\lambda$ multiplied by a given rational function. Algorithms that solve such problems are referred to as *modification algorithms*. When divisors are involved, one faces other instances where high-precision work is essential.

The text has five chapters, the first giving a brief introduction to the software packages `OPQ` and `SOPQ`, and the remaining four being organized according to the four subject matters of [27] subdivided into topical subsections. Each of these chapters will start out with a brief summary of notation and content and is followed by the exercises, each being solved immediately after it has been formulated. Those marked by an asterisk are more advanced. The exercises taken from [27] are so identified. Equations are numbered consecutively in each chapter, Eq. (i,k) being the kth numbered equation in Chapter i, and so are the exercises, examples, and demos.

The majority of the MATLAB scripts in this book were run with MATLAB Release R2011a; a few others used Releases R2012b, R2014a, and R2015a. In general, it does not matter all that much what release is being used. There are, however, a few instances where it does: The scripts in Exs. 2.34(c) and 5.1(d) work properly in Release R2011a, but not in Release R2014a, because of a problem with the MATLAB routine `eig.m`, which in the latter cases produces eigenvectors of insufficient accuracy (see also Demo 1.17). In Ex. 5.2(b), the MATLAB function `erfc(z)` accepts symbolic arguments `z` in MATLAB Release R2014a, but not in the earlier Release R2011a.

In variable-precision work, a basic tool is loading multiprecision arrays from a source text file into the MATLAB working window. Since the MATLAB Symbolic Toolbox does not currently provide such a loading routine, one named `loadvpa.m` (cf. Appendix B.3) has been developed in collaboration with MathWorks, Inc. It is shown below.

```
function a=loadvpa(filename,nheader_lines,dig,n,m)
%LOADVPA Loading a multiprecision source file.
fid=fopen(filename,'r');
%Skip first nheader_lines of header
for i =  1:nheader_lines
fgetl(fid);
end
C = textscan(fid,'%s');
C = C{:};
fclose(fid);
for k=1:numel(C)
  a(k)=vpa(C{k},dig);
end
a=reshape(a,m,n)';
```

For example, the commands (cf. answer to Ex. 5.1(c))

```
ab32=loadvpa('ab_hrhermite.txt',3,32,200,2);
ab=vpa(ab32(1:5,:),25)
```

display the first five recurrence coefficients to 25 decimal digits

```
ab =

[ 0.5641895835477562869480795, 0.8862269254527580136490837]
[ 0.9884253928468002854870634, 0.1816901138162093284622325]
[  1.285967619363939960282789, 0.3413251289594391985641718]
[  1.524720844080115303513002, 0.5049621529880016319357512]
[  1.730192274309439256771561, 0.6702641946396190856785084]
```

of the half-range Hermite measure $d\lambda(t) = e^{-t^2}dt$ on $\mathbb{R}_+$ as type `sym` by loading the file `ab_hrhermite.txt` containing three comment lines followed by the first 200 recurrence coefficients for $d\lambda$, accurate to 32 digits. In contrast, the MATLAB commands

```
load -ascii ab_hrhermite.txt;
ab_hrhermite
```

load and display the file `ab_hrhermite.txt` as type `double` to double-precision accuracy.

Chapter 1

A Guide to the Software Packages OPQ and SOPQ

In this introductory chapter we briefly describe and illustrate the use of the principal MATLAB functions contained in the software packages *Orthogonal Polynomials and Quadrature* (OPQ), located at

http://dx.doi.org/10.4231/R7959FHP,

and its symbolic counterpart (SOPQ), located at

http://dx.doi.org/10.4231/R7ZG6Q6T.

(For additional details see also [27], and consult Appendices A and B for the complete content of these packages.) As a general rule, a double-precision function and its symbolic analogue have the same name, except that in the latter the letter "s" (for symbolic) is inserted at the beginning of the name, and "dig," the number of digits to be employed, is inserted as the first input argument. Moreover, at the beginning of each symbolic function, all numerical input variables are converted to type "sym," even if they are already of this type, but especially if they are not.

1.1 ▪ Recurrence relation

Monic orthogonal polynomials $\pi_k(\cdot;\mathrm{d}\lambda)$ relative to a positive measure $\mathrm{d}\lambda$ supported on the real line $\mathbb{R}$ are known to satisfy a *three-term recurrence relation*

$$\begin{gathered}\pi_{k+1}(t)=(t-\alpha_k)\pi_k(t)-\beta_k\pi_{k-1}(t), \quad k=0,1,2,\ldots,\\ \pi_0(t)=1, \;\; \pi_{-1}(t)=0,\end{gathered} \tag{1.1}$$

where $\alpha_k=\alpha_k(\mathrm{d}\lambda)$, $\beta_k=\beta_k(\mathrm{d}\lambda)$ are real, resp., positive, coefficients depending on $\mathrm{d}\lambda$. Indeed (cf. Ex. 2.3(a)),

$$\begin{aligned}\alpha_k(\mathrm{d}\lambda)&=\frac{(t\pi_k,\pi_k)_{\mathrm{d}\lambda}}{(\pi_k,\pi_k)_{\mathrm{d}\lambda}}, \quad k=0,1,2,\ldots,\\ \beta_k(\mathrm{d}\lambda)&=\frac{(\pi_k,\pi_k)_{\mathrm{d}\lambda}}{(\pi_{k-1},\pi_{k-1})_{\mathrm{d}\lambda}}, \quad k=1,2,3,\ldots,\end{aligned} \tag{1.2}$$

where the inner products in the numerators and denominators are given by $(u,v)_{\mathrm{d}\lambda}=\int_{\mathbb{R}}u(t)v(t)\mathrm{d}\lambda(t)$. The coefficient β_0 is conveniently defined to be $\beta_0=\int_{\mathbb{R}}\mathrm{d}\lambda(t)$.

The *Jacobi matrix* J is the (in general) infinite symmetric tridiagonal matrix

$$J(d\lambda)=\begin{bmatrix} \alpha_0 & \sqrt{\beta_1} & & & 0 \\ \sqrt{\beta_1} & \alpha_1 & \sqrt{\beta_2} & & \\ & \sqrt{\beta_2} & \alpha_2 & \sqrt{\beta_3} & \\ & & \ddots & \ddots & \ddots \\ 0 & & & & \end{bmatrix}, \tag{1.3}$$

formed with the recurrence coefficients (1.2), and J_n the principal minor matrix of order n,

$$J_n(d\lambda)=J(d\lambda)_{[1:n,1:n]}. \tag{1.4}$$

For some special classes of orthogonal polynomials, for example the classical orthogonal polynomials, the recurrence coefficients are known explicitly. For these, the package OPQ provides routines for generating the first N of them. The syntax of these routines is

$$\texttt{ab} = \texttt{r\_}\textit{name}(\texttt{N}), \tag{1.5}$$

where *name* identifies the name of the orthogonal polynomials and ab is the N×2 array

$$\texttt{ab}=\begin{bmatrix} \alpha_0 & \beta_0 \\ \alpha_1 & \beta_1 \\ \vdots & \vdots \\ \alpha_{N-1} & \beta_{N-1} \end{bmatrix}$$

of the first N recurrence coefficients. There may be additional input parameters following N in (1.5). The corresponding symbolic routine is ab =sr_*name*(dig,N), where dig is the number of decimal digits used.

Table 1.1 provides some details for these special routines.

Table 1.1. *Routines for classical orthogonal polynomials*

Name	Input parameters	Weight function $w(t)$	Support
jacobi	N,a,b	$(1-t)^{\alpha}(1+t)^{\beta}$, $\alpha,\beta>-1$	$t\in[-1,1]$
jacobi01	N,a,b	$(1-t)^{\alpha}t^{\beta}$, $\alpha,\beta>-1$	$t\in[0,1]$
hermite	N,mu	$\lvert t\rvert^{2\mu}e^{-t^2}$, $\mu>-1/2$	$t\in[-\infty,\infty]$
laguerre	N,a	$t^{\alpha}e^{-t}$, $\alpha>-1$	$t\in[0,\infty]$
hahn	N,a,b	$\binom{\alpha+k}{k}\binom{\beta+N-k}{N-k}$, $\alpha,\beta\geq 0$	$k\in\{0,1,2,\ldots,N\}$

In the two jacobi routines, if b=0, the input parameter b may be omitted, and if a=b=0, both input parameters a, b may be omitted. Likewise for mu and a in the routines hermite, resp., laguerre, and for a, b in the routine hahn. The latter, if a=b=0, yields the classical Chebyshev discrete orthogonal polynomials on $[0,N]$.

Demo 1.1. The first ten recurrence coefficients for the Jacobi polynomial with parameters $\alpha = -1/2$, $\beta = 3/2$.

In MATLAB double precision, they are produced by the routine `r_jacobi.m` (cf. Appendix A.1)

```
>> ab=r_jacobi(10,-1/2,3/2)
ab =
          6.666666666666666e-01      4.712388980384690e+00
          1.333333333333333e-01      1.388888888888889e-01
          5.714285714285714e-02      2.100000000000000e-01
          3.174603174603174e-02      2.295918367346939e-01
          2.020202020202020e-02      2.376543209876543e-01
          1.398601398601399e-02      2.417355371900826e-01
          1.025641025641026e-02      2.440828402366864e-01
          7.843137254901961e-03      2.455555555555556e-01
          6.191950464396285e-03      2.465397923875433e-01
          5.012531328320802e-03      2.472299168975069e-01
>>
```

and in 32-digit precision by the routine `sr_jacobi.m` (cf. Appendix B.1)

```
>> ab=sr_jacobi(32,10,-1/2,3/2)
ab =
[   0.66666666666666666666666666666667,  4.7123889803846898576939650749193]
[   0.13333333333333333333333333333333, 0.13888888888888888888888888888889]
[  0.057142857142857142857142857142857,                               0.21]
[  0.031746031746031746031746031746032, 0.22959183673469387755102040816327]
[   0.02020202020202020202020202020202, 0.23765432098765432098765432098765]
[  0.013986013986013986013986013986014, 0.24173553719008264462809917355372]
[   0.01025641025641025641025641025641, 0.24408284023668639053254437869822]
[ 0.0078431372549019607843137254901961, 0.24555555555555555555555555555556]
[ 0.0061919504643962848297213622291022, 0.24653979238754325259515570934256]
[ 0.0050125313283208020050125313283208, 0.24722991689750692520775623268698]
>>
```

1.2 ▪ Chebyshev and modified Chebyshev algorithm

These are algorithms generating the first N recurrence coefficients from the first $2N$ *moments* of $d\lambda$,

$$\mu_k = \int_{\mathbb{R}} t^k d\lambda(t), \quad k = 0, 1, \ldots, 2N-1, \tag{1.6}$$

or from the first $2N$ *modified moments*,

$$m_k = \int_{\mathbb{R}} p_k(t) d\lambda, \quad k = 0, 1, \ldots, 2N-1, \tag{1.7}$$

where $\{p_k\}$ is a system of $2N$ monic polynomials satisfying a three-term recurrence relation

$$p_{k+1}(t) = (t - a_k)p_k(t) - b_k p_{k-1}(t), \quad k = 0, 1, 2, \ldots,$$
$$p_0(t) = 1, \quad p_{-1}(t) = 0$$

with known coefficients $a_k \in \mathbb{R}$, $b_k \geq 0$. In the case $a_k = b_k = 0$, all $k \geq 0$, the modified moments (1.7) become the ordinary power moments (1.6).

In the package OPQ, both algorithms have the same syntax (cf. Appendix A.1),

$$\texttt{ab} = \texttt{chebyshev(N,mom,abm)},$$

where $\texttt{mom} = [m_0 \; m_1 \; \ldots \; m_{2N-1}]$ is the $1 \times 2N$ array of modified moments and abm the $(2N-1) \times 2$ array of known recurrence coefficients

$$\texttt{abm} = \begin{bmatrix} a_0 & b_0 \\ a_1 & b_1 \\ \vdots & \vdots \\ a_{2N-2} & b_{2N-2} \end{bmatrix}.$$

When abm=0, this implements the classical Chebyshev algorithm. The third input argument, abm, in this case, may be omitted. The corresponding symbolic routine in SOPQ is (cf. Appendix B.1)

$$\texttt{ab} = \texttt{schebyshev(dig,N,mom,abm)}.$$

Demo 1.2. *Elliptic Chebyshev polynomials.*

These are orthogonal with respect to the measure

$$\mathrm{d}\lambda(t;\omega) = [(1-\omega^2 t^2)(1-t^2)]^{-1/2}\mathrm{d}t \quad \text{on } [-1,1],\ 0 < \omega < 1.$$

In the limit $\omega \downarrow 0$, we get the Chebyshev polynomials of the first kind, otherwise an "elliptic" modification thereof. It seems natural, therefore, to use modified moments relative to the monic Chebyshev polynomials,

$$m_0 = \int_{-1}^{1} \mathrm{d}\lambda(t;\omega), \quad m_k = \frac{1}{2^{k-1}} \int_{-1}^{1} T_k(t)\mathrm{d}\lambda(t;\omega), \quad k = 1,2,\ldots.$$

Their stable computation is possible (cf. [24, Example 2.29]), though not entirely trivial. The first $2N$ of them are generated by the OPQ routine mm_ellcheb.m (cf. Appendix A.5).

The modified Chebyshev algorithm for the first N recurrence coefficients $\alpha_k(\mathrm{d}\lambda)$, $\beta_k(\mathrm{d}\lambda)$ can thus be implemented by the routine (cf. Appendix A.1)

```
function ab=r_ellcheb(N,om2)
%R_ELLCHEB Recurrence coefficients for the elliptic Chebyshev
%weight function.

if((N<=0)|(om2<0)|(om2>=1))
  error('parameter(s) out of range');
end
abm=r_jacobi(2*N-1,-1/2);
mom=mm_ellcheb(N,om2);
ab=chebyshev(N,mom,abm);
```

Run with $\texttt{om2} = \omega^2 = .999$ and $N = 40$, it yields

```
>> ab=r_ellcheb(40,.999)
ab =
        0      9.682265121100620e+00
        0      7.937821421385184e-01
        0      1.198676724605757e-01
        0      2.270401183698990e-01
        0      2.410608787266061e-01
        0      2.454285325203698e-01
        0      2.473016530297635e-01
        0      2.482587060199245e-01
      ...      ...

        0      2.499915376529289e-01
        0      2.499924312667191e-01
        0      2.499932210069769e-01
>>
```

By symmetry, all $\alpha_k = 0$, and, evidently, $\beta_k \to 1/4$ as $k \to \infty$.

1.3 ▪ Discretization procedures

1.3.1 ▪ Stieltjes procedure

In [59, p. 383], Stieltjes wrote (translated from the French, and using our notation and equation numbers), "The relations (1.1) and (1.2) allow us to compute, one by one, all the polynomials $\pi_1(t), \pi_2(t), \ldots$. To begin with, one has

$$\alpha_0 = \int_{\mathbb{R}} t\,\mathrm{d}\lambda(t) : \int_{\mathbb{R}} \mathrm{d}\lambda(t),$$

which lets us obtain $\pi_1(t)$. The formulae (1.2) then give α_1, β_1, which let us obtain $\pi_2(t)$, ..." In [15, §2], therefore, we called this procedure of bootstrapping oneself up from degree zero to any given degree, using alternately (1.2) and (1.1), the *Stieltjes procedure*, a nomenclature that has become widely adopted. The main problem, of course, is the computation of the integrals in the inner products of (1.2), for which Gaussian quadrature relative to the measure $\mathrm{d}\lambda$ would be ideal, but not possible, since this presupposes knowledge of the respective orthogonal polynomials, something we want to obtain in the first place! Therefore, we need to look for other ways of implementing Stieltjes's procedure, which leads to what we have called *discretized Stieltjes procedures.*

1.3.2 ▪ Discretized Stieltjes procedure (also referred to in the literature as the Gautschi–Stieltjes procedure [58])

Discrete measures

The simplest case is a discrete measure (0.3), $\mathrm{d}\lambda = \mathrm{d}\lambda_N$, in which case all inner products in (1.2) are finite sums, and therefore can be evaluated without difficulty, once the respective orthogonal polynomials π_k (or their recurrence coefficients) have been obtained. The Stieltjes procedure, in this case, is implemented in the `OPQ` routine (cf. Appendix A.1)

$$\texttt{ab} = \texttt{stieltjes(N,xw)},$$

where `xw` is an N×2 array, the first column of which contains the abscissae x_k, and the second column the jumps w_k of $d\lambda_N$. An alternative routine, accomplishing the same thing, is (cf. Appendix A.1)

$$\mathtt{ab} = \mathtt{lanczos(N,xw)},$$

which is based on an orthogonal similarity transformation by Lanczos's method of the symmetric matrix of order $N+1$,

$$\begin{bmatrix} 1 & \sqrt{\boldsymbol{w}}^{\mathrm{T}} \\ \sqrt{\boldsymbol{w}} & \boldsymbol{D_x} \end{bmatrix}, \quad \boldsymbol{w}^{\mathrm{T}} = [w_1, w_2, \ldots, w_N],\ \boldsymbol{D_x} = \mathrm{diag}(x_1, x_2, \ldots, x_N),$$

to the extended Jacobi matrix

$$\begin{bmatrix} 1 & \sqrt{\beta_0}\boldsymbol{e}_1^{\mathrm{T}} \\ \sqrt{\beta_0}\boldsymbol{e}_1 & J_N(d\lambda_N) \end{bmatrix}, \quad \boldsymbol{e}_1^{\mathrm{T}} = [1, 0, \ldots, 0] \in \mathbb{R}^N.$$

Example 1.1. *Discrete Chebyshev polynomials on* $[0,1)$.

These are orthogonal with respect to the discrete N-point measure

$$d\lambda_N(t) = \frac{1}{N}\sum_{k=0}^{N-1}\delta(t-k/N)dt.$$

Their recurrence coefficients are known explicitly, being essentially the recurrence coefficients of the N-point Hahn polynomial with parameters $\alpha=\beta=0$ (cf. Table 1.1), except for a linear transformation of the interval $[0, N-1]$ to the interval $[0, 1-1/N]$, which requires division of the Hahn α-coefficients by N and of the β-coefficients by N^2. ∎

A typical difficulty with Stieltjes's procedure, when the support points are equally spaced, as in Example 1.1, or nearly so, is the deterioration of accuracy of the recurrence coefficients α_n, β_n as n approaches N. This is illustrated by the next demo.

Demo 1.3. *Discrete Chebyshev polynomials on* $[0,1)$ *generated by* `stieltjes.m`*, resp.,* `lanczos.m`.

The script

```
%DEMO1_3  Discrete Chebyshev polynomials on [0,1)
%
f2='%10.0f    <=%3.0f        <=1e-14         <=1e-14\n';
f3='%17.0f %13.3e %13.3e\n';
f4='%10.0f         %13.3e %13.3e\n';
f5='                         %6.3f\n';
disp('        N      n        erra            errb')
for N=80
  xw=zeros(N,2);
  xw(:,1)=linspace(0,1-1/N,N)'; xw(:,2)=ones(N,1)/N;
  abex=r_hahn(N-1);
  abex(:,1)=abex(:,1)/N; abex(:,2)=abex(:,2)/(N^2);
  abex(1,2)=1;
  ab=stieltjes(N,xw);
```

```
    %ab=lanczos(N,xw);
    errab=zeros(N,2);
    errab(:,1)=abs(ab(:,1)-abex(:,1));
    errab(:,2)=abs(ab(:,2)-abex(:,2));
    erram=max(errab(:,1)); errbm=max(errab(:,2));
    f=find(errab(:,1)>1e-14 | errab(:,2)>1e-14);
    if ~isempty(f)
      errab=errab(f,:);
      fprintf(f2,N,f(1)-2)
      fprintf(f3,[f(1:4:13)-1 errab(1:4:13,:)]')
    end
    fprintf(f4,N,[erram errbm]')
end
```

determines those values of n, if any, for which the absolute error either of α_n or of β_n is larger than 10^{-14} and prints a selected number of them. It also prints the largest absolute errors of α_n and β_n.

The results (for $N = 80$) are shown below.

```
>> Demo1_3              Stieltjes procedure
         N       n          erra              errb
        80     <= 59      <=1e-14           <=1e-14
                  60      2.720e-15         1.283e-14
                  64      6.457e-12         3.996e-11
                  68      7.295e-08         3.725e-07
                  72      1.475e-03         1.333e-02
        80                3.114e-01         1.092e-01
>>
```

It is seen that the accuracy of the α_n, β_n rapidly deteriorates as $n \to N$. In contrast, the routine `lanczos.m` has no problem with accuracy, the maximum error being of the order 10^{-15}:

```
>> Demo1_3                Lanczos procedure
         N       n          erra              errb
        80                8.882e-16         1.554e-15
>>
```

It is also faster than `stieltjes.m`, taking .474 sec to run the script a thousand times as opposed to 2.183 sec for `stieltjes.m`.

For N much larger than 80, Stieltjes's procedure may also be subject to underflow problems, which must be dealt with by appropriate scalings (cf. [24, Example 2.35]).

However, for nodes x_k distributed more like zeros of orthogonal polynomials, Stieltjes's procedure is quite competitive, as is shown in the next demo.

Demo 1.4. *Discrete polynomials orthogonal with respect to Fejér nodes and weights, generated by* `stieltjes.m`*, resp.,* `lanczos.m`*.*

Here, the support points of the discrete measure $d\lambda_N$ are the N Fejér quadrature nodes, and the jumps the Fejér quadrature weights (cf. [24, §2.2.2]). These are generated by the routine `fejer.m` (cf. Appendix A.1). The script below computes, either by Stieltjes's or by Lanczos's procedure, the $2N$ recurrence coefficients of the corresponding discrete orthogonal polynomials, for $N = 20$, 40, 80, 160, and 320. It also measures the time consumed by each procedure.

```
%DEMO1_4 Discrete polynomials orthogonal with respect to the
%Fejer nodes and weights.

f0='%10.0f %12.4e %12.4e\n';
tic
dig=32; digits(dig);
for N=[20 40 80 160 320]
  xw=fejer(N); sxw=sfejer(dig,N);
  ab=stieltjes(N,xw); sab=sstieltjes(dig,N,sxw);
%  ab=lanczos(N,xw); sab=slanczos(dig,N,sxw);
  errab=double(abs(ab-sab)); merrab=max(errab);
  fprintf(f0,N,merrab(1),merrab(2))
end
toc
```

From the results produced, shown in Table 1.2, it can be seen that Stieltjes's procedure is generally at least as accurate as Lanczos's, but considerably (about 25 times) faster.

Table 1.2. *Stieltjes vs Lanczos procedure for the discrete Fejér measure*

	Stieltjes		Lanczos	
N	erra	errb	erra	errb
20	1.3220e−15	1.4051e−16	1.1102e−15	4.5293e−16
40	2.8673e−15	1.0774e−15	2.1094e−15	1.0219e−15
80	3.5790e−15	1.3173e−15	4.5103e−15	2.0040e−15
160	5.3124e−15	2.3426e−15	1.1546e−14	4.8568e−15
320	1.2500e−14	4.7345e−15	1.4877e−14	6.1778e−15
	Timing		Timing	
	127.621549 sec		3151.102663 sec	

Multicomponent discretization

For a more general measure $\mathrm{d}\lambda$ consisting of an absolutely continuous measure plus a discrete measure,

$$\mathrm{d}\lambda(t) = w(t)\mathrm{d}t + \sum_{\kappa=1}^{p} \omega_\kappa \delta(t-\xi_\kappa)\mathrm{d}t, \tag{1.8}$$

the basic idea behind discretization methods is to approximate $\mathrm{d}\lambda$ by a discrete measure

$$\mathrm{d}\lambda(t) \approx \mathrm{d}\lambda_{K+p}(t), \tag{1.9}$$

and then, as in the preceding subsection, applying either the routine `stieltjes.m` or `lanczos.m` to $\mathrm{d}\lambda_{K+p}$. This basically requires the approximation of the absolutely continuous part $w(t)\mathrm{d}t$ in (1.8) by a discrete K-point measure, to which the discrete component in (1.8) can then simply be added on to obtain (1.9). It suffices therefore to find approximations of the type

$$\int_{\mathbb{R}} p(t)w(t)\mathrm{d}t \approx \sum_{k=1}^{K} w_k p(x_k),$$

that is,

$$w(t)\mathrm{d}t \approx \sum_{k=1}^{K} w_k \delta(t - x_k)\mathrm{d}t. \tag{1.10}$$

A general *m-component discretization* can be obtained by finding an identity

$$\int_{\mathbb{R}} p(t)w(t)\mathrm{d}t = \sum_{i=1}^{m} \int_{a_i}^{b_i} h_i(t)p(c_i t + d_i)w_i(t)\mathrm{d}t, \tag{1.11}$$

where w_i are positive weight functions on $[a_i, b_i]$, and then applying M-point Gaussian quadrature to each component integral on the right of (1.11),

$$\int_{\mathbb{R}} p(t)w(t)\mathrm{d}t \approx \sum_{i=1}^{m} \sum_{\mu=1}^{M} w_\mu^{(i)} h_i(x_\mu^{(i)}) p(c_i x_\mu^{(i)} + d_i), \tag{1.12}$$

where $\{x_\mu^{(i)}, w_\mu^{(i)}\}$ are the nodes and weights of the Gaussian quadrature rule for the weight function w_i on $[a_i, b_i]$. The approximation (1.12) evidently is an approximation of the desired type (1.10) with $K = mM$.

If h_i are polynomials of degree $\leq d$, and if we select $M = N + \lfloor (d+1)/2 \rfloor$, then the procedure will produce exact results for the first N recurrence coefficients, since in (1.11)—where, according to (1.2), the polynomial $p(t)$ is either $t\pi_k^2(t)$ or $\pi_k^2(t)$ (for α_k) and $\pi_k^2(t)$ or $\pi_{k-1}^2(t)$ (for β_k), $k \leq N-1$, hence $p \in \mathbb{P}_{2N-1}$ in all cases—the integrals are evaluated exactly by the $(N + \lfloor (d+1)/2 \rfloor)$-point Gauss quadrature formulae.

The approximation (1.12) may look complicated and a bit mysterious but, if constructed skillfully, provides a powerful tool for obtaining interesting and difficult types of orthogonal polynomials. All else failing, one can always resort to using the generally applicable procedure described below on p. 17 in the next subsection.

The syntax of the multicomponent discretization routine in `OPQ` is (cf. Appendix A.1)

```
[ab,Mcap,kount]=mcdis(N,eps0,quad,Mmax).
```

The routine generates the approximations (1.12) using an increasing sequence of M values (which the user can change to suit his or her own purposes) and terminates as soon as two successive approximations to the coefficients $\beta_k(\mathrm{d}\lambda)$, $k = 0, 1, 2, \ldots, N$, agree to within a relative error `eps0`. If this does not happen for any `M`$\leq$`Mmax`, the routine exits with an error message "`Mcap exceeds Mmax`".

The all-important input argument `quad` is a routine that constructs the approximation (1.12). It must have the syntax

```
xw=quad(M,i)                                                    (1.13)
```

and obtains the Gaussian quadrature approximation for the ith integral on the right of (1.11). The routine has to be provided, and named, by the user and prefixed by "@" when calling the routine `mcdis.m`.

There are several global parameters at the beginning of the routine `mcdis.m`, which the user must set before calling the routine. These are

`mc`	the number m of component intervals
`mp`	the number p of points in the discrete part of $d\lambda$ (`mp`$=0$ if there is none)
`iq`	to be set equal to 1 if the user provides his or her own quadrature routine, and different from 1 otherwise
`idelta`	has the default value 1, but is preferably set equal to 2 if `iq`$=1$ in order to speed up convergence
`irout`	to be set equal to 1 if the routine `stieltjes.m` is to be used, otherwise different from 1
`DM`	is a $p\times 2$ array $[\,[\xi_1\ \omega_1];[\xi_2\ \omega_2];\cdots;[\xi_p\ \omega_p]\,]$ specifying the abscissae and jumps of the discrete part of $d\lambda$, if any
`AB`	(used only if `iq`$\neq 1$) is an $m\times 2$ array $[\,[a_1\ b_1];[a_2\ b_2];\cdots;[a_m\ b_m]\,]$ specifying (from left to right) the component intervals, where $a_1=-$`Inf` if the first interval extends to $-\infty$, and $b_m=$ `Inf` if the last interval extends to ∞

There is an additional global variable uv for holding Fejér nodes and weights, which is set by the routine if `iq`$\neq 1$ and used in `quad.m`, where it also must appear as a global variable.

The output variable `ab` is the `N`×2 array of recurrence coefficients $\alpha_k(d\lambda)$, $\beta_k(d\lambda)$, $k\leq N-1$, the variable `Mcap` the value of M in (1.12) achieving the desired accuracy, and `kount` the number of times M had to be incremented.

The workings of the routine `mcdis.m` are best explained by examples.

Example 1.2. *The measure*

$$d\lambda(t;\sigma)=w(t;\sigma)dt,\quad w(t;\sigma)=\sigma(1-t^2)^{-1/2}+1-\sigma\ \ on\,[-1,1],\ 0\leq\sigma\leq 1.$$

This may be thought of as interpolating linearly between the Chebyshev and the Legendre measure.

Here,

$$\int_{-1}^{1}p(t)w(t;d\sigma)dt=\sigma\int_{-1}^{1}p(t)(1-t^2)^{-1/2}dt+(1-\sigma)\int_{-1}^{1}p(t)dt, \tag{1.14}$$

which suggests a two-component discretization, both components being supported on $[-1,1]$, that is, $m=2$ and $[a_1,b_1]=[a_2,b_2]=[-1,1]$ in (1.11); moreover, $h_1=\sigma$, $h_2=1-\sigma$, and N-point Gauss–Chebyshev quadrature is appropriate in the first integral on the right of (1.14) and Gauss–Legendre quadrature in the second integral. As noted above, the polynomials h_1, h_2 being of degree 0, this will produce exact answers for the first N recurrence coefficients.

The quadrature routine (1.13), in this example, named `quad_chebleg.m`, will have the following form where the global variable `sig` holds the value of σ:

```
function xw=quad_chebleg(N,i)
%QUAD_CHEBLEG A quadrature routine used by r_chebleg.m in
%mcdis.m.

global sig
ab1=r_jacobi(N,-1/2); ab2=r_jacobi(N);
xw1=gauss(N,ab1); xw2=gauss(N,ab2);
```

```
if i==1
  xw(:,1)=xw1(:,1); xw(:,2)=sig*xw1(:,2);
else
  xw(:,1)=xw2(:,1); xw(:,2)=(1-sig)*xw2(:,2);
end
```

Used in

```
function ab=r_chebleg(N)
%R_CHEBLEG  First N recurrence coefficients for the
%linearly interpolated Chebyshev and Legendre measures.

global sig mc mp iq idelta irout
mc=2; mp=0; iq=1; idelta=2; irout=1;
Mmax=N+1; eps0=.5e-12;
[ab,Mcap,kount]=mcdis(N,eps0,@quad_chebleg,Mmax);
```

it generates the desired recurrence coefficients to a relative accuracy eps0. Note that we set Mmax equal to $N+1$, anticipating exact answers. (This is confirmed by the output variable kount to consistently have the value 1.) ∎

Demo 1.5. *Gauss quadrature nodes and weights of* $\mathrm{d}\lambda(\cdot\,;\sigma)$.

Since the measure $\mathrm{d}\lambda$ is symmetric, it suffices to look at the $\lfloor N/2 \rfloor$ positive nodes and corresponding weights. The driver routine

```
%RUN_CHEBLEG

global sig
N=40;
s=0;
for sig=0:.01:1
  s=s+1;
  ab=r_chebleg(N);
  xw=gauss(N,ab);
  for k=floor((N+3)/2):N
    x(s,k)=xw(k,1);
%     w(s,k)=xw(k,2);
  end
end
s=(0:.01:1)';
hold on
for k=floor((N+3)/2):N
  plot(s,x(:,k));set(gca,'FontSize',14)
  xlabel('sigma'); ylabel('nodes')
  axis([0 1.1 0 1.1])
%  plot(s,w(:,k));set(gca,'FontSize',14)
%  xlabel('sigma'); ylabel('weights')
%  axis([0 1.1 0 .085])
end
```

computes them for $N=40$ on the interval $\sigma \in [0,1]$ and produces the plots in Fig. 1.1. The nodes seemingly move along straight lines from the Legendre nodes to the Chebyshev

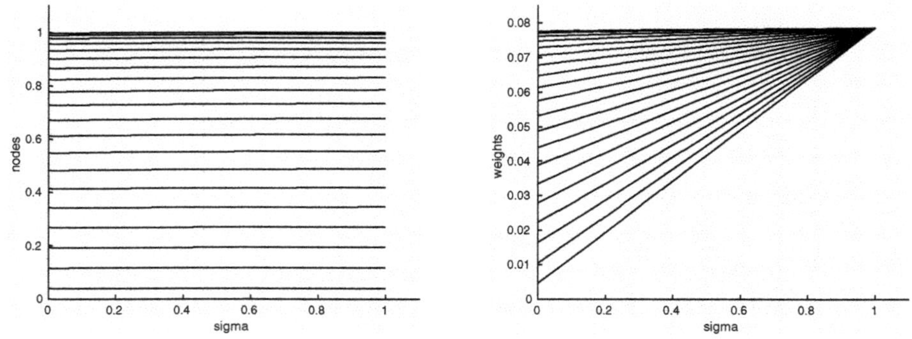

Figure 1.1. *Gauss quadrature nodes and weights for the measure* $\mathrm{d}\lambda(\cdot\,;\sigma)$ *of Example* 1.2

nodes, but in reality lie always slightly above these lines. Similarly for the weights, which seemingly move along straight lines from the Legendre weights to the constant Chebyshev weight $\pi/40 = .078539816\ldots$, but the first nine actually lie slightly above the respective straight lines, the next four initially below and then above, the next six always below, and the last weight first below, and then above the respective lines.

Example 1.3. *Normalized Jacobi measure plus a discrete measure.*
Here,

$$\begin{aligned} \mathrm{d}\lambda(t) &= w(t)\mathrm{d}t + \sum_{\kappa=1}^{p} \omega_\kappa \delta(t-\xi_\kappa)\mathrm{d}t, \\ w(t) &= (\beta_0^J)^{-1}(1-t)^\alpha(1+t)^\beta. \end{aligned} \tag{1.15}$$

Explicit formulae for the recurrence coefficients $\alpha_k(\mathrm{d}\lambda)$, $\beta_k(\mathrm{d}\lambda)$ are known only in very special cases, for example, in the case $p=1$, $\xi_1=-1$, $\omega_1=y$, and even then are quite complicated (see, e.g., [19, Eqs. (4.19)–(4.21)]). Adding more discrete support points makes an analytic treatment virtually intractable. Numerically, however, the matter is straightforward, regardless of how many support points there are. One simply applies a one-component discretization, using Gauss–Jacobi quadrature for the absolute continuous part of the measure, and adds on the discrete part of the measure, as described above. ∎

Demo 1.6. *Normalized Jacobi measure plus a 1-point atomic measure.*
We let $p=1$, $\omega_1 = y = [0,.1,.2,.5,1,2,5]$ in Example 1.3 and consider the case $\xi_1 = -1$. The routines required are (cf. Appendix A.1)

```
function xw=quad_jacplusdis(N,i)
%QUAD_JACPLUSDIS Quadrature routine used in r_jacplusdis.m.

global abj
xw=gauss(N,abj);
```

and

```
function ab=r_jacplusdis(N,alpha,beta,ty)
%R_JACPLUSDIS  Recurrence coefficients for the Jacobi measure
%plus a discrete measure.
```

```
global mc mp iq idelta irout DM abj
mc=1; mp=size(ty,1); iq=1; idelta=2; irout=1; DM=ty;
eps0=.5e-12; Mmax=N+1;
[ab,Mcap,kount]=mcdis(N,eps0,@quad_jacplusdis,Mmax);
```

We anticipate exact answers with N-point Gauss quadrature, and therefore take `Mmax` $= N+1$ for verification.

The driver routine

```
%RUN_JACPLUSDIS

global alpha beta abj
f0='%8.1f %12.4e %12.4e\n';
N=40;
alpha=-1/2; beta=3/2;
abj=r_jacobi(N+1,alpha,beta); abj(1,2)=1;
fprintf('\n')
disp('      y       erra          errb')
for y=[0 .1 .5 1 2 5 10]
  ty=[-1,y];
  ab=r_jacplusdis(N,alpha,beta,ty);
  aex=zeros(N,1); bex=zeros(N,1);
  c=zeros(N,1); d=zeros(N,1);
  c(1)=1+y; d(1)=0; d(2)=1;
  for k=2:N
    d(k+1)=(beta+k)*(alpha+beta+k)*d(k)/((alpha+k-1)*(k-1));
    c(k)=(1+(beta+k)*(alpha+beta+k)*y*d(k)/((k-1)*(alpha ...
    +k-1)))/(1+y*d(k));
  end
  aex(1)=(abj(1,1)-y)/(1+y); bex(1)=1+y;
  for k=2:N
    aex(k)=abj(k,1)+2*(k-1)*(alpha+k-1)*(c(k)-1)/((alpha+beta ...
    +2*k-2)*(alpha+beta+2*k-1))+2*(beta+k)*(alpha+beta+k) ...
    *(1/c(k)-1)/((alpha+beta+2*k-1)*(alpha+beta+2*k));
    bex(k)=(c(k)/c(k-1))*abj(k,2);
  end
  errab=[abs((ab(:,1)-aex)./aex) abs((ab(:,2)-bex)./bex)];
  merrab=max(errab);
  fprintf(f0,y,merrab(1),merrab(2))
end
```

generates the first 40 recurrence coefficients for $\alpha=-1/2$ and $\beta=3/2$ along with the maximum relative error in the α- and β-coefficients determined by Eqs. (4.19)–(4.21) of [19]. The results, using the routine `stieltjes.m` in `mcdis.m`, are as follows:

```
>> run_jacplusdis
      y       erra          errb
     0.0   1.3058e-11   5.6724e-15
     0.1   1.9969e-12   5.1630e-15
     0.5   1.9969e-12   5.4997e-15
     1.0   2.0236e-12   5.0508e-15
```

```
   2.0   2.0170e-12   5.1630e-15
   5.0   2.0303e-12   5.1630e-15
  10.0   1.9702e-12   5.2752e-15
```

The maximum relative errors are, as expected, somewhat larger for the α-coefficients since they become smaller, by about 2–3 decimal orders, than the β-coefficients when their indices approach N. The results are similar when using `lanczos.m` in place of `stieltjes.m`.

Demo 1.7. *Same as Demo* 1.6, *but with* $\xi_1 = [-1.1, -1.5, -2, -5 - 10]$ *and* $y = [.1, .5, 2, 5, 10]$.

Explicit formulae for the α and β-coefficients are no longer known, but N-point Gauss quadrature, in theory, should still yield exact answers. In practice, however, when using `stieltjes.m`, this is only the case if $\xi_1 = -1.1$; otherwise, numerical instabilities prevent any answers from being produced, even if the error tolerance `eps0` in the routine `r_jacplusdis.m` is lowered to .5e–8. Not so, if one uses `lanczos.m` (`irout=0` in `r_jacplusdis.m`). Then meaningful answers, even with `eps0=.5e-12`, are successfully produced in all cases. Some selected results, obtained by the script

```
%RUN0_JACPLUSDIS

global alpha beta abj
f0='%8.1f %4.1f %19.12e %19.12e\n';
N=40;
alpha=-1/2; beta=3/2;
abj=r_jacobi(N+1,alpha,beta); abj(1,2)=1;
for xi1=[-1.1 -1.5 -2 -5 -10]
  for y=[.1 .5 1 2 5 10]
    ty=[xi1,y];
    ab=r_jacplusdis(N,alpha,beta,ty);
    for k=[20 40]
      fprintf(f0,xi1,y,ab(k,1),ab(k,2))
    end
  end
  fprintf('\n')
end
```

are shown below:

```
>> run0_jacplusdis

    xi1    y       ab(20:20:40,1)        ab(20:20:40,2)
   -1.1   0.1   2.105366396290e-03   2.488768339206e-01
                3.986783193559e-04   2.497949519668e-01
   -1.1   0.5   2.105367484214e-03   2.488768330665e-01
                3.986783193560e-04   2.497949519668e-01
   -1.1   1.0   2.105367620205e-03   2.488768329597e-01
                3.986783193560e-04   2.497949519668e-01
   -1.1   2.0   2.105367688200e-03   2.488768329063e-01
                3.986783193560e-04   2.497949519668e-01
   -1.1   5.0   2.105367728997e-03   2.488768328743e-01
```

		3.986783193560e-04	2.497949519668e-01
-1.1	10.0	2.105367742596e-03	2.488768328636e-01
		3.986783193560e-04	2.497949519668e-01
-1.5	0.1	1.585754691649e-03	2.491611958430e-01
		3.506644413496e-04	2.498199626339e-01
-1.5	0.5	1.585754691649e-03	2.491611958430e-01
		3.506644413496e-04	2.498199626339e-01
-1.5	1.0	1.585754691650e-03	2.491611958430e-01
		3.506644413496e-04	2.498199626339e-01
-1.5	2.0	1.585754691650e-03	2.491611958430e-01
		3.506644413496e-04	2.498199626339e-01
-1.5	5.0	1.585754691650e-03	2.491611958430e-01
		3.506644413496e-04	2.498199626339e-01
-1.5	10.0	1.585754691650e-03	2.491611958430e-01
		3.506644413496e-04	2.498199626339e-01
-2.0	0.1	1.499394389970e-03	2.492081090106e-01
		3.414866841140e-04	2.498247358051e-01
-2.0	0.5	1.499394389970e-03	2.492081090106e-01
		3.414866841140e-04	2.498247358051e-01
-2.0	1.0	1.499394389970e-03	2.492081090106e-01
		3.414866841140e-04	2.498247358051e-01
-2.0	2.0	1.499394389970e-03	2.492081090106e-01
		3.414866841140e-04	2.498247358051e-01
-2.0	5.0	1.499394389970e-03	2.492081090106e-01
		3.414866841140e-04	2.498247358051e-01
-2.0	10.0	1.499394389970e-03	2.492081090106e-01
		3.414866841140e-04	2.498247358051e-01
-5.0	0.1	1.419405293866e-03	2.492514632770e-01
		3.326081565771e-04	2.498293509715e-01
-5.0	0.5	1.419405293866e-03	2.492514632770e-01
		3.326081565771e-04	2.498293509715e-01
-5.0	1.0	1.419405293866e-03	2.492514632770e-01
		3.326081565771e-04	2.498293509715e-01
-5.0	2.0	1.419405293866e-03	2.492514632770e-01
		3.326081565771e-04	2.498293509715e-01
-5.0	5.0	1.419405293866e-03	2.492514632770e-01
		3.326081565771e-04	2.498293509715e-01
-5.0	10.0	1.419405293866e-03	2.492514632770e-01
		3.326081565771e-04	2.498293509715e-01
-10.0	0.1	1.401539502313e-03	2.492611334625e-01
		3.305721972151e-04	2.498304089546e-01
-10.0	0.5	1.401539502313e-03	2.492611334625e-01
		3.305721972151e-04	2.498304089546e-01
-10.0	1.0	1.401539502313e-03	2.492611334625e-01
		3.305721972151e-04	2.498304089546e-01
-10.0	2.0	1.401539502313e-03	2.492611334625e-01

```
                 3.305721972151e-04  2.498304089546e-01
   -10.0   5.0   1.401539502313e-03  2.492611334625e-01
                 3.305721972151e-04  2.498304089546e-01
   -10.0  10.0   1.401539502313e-03  2.492611334625e-01
                 3.305721972151e-04  2.498304089546e-01
>>
```

Example 1.4. *The logistic measure*

$$\mathrm{d}\lambda(t)=\frac{e^{-t}}{(1+e^{-t})^2}\,\mathrm{d}t \quad \text{on } \mathbb{R}.$$

This is a symmetric measure supported on the whole real line. It is natural to write

$$\int_{-\infty}^{\infty} f(t)\mathrm{d}\lambda(t)=\int_0^{\infty}\frac{f(-t)}{(1+e^{-t})^2}e^{-t}\mathrm{d}t+\int_0^{\infty}\frac{f(t)}{(1+e^{-t})^2}e^{-t}\mathrm{d}t$$

and discretize the integrals on the right by applying Gauss–Laguerre quadrature to the function $f(\mp t)(1+e^{-t})^{-2}$. The recurrence coefficients for $\mathrm{d}\lambda$ happen to be known explicitly: $\alpha_k(\mathrm{d}\lambda)=0$, $\beta_0(\mathrm{d}\lambda)=1$, $\beta_k(\mathrm{d}\lambda)=\pi^2k^4/(4k^2-1)$, $k\geq 1$ (cf. [4, Eq. (8.7) where $\lambda=0$, $x=t/\pi$]) so that the accuracy obtained by the discretization can be easily checked. ∎

Demo 1.8. *The first* 40 *recurrence coefficients* $\beta_k(\mathrm{d}\lambda)$ *of the logistic measure.*

For reasons of apparent instabilities, the discretization in Example 1.4 is implemented in 32-digit arithmetic by the routine `sr_logist.m` (cf. Appendix B.1),

```
function ab=sr_logist(dig,N)
%SR_LOGIST Recurrence coefficients for the logistic measure.

global mc mp iq idelta irout
mc=2; mp=0; iq=1; idelta=2; irout=1; Mmax=200;
eps0=.5e-12;
[ab,Mcap,kount]=smcdis(dig,N,eps0,@squad_logist,Mmax);
[Mcap,kount]
```

where the quadrature routine `squad_logist.m` is given by

```
function xw=squad_logist(N,i)
%SQUAD_LOGIST Quadrature routine used by sr_logist in smcdis.

global dig
ab=sr_laguerre(dig,N);
xw=sgauss(dig,N,ab);
xw(:,2)=xw(:,2)./((1+exp(-xw(:,1))).^2);
if i==1, xw(:,1)=-xw(:,1); end
```

The driver

```
%SRUN_LOGIST
```

```
f0='%8.0f %20.11e %12.4e %12.4e\n';
f1='%42.4e %12.4e\n';
global dig
dig=32; digits(dig);
for N=40
  ab=sr_logist(dig,N);
  ab0=double(ab);
  abex=zeros(N,2);
  abex(1,2)=1; k=(1:N-1)';
  abex(2:N,2)=pi^2*k.^4./(4*k.^2-1);
  errab=abs((ab0-abex)./moll(abex));
  erra=errab(:,1); errb=errab(:,2);
  merra=max(erra); merrb=max(errb);
  for n=[1 2 7 16 27 40]
    fprintf(f0,n-1,ab0(n,2),erra(n),errb(n))
  end
  fprintf(f1,merra,merrb)
end
```

produces the results shown in Table 1.3. It took `Mcap` $=121$ quadrature points and three iterations for the routine `smcdis.m` to converge.

Table 1.3. *Results for the two-component discretization in Example* 1.4

n	β_n	err α	err β
0	1.00000000000(0)	2.2652(−41)	0.0000(0)
1	3.28986813370(0)	3.5949(−40)	0.0000(0)
6	8.94476035232(0)	5.1236(−39)	0.0000(0)
15	5.55782783988(2)	3.2247(−39)	0.0000(0)
26	1.66858022227(3)	2.7681(−38)	0.0000(0)
39	3.75353402519(3)	4.1112(−38)	0.0000(0)
		8.0918(−38)	1.9030(−16)

General-purpose multicomponent discretization

Here we use again a representation of the type (1.11) for the integral $\int_{\mathbb{R}} p(t) w(t)\mathrm{d}t$, but in the special form with $h_i(t)=1$, $c_i=1$, $d_i=0$, $w_i(t)=w(t)$, that is,

$$\int_{\mathbb{R}} p(t)w(t)\mathrm{d}t = \sum_{i=1}^{m}\int_{a_i}^{b_i} p(t)w(t)\mathrm{d}t. \tag{1.16}$$

In order to discretize the component integrals on the right of (1.16), we now simply apply an M-point Fejér quadrature rule to the whole integrand, including the weight function w, but transformed from the standard interval $[-1,1]$ to the intervals $[a_i, b_i]$.

The procedure is implemented in the general-purpose quadrature routine `quadgp.m`, to which must be appended the function routine `wf.m` evaluating the weight function w for a vector-valued argument—the M nodes of the Fejér quadrature rule. These, along with the corresponding weights, are generated within the routine `mcdis.m`. The routine must be called with `@quadgp` as the third input variable and with the global variables `mc` $=m$,

AB = [[a_1 b_1];[a_2 b_2];$\cdots$;[a_{mc} b_{mc}]], mp = 0, iq $\neq$ 1, idelta = 1, and irout = 1 or $\neq 1$ depending on whether Stieltjes's or Lanczos's method is to be used.

Demo 1.9. *Half-range Hermite measure.*

This is the measure $d\lambda(t) = e^{-t^2}dt$ supported on the positive line $\mathbb{R}_+$. A satisfactory decomposition of the interval $[0,\infty]$ has been found, by trial and error, to be $[0,\infty] = [0,3] \cup [3,6] \cup [6,9] \cup [9,\infty]$. This is used in the script

```
%DEMO1_9

global mc AB mp iq idelta irout
f0='%2.0f %16.12f %16.12f %12.4e %12.4e\n';
f1='%49.4e %12.4e\n';
ab1=loadvpa('ab_hrhermite.txt',3,32,200,2);
ab1=double(ab1(1:40,:));
mc=4; AB=[[0 3];[3 6];[6 9];[9 Inf]];
mp=0; iq=0; idelta=1; irout=1;
N=40; eps0=.5e-12; Mmax=100;
[ab,Mcap,kount]=mcdis(N,eps0,@quadgp,Mmax);
err=abs((ab-ab1)./ab1); erra=err(:,1);
errb=err(:,2); merra=max(erra); merrb=max(errb);
fprintf('\n')
disp( ...
' n       alpha_n              beta_n              erra              errb')
for n=[1 2 7 16 27 40]
  fprintf(f0,n-1,ab(n,1),ab(n,2),erra(n,1),errb(n,1))
end
fprintf(f1,merra,merrb)
fprintf('       Mcap=%2.0f  kount=%1.0f\n',Mcap,kount)
```

which employs Stieltjes's method (irout = 1) to generate the first forty recurrence coefficients for the orthogonal polynomials $\pi_k(\cdot; d\lambda)$. It also determines the respective relative errors, using as reference the 32-digit values in the file ab_hrhermite obtained in Ex. 2.17 by a different method. The output

```
>> Demo1_9

 n      alpha_n             beta_n              erra          errb
 0   0.564189583548    0.886226925453    0.0000e+00    1.2528e-16
 1   0.988425392847    0.181690113816    1.1232e-16    0.0000e+00
 6   2.080620336401    1.002347851011    0.0000e+00    0.0000e+00
15   3.214270636071    2.500927917134    2.7632e-16    1.7757e-16
26   4.203048578872    4.333867901230    6.3395e-16    2.0494e-16
39   5.131532886894    6.500356237707    2.0770e-15    1.3664e-16
                                         2.0770e-15    7.6124e-16
     Mcap=81   kount=1
>>
```

shows that results accurate to machine precision are obtained with Mcap = 81 after just one iteration.

1.4 ▪ Modification algorithms

The task here is to generate the recurrence coefficients $\hat{\alpha}_k$, $\hat{\beta}_k$ for the polynomials orthogonal relative to a modified measure

$$d\hat{\lambda}(t) = r(t)d\lambda(t), \tag{1.17}$$

where r is a rational function (usually positive on the support of $d\lambda$), and $d\lambda$ is given by its recurrence coefficients α_k, β_k. In principle, the task can be accomplished if we can deal with the special cases of linear and quadratic factors, resp., divisors. The extension to more general rational functions can then be obtained by a repeated application of these special cases.

The relevant OPQ routines were originally given names beginning with chri (for Christoffel; cf. Appendix A.1), but are renamed here in a mnemonically more satisfactory manner, and, in the case of divisors, also organized differently.

1.4.1 ▪ Linear factors

We begin with the case of a linear factor $r(t) = t - z$, where $z \in \mathbb{C}$ is outside the support of $d\lambda$. The routine generating the first N recurrence coefficients $\hat{\alpha}_k = \alpha_k(d\hat{\lambda})$, $\hat{\beta}_k = \beta_k(d\hat{\lambda})$ requires $N+1$ coefficients $\alpha_k = \alpha_k(d\lambda)$, $\beta_k = \beta_k(d\lambda)$, one more than the number of coefficients asked for; cf. Algorithm 2.5 of [24].

```
function ab=mod_linfac(N,ab0,z)
   %MOD_LINFAC Modification by a linear factor.
   %   AB=MOD_LINFAC(N,AB0,Z) generates the Nx2 array AB of the
   %   first N recurrence coefficients for polynomials orthogonal
   %   with respect to the modified weight function (T-Z)*W(T),
   %   where Z is real or complex, outside the support interval
   %   of the weight function W, and the latter is specified by
   %   the (N+1)x2 array AB0 of recurrence coefficients.

   N0=size(ab0,1);
   if N0<N+1, error('input array ab0 too short'), end
   r=zeros(N+1,1);
   r(1)=z-ab0(1,1); r(2)=z-ab0(2,1)-ab0(2,2)/r(1);
   ab(1,1)=ab0(2,1)+r(2)-r(1); ab(1,2)=-r(1)*ab0(1,2);
   if N==1, return, end
   for k=2:N
     r(k+1)=z-ab0(k+1,1)-ab0(k+1,2)/r(k);
     ab(k,1)=ab0(k+1,1)+r(k+1)-r(k);
     ab(k,2)=ab0(k,2)*r(k)/r(k-1);
   end
```

When z is nonreal, the polynomials so generated are complex-valued formal orthogonal polynomials.

Demo 1.10. *Recurrence coefficients for Jacobi measure* $d\lambda^{(-1/2,3/2)}$ *from those for* $d\lambda^{(-1/2,1/2)}$.

This is a matter of applying to $d\lambda^{(-1/2,1/2)}$ a modification algorithm with the linear factor $r(t) = t - z$, $z = -1$. The results are checked against a direct evaluation of the recurrence coefficients of $d\lambda^{(-1/2,3/2)}$ (cf. also Demo 1.1).

```
%DEMO1_10

f0='%2.0f %19.15f %19.15f %12.4e %12.4e\n';
f1='%55.4e %12.4e\n';
disp( ...
' k            alpha_k                    beta_k               erra             errb')
N=40; alpha=-1/2; beta=1/2;
ab0=r_jacobi(N+1,alpha,beta);
ab1=r_jacobi(N,alpha,beta+1);
ab=mod_linfac(N,ab0,-1);
err=abs((ab-ab1)./ab1); merr=max(err);
for k=1:N
  fprintf(f0,k-1,ab(k,1),ab(k,2),err(k,1),err(k,2))
end
fprintf(f1,merr(1),merr(2))
```

Output:

```
>> Demo1_10
 k        alpha_k              beta_k             erra         errb
 0   0.666666666666667   4.712388980384691   0.0000e+00   1.8848e-16
 1   0.133333333333333   0.138888888888889   6.2450e-16   0.0000e+00
 2   0.057142857142857   0.210000000000000   3.6429e-16   1.3217e-16
 3   0.031746031746032   0.229591836734694   0.0000e+00   0.0000e+00
 4   0.020202020202020   0.237654320987654   8.5869e-16   0.0000e+00
 5   0.013986013986014   0.241735537190083   2.1086e-15   0.0000e+00
 6   0.010256410256410   0.244082840236686   3.5518e-15   1.1371e-16
 7   0.007843137254902   0.245555555555556   3.5388e-15   1.1303e-16
 8   0.006191950464396   0.246539792387543   1.1486e-14   0.0000e+00
 9   0.005012531328321   0.247229916897507   0.0000e+00   1.1227e-16
..          ...                 ...               ...          ...

36   0.000365296803653   0.249812347532370   8.5182e-14   1.1111e-16
37   0.000346320346320   0.249822222222222   8.5153e-14   1.1110e-16
38   0.000328785138912   0.249831337493675   1.4559e-13   1.1110e-16
39   0.000312548835756   0.249839769267746   4.0413e-14   1.1109e-16
                                             1.9050e-13   2.2253e-16
>>
```

The maximum relative error in the α_k is seen to be about three orders of magnitude larger than the one for the β_k. The reason is the relatively small magnitude of most of the α's, compared to the magnitude of the β's. Overall, however, the accuracy achieved is close to machine precision.

1.4.2 ▪ Quadratic factors

In the case of quadratic factors (cf. Ex. 2.39),

$$r(t) = (t-x)^2 + y^2, \quad x \in \mathbb{R},\ y > 0, \tag{1.18}$$

the relevant modification algorithm is implemented in the routine `mod_quadfac.m`,

```
function ab=mod_quadfac(N,ab0,x,y)
%MOD_QUADFAC Modification by a quadratic factor.
% AB=MOD_QUADFAC(N,AB0,X,Y) generates the Nx2 array AB of the
% first N>1 recurrence coefficients for polynomials orthogonal
% with respect to the modified weight function ((T-X)^2+Y^2)*W(T),
% where X and Y>0 are real and the weight function W is specified
% by the (N+1)x2 array AB0 of recurrence coefficients.

if N<2, error('N out of range'), end
N0=size(ab0,1);
if N0<N+1, error('input array ab0 too short'), end
s=x-ab0(1,1); t=y; eio=0;
for k=1:N
  d=s^2+t^2;
  er=-ab0(k+1,2)*s/d; ei=ab0(k+1,2)*t/d;
  s=x+er-ab0(k+1,1); t=y+ei;
  ab(k,1)=x+t*er/ei-s*ei/t;
  ab(k,2)=t*eio*(1+(er/ei)^2);
  eio=ei;
end
ab(1,2)=ab0(1,2)*(ab0(2,2)+(ab0(1,1)-x)^2+y^2);
```

The modification algorithm can be tested against a three-component discretization procedure (cf. Section 1.3.2) applied to

$$d\hat{\lambda}(t) = (t^2 - 2xt + x^2 + y^2)d\lambda(t). \tag{1.19}$$

We found, however, that this latter procedure is subject to considerable error accumulation when N is larger than about 20. We therefore implement it in variable-precision arithmetic.

Demo 1.11. *Laguerre measure modified by a quadratic factor,*

$$d\hat{\lambda}(t) = ((t-x)^2 + y^2)d\lambda(t), \quad d\lambda(t) = e^{-t}dt, \; t \in \mathbb{R}_+, \; x = 1, \; y = 2,$$

computed by a modification algorithm and tested by a three-component discretization procedure based on (1.19).

This requires a symbolic quadrature routine for use in `smcdis.m`, which is as follows:

```
function xw=squad_mod_lag(N,i)
%SQUAD_MOD_LAG  Symbolic quadrature routine used by Demo1.11
%in smcdis

global x0 y0 dig
x0=vpa(x0); y0=vpa(y0);
ab1=sr_laguerre(dig,N,2);
ab2=sr_laguerre(dig,N,1);
ab3=sr_laguerre(dig,N);
if i==1
  xw=sgauss(dig,N,ab1);
elseif i==2
```

```
    xw=sgauss(dig,N,ab2); xw(:,2)=-2*x0*xw(:,2);
  else
    xw=sgauss(dig,N,ab3); xw(:,2)=(x0^2+y0^2)*xw(:,2);
  end
```

The task stated in Demo 1.11 is implemented in the script `Demo1_11.m` for $N = 40$.

```
%DEMO1_11

f0='%2.0f %22.15e %22.15e %12.4e %12.4e\n';
f1='%61.4e %12.4e\n';
global mc mp iq idelta irout x0 y0 dig
dig=32; digits(dig);
N=40; x0=1; y0=2;
ab0=r_laguerre(N+1);
ab=mod_quadfac(N,ab0,x0,y0);
mc=3; mp=0; iq=1; idelta=2; irout=1; eps0=.5e-28; Mmax=100;
[ab1,Mcap,kount]=smcdis(dig,N,eps0,@squad_mod_lag,Mmax);
disp( ...
' k         alpha_k                    beta_k                    erra            errb')
err=abs((ab-double(ab1))./double(ab1)); merr=max(err);
for k=1:N
  fprintf(f0,k-1,ab(k,1),ab(k,2),err(k,1),err(k,2))
end
fprintf(f1,merr(1),merr(2))
```

It yields the following results:

```
>> Demo1_11
 k         alpha_k                    beta_k                 erra          errb
 0  1.400000000000000e+00  5.000000000000000e+00   0.0000e+00   0.0000e+00
 1  4.436065573770492e+00  2.440000000000000e+00   0.0000e+00   0.0000e+00
 2  6.516215652856284e+00  7.186240257995164e+00   0.0000e+00   1.2359e-16
 3  8.559966180045786e+00  1.424089419765055e+01   0.0000e+00   0.0000e+00
 4  1.061520081934103e+01  2.312518999110937e+01   0.0000e+00   1.5363e-16
 5  1.265947155885006e+01  3.393300095375606e+01   1.4032e-16   2.0940e-16
 6  1.468990241955731e+01  4.676021057721211e+01   1.2092e-16   1.5195e-16
 7  1.671150839573743e+01  6.162172821041541e+01   0.0000e+00   1.1531e-16
 8  1.872847509830807e+01  7.850249780460516e+01   0.0000e+00   0.0000e+00
 9  2.074285745916952e+01  9.738946039854773e+01   1.7127e-16   4.3775e-16
..          ...                    ...                 ...          ...

36  7.486963508886146e+01  1.364154958058302e+03   1.8981e-16   5.0003e-16
37  7.687139203023193e+01  1.439091610491662e+03   1.8487e-16   3.1600e-16
38  7.887307978006686e+01  1.516029069214873e+03   0.0000e+00   1.4998e-16
39  8.087470276316370e+01  1.594967304315068e+03   1.7571e-16   1.4256e-16
                                                   2.1254e-16   6.3468e-16
>>
```

As can be seen from the maximum relative errors, displayed at the lower right-hand corner of the output, quadratic modification, in this example, is accurate to essentially machine precision.

The special case $x = 0$ of (1.18) warrants a routine of its own, `mod_symmquad fac.m`. Another modification of interest,

$$d\hat{\lambda}(t) = (t-x)^2 d\lambda(t), \tag{1.20}$$

is dealt with by the routine `mod_quad0fac.m` (cf. [24, pp. 127–128]).

1.4.3 ▪ Linear divisors

Modification of a measure $d\lambda$ by a linear divisor $t-z$, where $z \in \mathbb{C}\backslash\text{supp}(d\lambda)$, requires ratios

$$r_k(z) = \frac{\rho_{k+1}(z; d\lambda)}{\rho_k(z; d\lambda)}, \quad k = -1, 0, 1, 2, \ldots, \tag{1.21}$$

of Cauchy integrals

$$\rho_n(z; d\lambda) = \int_{\mathbb{R}} \frac{\pi_n(t; d\lambda)}{z-t} d\lambda(t), \quad n = 0, 1, 2, \ldots, \ \rho_{-1}(z; d\lambda) = 1. \tag{1.22}$$

The first $N+1$ ratios $r_{-1}, r_0, \ldots, r_{N-1}$ can be generated recursively by

$$\begin{aligned} r_k(z) &= z - \alpha_k - \beta_k / r_{k-1}(z), \quad k = 0, 1, 2, \ldots, N-1, \\ r_{-1}(z) &= \rho_0(z), \end{aligned} \tag{1.23}$$

where $\alpha_k = \alpha_k(d\lambda)$, $\beta_k = \beta_k(d\lambda)$. The recursion, however, is unstable and must be executed in variable-precision arithmetic with as many digits as are required to obtain all $N+1$ ratios to full machine precision. Note that this requires knowledge of

$$\rho_0(z) = \int_{\mathbb{R}} \frac{d\lambda(t)}{z-t},$$

the Cauchy transform of the measure $d\lambda$.

In the case of the Legendre measure $d\lambda(t) = dt$ on $[-1, 1]$, for example, we have

$$\rho_0(z) = \log \frac{z+1}{z-1},$$

and the routine `cr_legendre.m` not only computes the first $N+1$ ratios r_k, but also the number of digits needed to obtain them to full machine precision.

```
function [r,dig]=cr_legendre(N,z)
%CR_LEGENDRE Cauchy integral ratios for Legendre measure.
%   [R,DIG]=CR_LEGENDRE(N,Z) generates to MATLAB double precision
%   the (N+1)x1 array R of the ratios r_n=rho_{n+1}(Z)/rho_n(Z),
%   n=-1,0,1,...,N-1, of Cauchy integrals rho_{-1}=1, rho_k(Z)
%   =INT pi_k(T)dT/(Z-T) for monic Legendre polynomials pi_k,
%   k=0,1,...,N. The output variable DIG is the number of digits
%   needed to obtain the ratios to MATLAB double precision.

r1=zeros(N+1,1); err=1; dig=12;
while err>eps
  r0=r1;
```

```
    dig=dig+10; digits(dig);
    z=vpa(z); r1=vpa(zeros(N+1,1));
    r1(1)=log((z+1)/(z-1)); ab0=sr_jacobi(dig,N);
    for k=1:N
      r1(k+1)=z-ab0(k,1)-ab0(k,2)/r1(k);
    end
    err=max(double(abs((r1-r0)./r1)));
  end
  r=double(r1);
```

Once the ratios $r_k = r_k(z)$, $k = -1, 0, 1, \ldots, N-1$, are available, the desired recurrence coefficients $\hat{\alpha}_k$, $\hat{\beta}_k$ can be generated by (cf. Algorithm 5 of Ex. 2.40)

$$\hat{\alpha}_0 = \alpha_0 + r_0(z), \quad \hat{\beta}_0 = -r_{-1}(z)$$

and, if $N > 1$, by

$$\left.\begin{aligned} \hat{\alpha}_k &= \alpha_k + r_k(z) - r_{k-1}(z) \\ \hat{\beta}_k &= \beta_{k-1} r_{k-1}(z)/r_{k-2}(z) \end{aligned}\right\} \quad k = 1, 2, \ldots, N-1.$$

This is implemented in the routine mod_lindiv.m:

```
function ab=mod_lindiv(N,ab0,z,r)
%MOD_LINDIV Modification by a linear divisor.
%   AB=MOD_LINDIV(N,AB0,Z,R) generates the Nx2 array AB of the
%   first N recurrence coefficients for polynomials orthogonal
%   with respect to the modified weight function W(T)/(T-Z),
%   where Z is real or complex, outside the support interval
%   of the weight function W, and the latter is specified by the
%   Nx2 array AB0 of recurrence coefficients. The input argument
%   R is the (N+1)x1 array of ratios r_n=rho_{n+1}(Z)/rho_n(Z),
%   n=-1,0,1,...,N-1, of Cauchy integrals rho_{-1}=1, rho_k(Z)
%   =INT pi_k(T)W(T)dT/(Z-T) for the monic orthogonal polynomials
%   pi_k, k=0,1,...,N, relative to the weight function W.

N0=size(ab0,1); r=zeros(N+1,1);
if N0<N, error('input array too short'), end
ab(1,1)=ab0(1,1)+r(2); ab(1,2)=-r(1);
if N==1, return, end
for k=2:N
  ab(k,1)=ab0(k,1)+r(k+1)-r(k);
  ab(k,2)=ab0(k-1,2)*r(k)/r(k-1);
end
```

Demo 1.12. *Test of the routines* mod_linfac.m *and* mod_lindiv.m.

Modify the Legendre measure $d\lambda$ by a linear divisor $t-z$, $z = 2e^{m\pi i/10}$, $m = 0, 1, \ldots, 10$, to obtain the complex measure $d\lambda_1$. Then modify $d\lambda_1$ multiplying by the same linear function $t - z$ to obtain $d\lambda_2$. Check to what extent $d\lambda_2 = d\lambda$. Use $N = 40$.

The test is carried out by the script

```
%DEMO1_12

f0='%8.0f %12.4e %12.4e\n';
N=40;
disp('        m      erra          errb')
for m=0:10
  z=2*exp(m*pi*i/10);
  ab=r_jacobi(N+1); r=cr_legendre(N+1,z);
  ab1=mod_lindiv(N+1,ab,z,r);
  ab2=mod_linfac(N,ab1,z);
  err=ab(1:N,:)-ab2;
  erra=max(abs(err(:,1)));
  errb=max(abs(err(:,2)./ab(1:N,2)));
  fprintf(f0,m,erra,errb)
end
```

The output displays the maximum absolute errors in the α-coefficients and the maximum relative errors in the β coefficients.

```
>> Demo1_12
       m      erra          errb
       0   4.4409e-16   4.4401e-16
       1   2.4825e-16   4.5732e-16
       2   3.1402e-16   4.7673e-16
       3   3.1402e-16   4.5256e-16
       4   4.5776e-16   4.5540e-16
       5   4.4409e-16   4.4400e-16
       6   4.5776e-16   6.7345e-16
       7   4.4409e-16   4.7993e-16
       8   4.4409e-16   4.7670e-16
       9   2.4825e-16   4.4668e-16
      10   4.4409e-16   4.4401e-16
>>
```

It is seen that the recurrence coefficients for $d\lambda_2$ equal those for $d\lambda$ to within machine precision.

1.4.4 ▪ Quadratic divisors

As in the case of linear divisors, modification by a quadratic divisor $(t-x)^2+y^2$, $x \in \mathbb{R}$, $y > 0$, requires the ratios (1.21), (1.22) of Cauchy integrals to be generated in sufficiently high precision by the recursion (1.23). The procedure is implemented in the routine mod_quaddiv.m (cf. Algorithm 6 of Ex. 2.41)

```
function ab=mod_quaddiv(N,ab0,x,y,r)
%MOD_QUADDIV Modification by a quadratic divisor.
%   AB=MOD_QUADDIV(N,AB0,X,Y,R) generates the Nx2 array AB of the
%   first N recurrence coefficients for polynomials orthogonal with
%   respect to the modified weight function W(T)/((T-X)^2+Y^2),
%   where X is real and Y>0 and the weight function W is specified
```

```
%   by the Nx2 array AB0 of recurrence coefficients. The input
%   argument R is the (N+1)x1 array of ratios r_n=rho_{n+1}(Z)
%   /rho_n(Z), n=-1,0,1,...,N-1, of Cauchy integrals rho_{-1}=1,
%   rho_k(Z)=INT pi_k(T)W(T)dT/(Z-T), Z=X+iY, k=0,1,...,N,
%   for the monic orthogonal polynomials pi_k, k=0,1,...,N,
%   relative to the weight function W.

if N<2, error('N out of range'), end
N0=size(ab0,1); r=zeros(N+1,1); s=zeros(N,1); t=zeros(N,1);
z=complex(x,y);
if N0<N, error('input array ab0 too short'), end
for k=1:N
  s(k)=-(real(r(k+1))+(imag(r(k+1))/imag(r(k)))*real(r(k)));
  t(k)=(imag(r(k+1))/imag(r(k)))*abs(r(k))^2;
end
ab(1,1)=x+y*real(r(1))/imag(r(1)); ab(1,2)=-imag(r(1))/y;
ab(2,1)=ab0(2,1)-s(2)+s(1);
ab(2,2)=ab0(2,2)+s(1)*(ab0(1,1)-ab(2,1))-t(2);
if N==2, return, end
ab(3,1)=ab0(3,1)-s(3)+s(2);
ab(3,2)=ab0(3,2)+s(2)*(ab0(2,1)-ab(3,1))-t(3)+t(2);
if N==3, return, end
for k=3:N-1
  ab(k+1,1)=ab0(k+1,1)-s(k+1)+s(k);
  ab(k+1,2)=ab0(k-1,2)*t(k)/t(k-1);
end
```

Multiplying the resulting measure again by $(t-x)^2+y^2$ should restore the original measure.

Demo 1.13. *Test of the routines* `mod_quaddiv.m` *and* `mod_quadfac.m`.
Modify the Legendre measure $d\lambda$ by the quadratic divisor $(t-x)^2+y^2$ and multiply the resulting measure by the same quadratic factor. Check to what extent the original Legendre measure is recovered.

This is done, for $x=1$, $y=2$, by the script

```
%DEMO1_13

f0='%2.0f %22.15e %22.15e %12.4e %12.4e\n';
f1='%61.4e %12.4e\n';
disp( ...
' k         alpha_k                  beta_k                 erra          errb')
x=1; y=2; z=complex(x,y);
N=40;
ab0=r_jacobi(N+1); r=cr_legendre(N+1,z);
ab1=mod_quaddiv(N+1,ab0,x,y,r);
ab2=mod_quadfac(N,ab1,x,y);
err=abs(ab0(1:N,:)-ab2);
erra=err(:,1);
errb=err(:,2)./ab0(1:N,2);
```

```
erram=max(erra); errbm=max(errb);
for k=1:N
  fprintf(f0,k-1,ab2(k,1),ab2(k,2),erra(k),errb(k))
end
fprintf(f1,erram,errbm)
```

producing the following output:

```
>> Demo1_13
 k          alpha_k                    beta_k                  erra          errb
 0  5.551115123125783e-17  2.000000000000000e+00   5.5511e-17   0.0000e+00
 1  6.938893903907228e-17  3.333333333333333e-01   6.9389e-17   0.0000e+00
 2  6.938893903907228e-18  2.666666666666667e-01   6.9389e-18   0.0000e+00
 3 -1.665334536937735e-16  2.571428571428572e-01   1.6653e-16   2.1588e-16
 4  2.012279232133096e-16  2.539682539682541e-01   2.0123e-16   6.5573e-16
 5  7.632783294297951e-17  2.525252525252524e-01   7.6328e-17   6.5947e-16
 6  6.938893903907228e-18  2.517482517482517e-01   6.9389e-18   2.2050e-16
 7  8.326672684688674e-17  2.512820512820513e-01   8.3267e-17   0.0000e+00
 8  2.914335439641036e-16  2.509803921568628e-01   2.9143e-16   2.2118e-16
 9  4.857225732735060e-17  2.507739938080495e-01   4.8572e-17   2.2136e-16
..          ...                      ...                   ...          ...

36  2.428612866367530e-16  2.500482346131584e-01   2.4286e-16   0.0000e+00
37 -2.220446049250313e-16  2.500456621004567e-01   2.2204e-16   2.2200e-16
38  3.261280134836397e-16  2.500432900432901e-01   3.2613e-16   0.0000e+00
39 -6.938893903907228e-18  2.500410981423640e-01   6.9389e-18   0.0000e+00
                                                        3.2613e-16   6.5947e-16
>>
```

The maximum discrepancy, shown at the lower right-hand corner of the output, demonstrates that we have agreement essentially to machine precision.

1.5 ▪ Sobolev orthogonal polynomials

Polynomials orthogonal with respect to an inner product

$$(u,v)_S = \int_{\mathbb{R}} u(t)v(t)\mathrm{d}\lambda_0(t) + \int_{\mathbb{R}} u'(t)v'(t)\mathrm{d}\lambda_1(t) \\ + \cdots + \int_{\mathbb{R}} u^{(s)}(t)v^{(s)}(t)\mathrm{d}\lambda_s(t), \quad s \geq 1,$$

involving derivatives of order up to s, each endowed with its own measure, are called *Sobolev orthogonal polynomials.* As with any sequence of monic polynomials of increasing degrees $k = 1,2,3,\ldots$, they must satisfy an extended recurrence relation of the form

$$\pi_{k+1}(t) = t\pi_k(t) - \sum_{j=0}^{k} \beta_j^k \pi_{k-j}(t), \quad k = 0,1,2,\ldots . \tag{1.24}$$

Associated with it is the upper Hessenberg matrix of recurrence coefficients

$$H_n = \begin{bmatrix} \beta_0^0 & \beta_1^1 & \beta_2^2 & \cdots & \beta_{n-2}^{n-2} & \beta_{n-1}^{n-1} \\ 1 & \beta_0^1 & \beta_1^2 & \cdots & \beta_{n-3}^{n-2} & \beta_{n-2}^{n-1} \\ 0 & 1 & \beta_0^2 & \cdots & \beta_{n-4}^{n-2} & \beta_{n-3}^{n-1} \\ \vdots & \vdots & \vdots & \vdots & \vdots & \vdots \\ 0 & 0 & 0 & \cdots & \beta_0^{n-2} & \beta_1^{n-1} \\ 0 & 0 & 0 & \cdots & 1 & \beta_0^{n-1} \end{bmatrix},$$

the coefficients β_j^k, $j = 0, 1, \ldots, k$, for each fixed k being located in the $(k+1)$st column of $\boldsymbol{H}_n$ from the diagonal element ($j = 0$) up to the first row ($j = k$). The computational task is to compute the matrix $\boldsymbol{H}_n$, given the $s+1$ measures $\mathrm{d}\lambda_0, \mathrm{d}\lambda_1, \ldots, \mathrm{d}\lambda_s$.

There are again two types of methods: one based on modified moments, the other on discretization. In the former, there are $s+1$ sets of modified moments,

$$m_k^{(\sigma)} = \int_{\mathbb{R}} p_k(t)\mathrm{d}\lambda_\sigma(t), \quad k = 0, 1, 2, \ldots; \ \sigma = 0, 1, \ldots, s,$$

the first $2n$ of each uniquely determining the matrix $\boldsymbol{H}_n$. The polynomials p_k, as in Section 1.2, are assumed to satisfy a three-term recurrence relation with known coefficients a_k, b_k. The relevant OPQ routine, implemented only for $s = 1$, is (cf. Appendix A.1)

B = chebyshev_sob(N, mom, abm).

It produces the $N \times N$ upper triangular matrix B_N of recurrence coefficients, the coefficient β_j^k, $0 \le j \le k$, $0 \le k \le N-1$, occupying the position $(j+1, k+1)$ in the matrix. The input argument mom is the $2 \times (2N)$ array of modified moments $m_k^{(\sigma)}$, $k = 0, 1, \ldots, 2N-1$, $\sigma = 0, 1$, of the measures $\mathrm{d}\lambda_0$ and $\mathrm{d}\lambda_1$, and abm the $(2N-1) \times 2$ array of coefficients a_k, b_k, $k = 0, 1, \ldots, 2N-2$, defining the polynomials p_k.

Demo 1.14. *Altheimer's polynomials.*

These are the Sobolev polynomials orthogonal with respect to the measures $\mathrm{d}\lambda_0(t) = \mathrm{d}t$, $\mathrm{d}\lambda_1(t) = \gamma \mathrm{d}t$ on $[-1, 1]$, $\gamma > 0$. *Determine the relevant upper Hessenberg matrix H_n. Take $n = 5$, $\gamma = 2$.*

A natural choice of modified moments are those relative to the monic Legendre polynomials, in which case, by orthogonality, all modified moments $m_k^{(0)}$, $m_k^{(1)}$ are zero when $k > 0$, while $m_0^{(0)} = 2$ and $m_0^{(1)} = 2\gamma$. The matrix B_n of recurrence coefficients, therefore, can be generated by the routine

```
function B=rsob_alth(N,g)
%RSOB_ALTH  Altheimer polynomials.
%   B=RSOB_ALTH(N,G) generates the NxN matrix B of recurrence
%   coefficients for Altheimer polynomials with parameter G,
%   using moments.
mom=zeros(2,2*N);
mom(1,1)=2; mom(2,1)=2*g;
abm=r_jacobi(2*N-1);
B=chebyshev_sob(N,mom,abm);
```

Since the elements h_{ik}, $i \le k$, of the matrix H_n are β_{k-i}^{k-1}, which are located in position $(k-i+1, k)$ of the matrix B_n, the desired upper Hessenberg matrix H_n is obtained by

```
function H=HfromB(B)
%HFROMB Conversion from B to H.
%   H=HFROMB(B) converts the upper triangular matrix B of
%   Sobolev recurrence coefficients to the upper Hessenberg
%   matrix H of rearranged coefficients.

N=size(B)*[1;0];
H=zeros(N);
for k=1:N
  for i=1:k
    H(i,k)=B(k-i+1,k);
  end
  if k<N, H(k+1,k)=1; end
end
```

Thus, for $n=5$, $\gamma=2$, running the two routines in succession, with $\mathtt{N}=5$ and $\mathtt{g}=2$, yields

$$H_5=\begin{bmatrix} 0 & 0.3333 & 0 & -0.1143 & 0 \\ 1.0000 & 0 & 0.6095 & 0 & -0.0063 \\ 0 & 1.0000 & 0 & 0.2461 & 0 \\ 0 & 0 & 1.0000 & 0 & 0.2353 \\ 0 & 0 & 0 & 1.0000 & 0 \end{bmatrix}. \tag{1.25}$$

It is seen that the coefficients β_j^k for fixed k and increasing j alternate between zero and nonzero values, which is a consequence of symmetry; cf. Ex. 3.4(b).

There are analogues to the Darboux formulae (1.2), for Sobolev orthogonal polynomials,

$$\beta_j^k=\frac{(t\pi_k,\pi_{k-j})_S}{(\pi_{k-j},\pi_{k-j})_S},\quad j=0,1,\dots,k;\ k=0,1,\dots,n-1,$$

which, combined with the recurrence relation (1.24), can be used to develop a discretized Stieltjes algorithm for Sobolev orthogonal polynomials. A basic underlying assumption is the availability, for each measure $d\lambda_\sigma$, of an n_σ-point quadrature rule

$$\int_{\mathbb{R}} p(t)d\lambda_\sigma(t)=\sum_{k=1}^{n_\sigma} w_k^{(\sigma)}p(x_k^{(\sigma)}),\quad p\in\mathbb{P}_{2(n-\sigma)-1},\ \sigma=0,1,\dots,s.$$

This is implemented, for arbitrary $s\geq 1$, by the routine (cf. Appendix A.1)

$$\mathtt{B}=\mathtt{stieltjes\_sob(N,s,nd,xw,a0,same)},$$

where $\mathtt{nd}=[n_0,n_1,\dots,n_s]$, and $\mathtt{xw}$ is the $\mathtt{md}\times(2s+2)$ array ($\mathtt{md}=\max(\mathrm{nd})$)

$$\mathtt{xw}=\left[\begin{array}{ccc|ccc} x_1^{(0)} & \cdots & x_1^{(s)} & w_1^{(0)} & \cdots & w_1^{(s)} \\ x_2^{(0)} & \cdots & x_2^{(s)} & w_2^{(0)} & \cdots & w_2^{(s)} \\ \vdots & & \vdots & \vdots & & \vdots \\ x_{\mathrm{md}}^{(0)} & \cdots & x_{\mathrm{md}}^{(s)} & w_{\mathrm{md}}^{(0)} & \cdots & w_{\mathrm{md}}^{(s)} \end{array}\right].$$

(In each column of $\mathtt{xw}$ the entries after $x_{n_\sigma}^{(\sigma)}$, resp., $w_{n_\sigma}^{(\sigma)}$, if any, are not used by the routine.) The input argument $\mathtt{a0}$ is $\alpha_0(d\lambda_0)$, and $\mathtt{same}$ is a logical variable equal to 1 if all quadrature rules have the same set of nodes, and equal to 0 otherwise.

Demo 1.15. *Altheimer's polynomials, revisited.*

Here, the obvious choice of the quadrature rule for $d\lambda_0$ and $d\lambda_1$ is the n-point Gauss–Legendre rule. Then B_n is now obtained by

```
function B=rsob_althrev(N,g)
%RSOB_ALTHREV  Altheimer polynomials.
%   B=ALTHREV(N,G) generates the NxN matrix B of recurrence
%   coefficients for Altheimer polynomials with parameter G,
%   using a discretized Stieltjes procedure.

s=1; nd=[N N];
a0=0; same=1;
ab=r_jacobi(N); zw=gauss(N,ab);
xw=[zw(:,1) zw(:,1) zw(:,2) g*zw(:,2)];
B=stieltjes_sob(N,s,nd,xw,a0,same);
```

and H_n as before in Demo 1.14. The result is identical with the one in (1.25).

For Sobolev orthogonal polynomials π_k, whose recurrence matrix is $B\in\mathbb{R}^{N\times N}$, the zeros of π_n, $n\le N$, can be found by the routine (cf. Appendix A.1)

$$\texttt{z}=\texttt{sobzeros(n,N,B)},$$

placing the n zeros, some of which could be complex, in the column vector $z\in\mathbb{C}^{n\times 1}$.

Example 1.5. *Sobolev orthogonal polynomials with only a few real zeros* (Meijer [51]).

The inner product, here, is

$$(u,v)_S=\int_{-1}^{3}u(t)v(t)\mathrm{d}t+\gamma\int_{-1}^{1}u'(t)v'(t)\mathrm{d}t+\int_{1}^{3}u'(t)v'(t)\mathrm{d}t,\quad \gamma>0,$$

and the measures involved, therefore,

$$\mathrm{d}\lambda_0(t)=\mathrm{d}t\ \text{ on }[-1,3],\quad \mathrm{d}\lambda_1(t)=\begin{cases}\gamma\mathrm{d}t & \text{on }[-1,1],\\ \mathrm{d}t & \text{on }(1,3].\end{cases}$$

The integrals required in Stieltjes's procedure,

$$\int_{-1}^{3}p(t)\mathrm{d}\lambda_0(t)=2\int_{-1}^{1}p(2x+1)\mathrm{d}x,$$
$$\int_{-1}^{3}p(t)\mathrm{d}\lambda_1(t)=\int_{-1}^{1}(\gamma p(x)+p(x+2))\,\mathrm{d}x,$$

where $p\in\mathbb{P}_{2n-1}$, can be evaluated exactly by n-point Gauss–Legendre quadrature. Stieltjes's procedure, so discretized, will produce the matrix H_n exactly.

The input arguments required for the routine `stieltjes_sob.m` are $\texttt{a0}=1$, $\texttt{same}=0$, $\texttt{nd}=[n,2n]$, and

$$\texttt{xw}=\begin{bmatrix}2\tau_1^G+1 & \tau_1^G & 2\lambda_1^G & \gamma\lambda_1^G\\ \vdots & \vdots & \vdots & \vdots\\ 2\tau_n^G+1 & \tau_n^G & 2\lambda_n^G & \gamma\lambda_n^G\\ & \tau_1^G+2 & & \lambda_1^G\\ & \vdots & & \vdots\\ & \tau_n^G+2 & & \lambda_n^G\end{bmatrix}\in\mathbb{R}^{2n\times 4}.\quad\blacksquare$$

Demo 1.16. *The zeros of Meijer's polynomial π_n for $n = 6$ and $\gamma = 44{,}000$.*
The zeros are produced by the script

```
%DEMO1_16

N=6; s=1; a0=1; same=0; g=44000; nd=[N 2*N];
ab=r_jacobi(N); zw=gauss(N,ab);
xw=zeros(2*N,2*(s+1));
xw(1:N,1)=2*zw(:,1)+1; xw(1:N,2)=zw(:,1);
xw(1:N,3)=2*zw(:,2); xw(1:N,4)=g*zw(:,2);
xw(N+1:2*N,2)=zw(:,1)+2; xw(N+1:2*N,4)=zw(:,2);
B=stieltjes_sob(N,s,nd,xw,a0,same);
z=sobzeros(N,N,B)
```

yielding

```
>> Demo1_16
z =
      -4.176763898909804e-01 - 1.703657992746572e-01i
      -4.176763898909804e-01 + 1.703657992746572e-01i
       8.453761089539273e-01 - 1.538233952529658e-01i
       8.453761089539273e-01 + 1.538233952529658e-01i
      -1.070135059563742e+00 + 0.000000000000000e+00i
       2.598402134930250e+00 + 0.000000000000000e+00i
>>
```

Thus, all but two zeros are complex.
More data and information on this example can be found in [24, Example 2.63].

1.6 ▪ Quadrature formulae

The principal routines, relating to numerical quadrature, are `gauss.m` (cf. Appendix A.2) and its symbolic analogue, `sgauss.m` (cf. Appendix B.2), which compute the Gauss quadrature formula

$$\int_{\mathbb{R}} f(x)\mathrm{d}\lambda(x) = \sum_{k=1}^{n} w_k f(x_k) + R_n(f), \quad R_n(\mathbb{P}_{2n-1}) = 0$$

for any given measure $\mathrm{d}\lambda$ identified by its array `ab` of recurrence coefficients. The method used is based on the fact that the nodes x_k are the eigenvalues of the Jacobi matrix (1.4), and the weights w_k equal to $\mu_0 v_{k,1}^2$, where $\mu_0 = \int_{\mathbb{R}} \mathrm{d}\lambda(x)$ and $v_{k,1}$ is the first component of the normalized eigenvector v_k corresponding to the eigenvalue x_k (cf. [24, Theorem 3.1] or Ex. 4.3). These are computed by the MATLAB function `eig.m`.[1]

Demo 1.17. *Gauss quadrature for the half-range Hermite measure.*
The 40-point Gauss quadrature rule for the measure $\mathrm{d}\lambda(t) = e^{-t^2}\mathrm{d}t$ on $[0,\infty]$, obtained first in MATLAB Release R2011a, and then in Release R2014a.

The array of the first 200 recurrence coefficients for the half-range Hermite measure has been computed to 32 digits by the Chebyshev algorithm in sufficiently high precision (see Ex. 2.17). They are available in the file `ab_hrhermite.txt` (cf. Appendix A.5) and can be downloaded by the routine `loadvpa.m` (see Preface). The script

[1] The routine `eig.m` of MATLAB Releases R2012–R2014 may produce inaccurate eigenvectors, hence inaccurate Gaussian weights. See Demo 1.17. Try an earlier or later release.

```
%DEMO1_17
%
f0='%2.0f %22.14e %22.14e %11.4e %11.4e\n';
disp( ...
' k              x                       w                 errx        errw')
dig=32;
sab=loadvpa('ab_hrhermite.txt',3,dig,200,2);
ab=double(sab);
N=40;
xw=gauss(N,ab); sxw=sgauss(dig,N,sab);
errx=double(abs((xw(:,1)-sxw(:,1))./sxw(:,1)));
errw=double(abs((xw(:,2)-sxw(:,2))./sxw(:,2)));
for k=1:N
  fprintf(f0,k,xw(k,1),xw(k,2),errx(k),errw(k))
end
```

produces the desired quadrature rule in MATLAB double precision and in 32-digit arithmetic. It prints the nodes and weights along with the respective relative errors. Run in MATLAB Release R2011a, they are

```
>> Demo1_17
 k            x                        w                  errx         errw
 1    5.15001035639217e-03   1.32096973126011e-02   2.6220e-15   1.5726e-14
 2    2.70780249482919e-02   3.05990537026153e-02   1.5170e-14   9.4384e-15
 3    6.62985802114275e-02   4.75471383164650e-02   5.5954e-15   3.7715e-15
 4    1.22439601598833e-01   6.34790918836327e-02   2.8599e-15   8.9579e-15
 5    1.94978680930686e-01   7.75289198733386e-02   4.4548e-15   2.6322e-14
 6    2.83275589505737e-01   8.85410175114734e-02   5.9474e-15   9.2448e-16
 7    3.86601842217663e-01   9.52301500661025e-02   3.7299e-15   1.5135e-14
 8    5.04170513432849e-01   9.64924486911256e-02   1.1442e-15   4.4846e-16
 9    6.35164593806001e-01   9.18076964714502e-02   6.6275e-16   9.6447e-16
..            ...                      ...                ...          ...

36    7.56638842907305e+00   5.51405346175267e-26   2.7707e-16   5.9699e-15
37    7.98377828755415e+00   9.01565188301616e-29   6.4936e-16   4.5860e-14
38    8.43831154135518e+00   5.69883374891866e-32   2.4736e-16   3.0997e-14
39    8.94984135501073e+00   9.01706601981072e-36   6.1566e-16   4.4237e-14
40    9.57048117889041e+00   1.19587866298430e-40   4.7281e-17   4.8185e-15
>>
```

which, with regard to accuracy, are as good as can be hoped for. Run in MATLAB Release R2014a, however, the output is rather different,

```
>> Demo1_17
 k            x                        w                  errx         errw
 1    5.15001035639240e-03   1.32096973126019e-02   4.2346e-14   3.9560e-14
 2    2.70780249482924e-02   3.05990537026152e-02   2.1275e-15   1.3293e-14
 3    6.62985802114287e-02   4.75471383164649e-02   1.2197e-14   2.8958e-15
 4    1.22439601598833e-01   6.34790918836319e-02   5.9305e-16   4.8152e-15
 5    1.94978680930687e-01   7.75289198733400e-02   3.2658e-16   7.3483e-15
 6    2.83275589505738e-01   8.85410175114743e-02   4.6046e-16   8.9500e-15
```

```
 7    3.86601842217664e-01    9.52301500661011e-02    7.1458e-16    2.7089e-16
 8    5.04170513432849e-01    9.64924486911256e-02    7.0383e-16    3.0464e-16
 9    6.35164593806001e-01    9.18076964714502e-02    3.8601e-16    1.2668e-15
..           ...                     ...                 ...           ...

36    7.56638842907305e+00    5.51356385016377e-26    1.5969e-16    8.8793e-05
37    7.98377828755415e+00    9.01021706331724e-29    1.2938e-16    6.0282e-04
38    8.43831154135518e+00    5.67622760875622e-32    1.7367e-16    3.9668e-03
39    8.94984135501072e+00    8.84790479913369e-36    2.0224e-17    1.8760e-02
40    9.57048117889041e+00    1.01916808839799e-40    1.3833e-16    1.4777e-01
>>
```

very much alike at the beginning, but, with regard to the quadrature weights w_k, dismally worse at the end. On the other hand, the 32-digit results for the weights are exactly the same in both releases, namely,

```
sw =

                                   0.013209697312601351772750775847755
                                   0.030599053702615630770812332934988
                                   0.047547138316464781760687618805642
                                   0.063479091883632155288530663069893
                                    0.077528919873340610577080841009O4
                                   0.088541017511473472909471020490714
                                   0.095230150066101070543322350216431
                                    0.096492448691125552476892237464 19
                                   0.091807696471450131730966057043738
              ...                               ...

          0.0000000000000000000000000551405346175270519011085670043 2
      0.00000000000000000000000000009015651883015744233747299091215
   0.00000000000000000000000000000005698833748918481924375901197454 3
                           9.0170660198111146529987770827383*10^(-36)
                           1.1958786629842989846688513729855*10^(-40)
```

This shows that the relative accuracy of the eigenvectors produced by the double-precision routine `eig.m` is not optimal in the more recent release.

Other frequently used quadrature formulae are those named Gauss–Radau and Gauss–Lobatto, which, like the Gauss formulae, are of maximum algebraic degree of exactness, but have fixed nodes at one, resp., both, endpoints of the interval of integration. They are implemented in the `OPQ` routines `radau.m`. resp., `lobatto.m`. In the case of Jacobi and Laguerre measures, the generation of Gauss–Radau formulae can be simplified (cf. [24, §3.1.1.2], giving rise to additional routines `radau_jacobi.m` and `radau_laguerre.m`. A similarly simplified, but also more robust, routine is `lobatto_jacobi.m` for Gauss–Lobatto formulae in the case of Jacobi measures. Extended quadrature routines of the type Gauss–Kronrod are implemented in `kronrod.m`, and Gauss-type quadrature rules involving derivatives in `turan.m` (cf. [24, §§3.1.2 and 3.1.3]). See Appendices A.2 and B.

Chapter 2

Answers to Exercises on Orthogonal Polynomials

In Section 2.1 we begin with a few elementary formal properties of orthogonal polynomials, with the all-important three-term recurrence relation (Exs. 2.3 and 2.4) occupying center stage. Equally important is the tridiagonal, symmetric *Jacobi matrix* of order n,

$$J_n(d\lambda)=\begin{bmatrix} \alpha_0 & \sqrt{\beta_1} & & & 0 \\ \sqrt{\beta_1} & \alpha_1 & \sqrt{\beta_2} & & \\ & \ddots & \ddots & \ddots & \\ & & \sqrt{\beta_{n-2}} & \alpha_{n-2} & \sqrt{\beta_{n-1}} \\ 0 & & & \sqrt{\beta_{n-1}} & \alpha_{n-1} \end{bmatrix}, \tag{2.1}$$

formed with the first n recurrence coefficients α_k, $0\le k\le n-1$, and the $n-1$ coefficients β_k, $1\le k\le n-1$. Its eigenvalues, for example, are the zeros of the nth-degree orthogonal polynomial $\pi_n(\cdot;d\lambda)$ (cf. Ex. 2.5(c)). Closely related to the three-term recurrence relation is the *Jacobi continued fraction*

$$\mathscr{J}(z;d\lambda)=\frac{\beta_0}{z-\alpha_0-}\,\frac{\beta_1}{z-\alpha_1-}\,\frac{\beta_2}{z-\alpha_2}-\cdots \tag{2.2}$$

(cf. Exs. 2.7 and 2.8), and an important special solution, the *Cauchy transform* of $\pi_n(\cdot;d\lambda)$,

$$\rho_n(z;d\lambda)=\int_{\mathbb{R}}\frac{\pi_n(t;d\lambda)}{z-t}\,d\lambda(t), \quad z\in\mathbb{C},\ z\notin \operatorname{supp}(d\lambda) \tag{2.3}$$

(cf. Exs. 2.9 and 2.10).

The first N polynomials π_n, $n=0,1,\dots,N-1$, orthogonal relative to an absolutely continuous measure $d\lambda$ are also orthogonal with respect to a discrete measure $d\lambda_N$ constructed by the N-point Gaussian quadrature formula for $d\lambda$; see Ex. 2.11.

Useful simplifications and implications of symmetry (Exs. 2.12 and 2.13*) are discussed in Section 2.2, and in Section 2.3 an example is given of a symmetric weight function supported on two separate intervals (Ex. 2.14*).

The classical theory of orthogonal polynomials rests solidly on the *moments*

$$\mu_k=\int_{\mathbb{R}} t^k d\lambda(t), \quad k=0,1,2,\dots, \tag{2.4}$$

of the underlying measure. They are also useful for computation, but require high-precision arithmetic. Indeed, there is a uniquely defined (but ill-conditioned) mapping from the first $2n$ moments μ_k, $k=0,1,2,\dots,2n-1$, to the first n recurrence coefficients α_k, β_k, $k=0,1,2,\dots,n-1$, which is implemented by the *Chebyshev algorithm*. Several applications of the algorithm are given in Exs. 2.17–2.23. Modified moments

$$m_k = \int_{\mathbb{R}} p_k(t)\mathrm{d}\lambda(t), \tag{2.5}$$

where p_k is a polynomial of degree k—typically orthogonal with respect to some other measure $\mathrm{d}\ell$—give rise to the *modified Chebyshev algorithm* implementing the mapping $\{m_k\}_{k=0}^{2n-1} \mapsto \{\alpha_k, \beta_k\}_{k=0}^{n-1}$ (cf. Exs. 2.15 and 2.16*).

Another classical approach to constructing orthogonal polynomials, or orthogonal functions in general, is the *Gram-Schmidt procedure*, which consists in starting with a linearly independent set of functions and successively orthogonalizing them. In the case of polynomials, we may start with the system of powers, $\{t^k\}$, which are certainly linearly independent, and thus proceed as follows:

$$\pi_0(t) = 1,$$
$$\pi_k(t) = t^k - \sum_{\ell=0}^{k-1} c_{k\ell}\pi_\ell(t), \quad k=1,2,3,\dots,$$

where orthogonality to all preceding polynomials $\pi_0, \pi_1, \dots, \pi_{k-1}$ is achieved by taking

$$c_{k\ell} = \frac{(t^k, \pi_\ell)_{\mathrm{d}\lambda}}{(\pi_\ell, \pi_\ell)_{\mathrm{d}\lambda}}, \quad \ell = 0,1,\dots,k-1,$$

where

$$(u,v)_{\mathrm{d}\lambda} = \int_{\mathbb{R}} u(t)v(t)\mathrm{d}\lambda(t).$$

All denominators are positive since the inner product is positive definite, that is, $(u,u)_{\mathrm{d}\lambda} > 0$ for all polynomials $u \in \mathbb{P}$, $u \not\equiv 0$, by assumption.

A similar, but computationally more attractive, approach is to generate the three-term recurrence relation (Exs. 2.3 and 2.4), using Darboux's formulae (cf. Ex. 2.3(a)) for the coefficients in tandem with the recurrence relation: since $\pi_0 = 1$, the Darboux formula for $k=0$ gives α_0, and this in turn π_1. Both Darboux formulae for $k=1$ then yield α_1 and β_1, and these in turn π_2. Evidently, we can continue this *ad infinitum* for $k=2,3,\dots$. This is an idea going back to Stieltjes (cf. Section 1.3.1) and forms the basis of what is now called the *Stieltjes procedure*. Approximating the inner products in Darboux's formulae by a discrete inner product gives rise to *discretized Stieltjes procedures* and can be extended into powerful *multicomponent discretization procedures*. These are implemented in the MATLAB function `mcdis.m` and its symbolic analogue `smcdis.m`. Such algorithms based on discretization constitute a second major class of methods for computing orthogonal polynomials. Section 2.5 illustrates their use in a number of examples (Exs. 2.34–2.38).

Algorithms for modifying a measure $\mathrm{d}\lambda$ by a quadratic factor (Ex. 2.39) or by polynomial divisors (Exs. 2.40–2.43*), finally, are the subject of Section 2.6.

2.1 ▪ Elementary properties

Exercise 2.1. (Ex. 1 of [27, §2].) *Explain why, under the assumptions made in the preface, the inner product $(p,q)_{\mathrm{d}\lambda} = \int_{\mathbb{R}} p(t)q(t)\mathrm{d}\lambda(t)$ of two polynomials p, q is well defined.*

Answer. Using the linearity of inner products, the inner product $(p,q)_{d\lambda}$ is readily expressed in terms of the coefficients of p and q, and in terms of the moments μ_r,

$$\begin{aligned}(p,q)_{d\lambda} &= \left(\sum_{k=0}^{n} c_k t^k, \sum_{\ell=0}^{m} d_\ell t^\ell\right) \\ &= \sum_{k=0}^{n}\sum_{\ell=0}^{m} c_k c_\ell (t^k, t^\ell) = \sum_{k=0}^{n}\sum_{\ell=0}^{m} c_k c_\ell \int_{\mathbb{R}} t^{k+\ell}\, d\lambda(t) \\ &= \sum_{k=0}^{n}\sum_{\ell=0}^{m} c_k c_\ell\, \mu_{k+\ell}.\end{aligned}$$

Exercise 2.2. (Ex. 2 of [27, §2].) *Show that monic orthogonal polynomials relative to an absolutely continuous measure are uniquely defined.* {Hint: Use Gram–Schmidt orthogonalization.} *Discuss the uniqueness in the case of discrete measures.*

Answer. An absolutely continuous measure is positive definite on $\mathbb{P}$; hence there exists a unique sequence of monic orthogonal polynomials that can be obtained by Gram–Schmidt orthogonalization of the sequence of powers t^k, $k = 0, 1, 2, \ldots$ (cf. [24, Theorem 1.6]). A discrete N-point measure is positive definite on $\mathbb{P}_{N-1}$, but not on $\mathbb{P}_N$; hence there exists a unique sequence of N, but not more, monic orthogonal polynomials (cf. [24, Theorem 1.7]).

Exercise 2.3. (Ex. 3 of [27, §2].) Three-term recurrence relation for the monic orthogonal polynomials $\pi_k(t) = \pi_k(t; d\lambda)$.

(a) *Derive the three-term recurrence relation*

$$\begin{gathered}\pi_{k+1}(t) = (t-\alpha_k)\pi_k(t) - \beta_k \pi_{k-1}(t), \quad k = 0, 1, \ldots, n-1, \\ \pi_{-1}(t) = 0, \quad \pi_0(t) = 1.\end{gathered} \tag{2.6}$$

In particular, derive Darboux's formula *for the coefficients* $\alpha_k = \alpha_k(d\lambda)$, $\beta_k = \beta_k(d\lambda)$,

$$\begin{aligned}\alpha_k(d\lambda) &= \frac{(t\pi_k, \pi_k)_{d\lambda}}{(\pi_k, \pi_k)_{d\lambda}}, \quad k = 0, 1, 2, \ldots, \\ \beta_k(d\lambda) &= \frac{(\pi_k, \pi_k)_{d\lambda}}{(\pi_{k-1}, \pi_{k-1})_{d\lambda}}, \quad k = 1, 2, \ldots.\end{aligned} \tag{2.7}$$

(b) *Express* $\|\pi_k\|^2_{d\lambda}$ *in terms of the* β*-coefficients in* (a).

Answer

(a) Since $\pi_{k+1}(t) - t\pi_k(t)$ is a polynomials of degree $\le k$ and the polynomials $\pi_0, \pi_1, \ldots, \pi_k$ are linearly independent (cf. the answer to Ex. 3.2(a)), it can be written, expanded in the orthogonal polynomials π_j, as

$$\pi_{k+1}(t) - t\pi_k(t) = -\alpha_k \pi_k(t) - \beta_k \pi_{k-1}(t) + \sum_{j=0}^{k-2} \gamma_{kj} \pi_j(t) \tag{2.8}$$

for certain constants α_k, β_k, and γ_{kj}, where $\pi_{-1}(t) = 0$ (for $k = 0$) and empty sums (for $k = 0$ and 1) are understood to be zero. Taking the inner product of both sides with π_k

yields, by orthogonality,

$$-(t\pi_k,\pi_k) = -\alpha_k(\pi_k,\pi_k),$$

that is,

$$\alpha_k = \frac{(t\pi_k,\pi_k)}{(\pi_k,\pi_k)}.$$

This proves the first relation in (2.7). The second is obtained by taking the inner product with π_{k-1} ($k \geq 1$),

$$-(t\pi_k,\pi_{k-1}) = -\beta_k(\pi_{k-1},\pi_{k-1}),$$

and noting that $(t\pi_k,\pi_{k-1}) = (\pi_k, t\pi_{k-1}) = (\pi_k, \pi_k + \cdots)$, where dots stand for a polynomial of degree $< k$. Thus, again by orthogonality, $(t\pi_k,\pi_{k-1}) = (\pi_k,\pi_k)$, giving

$$\beta_k = \frac{(\pi_k,\pi_k)}{(\pi_{k-1},\pi_{k-1})}.$$

Finally, taking the inner product of both sides of (2.8) with π_i, $i < k-1$, yields

$$-(t\pi_k,\pi_i) = \gamma_{ki}(\pi_i,\pi_i).$$

It is at this point where the property $(tp,q) = (p,tq)$ plays out its crucial role: one has $(t\pi_k,\pi_i) = (\pi_k, t\pi_i)$, which vanishes by orthogonality, since $t\pi_i \in \mathbb{P}_{k-1}$. Thus, $\gamma_{ki} = 0$ for $i < k-1$. The coefficient β_0 is arbitrary because it multiplies $\pi_{-1}(t) = 0$ in (2.6). It is conveniently defined, however, by $\beta_0 = \int_{\mathbb{R}} \mathrm{d}\lambda(t)$ (cf., e.g., part (b) of this exercise and Ex. 2.7).

(b) By the second relation in (2.7) we have

$$\|\pi_k\|_{\mathrm{d}\lambda}^2 = \beta_k \|\pi_{k-1}\|_{\mathrm{d}\lambda}^2, \quad k = 1,2,\ldots.$$

Repeated application of this relation then yields

$$\|\pi_k\|_{\mathrm{d}\lambda}^2 = \beta_k\beta_{k-1}\cdots\beta_1 \|\pi_0\|_{\mathrm{d}\lambda}^2.$$

If we define $\beta_0 = \int_{\mathbb{R}} \mathrm{d}\lambda(t) = (1,1)_{\mathrm{d}\lambda} = \|\pi_0\|_{\mathrm{d}\lambda}^2$, we get

$$\|\pi_k\|_{\mathrm{d}\lambda}^2 = \beta_0\beta_1\cdots\beta_k. \tag{2.9}$$

Exercise 2.4. (Ex. 4 of [27, §2].) *Derive the three-term recurrence relation for the orthonormal polynomials,*

$$\begin{gathered}\sqrt{\beta_{k+1}}\,\tilde{\pi}_{k+1}(t) = (t-\alpha_k)\tilde{\pi}_k(t) - \sqrt{\beta_k}\,\tilde{\pi}_{k-1}(t), \quad k = 0,1,2,\ldots, \\ \tilde{\pi}_{-1}(t) = 0, \quad \tilde{\pi}_0(t) = 1/\sqrt{\beta_0}.\end{gathered} \tag{2.10}$$

Answer. With $\tilde{\pi}_k$ denoting the orthonormal polynomials, one has

$$\pi_k(t) = \|\pi_k\|\tilde{\pi}_k(t), \quad k = 0,1,2,\ldots. \tag{2.11}$$

Inserting this into the three-term recurrence relation (2.6) of Ex. 2.3, dividing by $||\pi_{k+1}||$, and noting the second relation in (2.7) yields

$$\tilde{\pi}_{k+1}(t) = (t-\alpha_k)\frac{\tilde{\pi}_k(t)}{\sqrt{\beta_{k+1}}} - \beta_k \frac{\tilde{\pi}_{k-1}(t)}{\sqrt{\beta_{k+1}\beta_k}}.$$

Multiplying through by $\sqrt{\beta_{k+1}}$ gives the three-term recurrence relation (2.10). The starting values follow from those in (2.6), noting from (2.11) and (2.9) with $k=0$ that $1 = \sqrt{\beta_0}\tilde{\pi}_0$.

Exercise 2.5. (Ex. 5 of [27, §2].) (a) *With $\tilde{\pi}_k$ denoting the orthonormal polynomials relative to a measure* $\mathrm{d}\lambda$, *show that for integers $\ell \geq k > 0$*

$$\int_{\mathbb{R}} t\tilde{\pi}_k(t)\tilde{\pi}_\ell(t)\mathrm{d}\lambda(t) = \begin{cases} 0 & \text{if } \ell > k+1, \\ \sqrt{\beta_{k+1}} & \text{if } \ell = k+1, \\ \alpha_k & \text{if } \ell = k, \end{cases}$$

where $\alpha_k = \alpha_k(\mathrm{d}\lambda)$, $\beta_k = \beta_k(\mathrm{d}\lambda)$.

(b) *Use* (a) *to prove that the Jacobi matrix* (2.1) *is*

$$J_n = J_n(\mathrm{d}\lambda) = \int_{\mathbb{R}} t\boldsymbol{p}(t)\boldsymbol{p}^{\mathrm{T}}(t)\mathrm{d}\lambda(t),$$

where $\boldsymbol{p}(t)^{\mathrm{T}} = [\tilde{\pi}_0(t), \tilde{\pi}_1(t), \ldots, \tilde{\pi}_{n-1}(t)]$.

(c) *With notation as in* (b), *prove*

$$t\boldsymbol{p}(t) = J_n\boldsymbol{p}(t) + \sqrt{\beta_n}\tilde{\pi}_n(t)\boldsymbol{e}_n,$$

where $\boldsymbol{e}_n = [0,0,\ldots,1]^{\mathrm{T}} \in \mathbb{R}^n$. In particular, the zeros τ_ν of π_n are the eigenvalues of J_n, and $\boldsymbol{p}(\tau_\nu)$ corresponding eigenvectors.

Answer

(a) The proof follows from (cf. Ex. 2.4)

$$t\tilde{\pi}_k(t) = \sqrt{\beta_{k+1}}\,\tilde{\pi}_{k+1}(t) + \alpha_k\tilde{\pi}_k(t) + \sqrt{\beta_k}\,\tilde{\pi}_{k-1}(t), \quad k \geq 0,$$

multiplying by $\tilde{\pi}_\ell(t)$, integrating, and using orthonormality.

(b) We have

$$\int_{\mathbb{R}} t\boldsymbol{p}(t)\boldsymbol{p}^{\mathrm{T}}(t)\mathrm{d}\lambda(t) = \left[\int_{\mathbb{R}} t\tilde{\pi}_k(t)\tilde{\pi}_\ell(t)\mathrm{d}\lambda(t)\right]_{k,\ell=0,1,\ldots,n-1}. \tag{2.12}$$

By (a), the upper triangular part of the matrix on the right ($\ell \geq k$) is

$$\begin{bmatrix} \alpha_0 & \sqrt{\beta_1} & & & 0 \\ & \alpha_1 & \sqrt{\beta_2} & & \\ & & \ddots & \ddots & \\ & & & \alpha_{n-2} & \sqrt{\beta_{n-1}} \\ 0 & & & & \alpha_{n-1} \end{bmatrix}.$$

Since the matrix on the left in (2.12) is symmetric, so is the matrix on the right, and hence it is equal to $\boldsymbol{J}_n(d\lambda)$.

(c) The first n recurrence relations in (2.10) of Ex. 2.4, solving for $t\tilde{\pi}_k(t)$, $k=0,1,\dots,n-1$, can be given the form

$$t\boldsymbol{p}(t)=\boldsymbol{J}_n\boldsymbol{p}(t)+\sqrt{\beta_n}\tilde{\pi}_n(t)\boldsymbol{e}_n,$$

where the last term on the right arises only when $k=n-1$; hence $\boldsymbol{e}_n=[0,0,\dots,1]^{\mathrm{T}}\in\mathbb{R}^n$.

Exercise 2.6. (Ex. 8 of [27, §2].) The recurrence relation, in MATLAB, of the Chebyshev polynomials U_k of the second kind.

(a) *Using MATLAB, compute* $U_k(x)$ *for* $1\le k\le N$ *either by means of the three-term recurrence relation* $U_{n+1}(x)=2xU_n(x)-U_{n-1}(x)$ *for* $n=0,1,\dots,N-1$ *(where* $U_{-1}(x)=0$, $U_0(x)=1$*), or else by putting* $n=1:N$ *in the explicit formula* $U_n(\cos\theta)=\sin(n+1)\theta/\sin\theta$, *where* $x=\cos\theta$. *For selected values of* x *and* N, *determine which of the two methods, by timing each, is more efficient.*

(b) *Using MATLAB, compute the single value* $U_N(x)$ *either by use of the three-term recurrence relation, or by direct computation based on the trigonometric formula for* $U_N(\cos\theta)$. *For selected values of* x *and* N, *determine which of the two methods, by timing each, is more efficient.*

Answer

(a) The tasks indicated in (a) are carried out by the script `cheb2a.m`,

```
%CHEB2A

tic
for i=1:10000
  for x=[-.9 -.5 -.1 .1 .5 .9]
    for N=[10 25 50 100 250 500]
      Uk=0; U=1;
      for k=1:N
        Ukm1=Uk; Uk=U;
        U=2*x*Uk-Ukm1;
      end
    end
  end
end
trec=toc
tic
for i=1:10000
  for x=[-.9 -.5 -.1 .1 .5 .9]
    for N=[10 25 50 100 250 500]
      th=acos(x); n=1:N;
      U=sin((n+1)*th)/sin(th);
    end
  end
end
texpl=toc
```

The output of CHEB2A,

```
>> cheb2a
trec =
    1.886697
texpl =
    7.988459
```

shows that, for a sequence of Chebyshev polynomials, the recurrence relation is about four times faster.

(b) The tasks indicated in (b) are carried out by the script cheb2b.m,

```
%CHEB2B

tic
for i=1:10000
  for x=[-.9 -.5 -.1 .1 .5 .9]
    for N=[10 25 50 100 250 500]
      Uk=0; U=1;
      for k=1:N
        Ukm1=Uk; Uk=U;
        U=2*x*Uk-Ukm1;
      end
    end
  end
end
trec=toc
tic
for i=1:10000
  for x=[-.9 -.5 -.1 .1 .5 .9]
    for N=[10 25 50 100 250 500]
      th=acos(x);
      U=sin((N+1)*th)/sin(th);
   end
  end
end
texpl=toc
```

The output of CHEB2B,

```
>> cheb2b
trec =
    1.880321
texpl =
    1.032935
>>
```

shows that, for a single value of the Chebyshev polynomial, the recurrence relation is about two times slower.

Exercise 2.7. (Ex. 13 of [27, §2].) The Jacobi continued fraction (2.2).

Use the theory of continued fractions to prove that the nth convergent of $\mathcal{J}(z;d\lambda)$ *is*

$$\frac{\beta_0}{z-\alpha_0-}\,\frac{\beta_1}{z-\alpha_1-}\cdots\frac{\beta_{n-1}}{z-\alpha_{n-1}}=\frac{\sigma_n(z;d\lambda)}{\pi_n(z;d\lambda)},\quad n=1,2,3,\ldots, \tag{2.13}$$

where π_n *is the monic orthogonal polynomial of degree n, and* σ_n *a polynomial of degree* $n-1$ *satisfying the same basic three-term recurrence relation as the one for the* π_k*, but with different starting values,*

$$\begin{gathered}\sigma_{k+1}(z)=(z-\alpha_k)\sigma_k(z)-\beta_k\sigma_{k-1}(z),\quad k=1,2,3,\ldots,\\ \sigma_0(z)=0,\quad \sigma_1(z)=\beta_0.\end{gathered} \tag{2.14}$$

Answer. Let

$$C=b_0+\frac{a_1}{b_1+}\,\frac{a_2}{b_2+}\,\frac{a_3}{b_3+}\cdots$$

be an infinite continued fraction, and

$$C_n=b_0+\frac{a_1}{b_1+}\cdots\frac{a_n}{b_n}=\frac{A_n}{B_n}$$

its nth convergent. From the elementary theory of continued fractions, it is known that

$$\left.\begin{aligned}A_n&=b_nA_{n-1}+a_nA_{n-2}\\ B_n&=b_nB_{n-1}+a_nB_{n-2}\end{aligned}\right\}\quad n=1,2,3,\ldots,$$

where

$$A_{-1}=1,\ A_0=b_0;\quad B_{-1}=0,\ B_0=1.$$

The continued fraction C becomes the Jacobi continued fraction if we identify

$$\begin{aligned}a_1&=\beta_0,\quad a_\nu=-\beta_{\nu-1},\quad \nu\geq 2,\\ b_0&=0,\quad b_\nu=z-\alpha_{\nu-1},\quad \nu\geq 1.\end{aligned}$$

If we write $A_n=\sigma_n$ and $B_n=\tau_n$, then by the relations above we have

$$\sigma_{-1}=1,\ \sigma_0=0;\quad \tau_{-1}=0,\ \tau_0=1 \tag{2.15}$$

and

$$\sigma_1=(z-\alpha_0)b_0+\beta_0=\beta_0,\ \tau_1=(z-\alpha_0)\tau_0+\beta_0\tau_{-1}=z-\alpha_0,$$

$$\left.\begin{aligned}\sigma_n&=(z-\alpha_{n-1})\sigma_{n-1}-\beta_{n-1}\sigma_{n-2}\\ \tau_n&=(z-\alpha_{n-1})\tau_{n-1}-\beta_{n-1}\tau_{n-2}\end{aligned}\right\}\quad n=2,3,\ldots.$$

Letting $n=k+1$, the last pair of relations becomes

$$\left.\begin{aligned}\sigma_{k+1}&=(z-\alpha_k)\sigma_k-\beta_k\sigma_{k-1}\\ \tau_{k+1}&=(z-\alpha_k)\tau_k-\beta_k\tau_{k-1}\end{aligned}\right\}\quad k=1,2,3,\ldots, \tag{2.16}$$

while, for $k=0$,

$$\sigma_1=\beta_0,\quad \tau_1=z-\alpha_0. \tag{2.17}$$

Since $\tau_{-1}=0$, the second relation in (2.16) also holds for $k=0$. This shows that $\tau_k=\pi_k(z;\mathrm{d}\lambda)$ for $k=-1,0,1,2,\ldots$, and, if we denote $\sigma_k=\sigma_k(z;\mathrm{d}\lambda)$, the relations derived for σ_k are exactly those for $\sigma_k(z;\mathrm{d}\lambda)$ claimed in the exercise.

The first relation in (2.16), in view of (2.15) and (2.17), holds also for $k=0$, but only if we redefine $\sigma_{-1}=-1$.

Exercise 2.8. (Ex. 14 of [27, §2].) *Show that the numerator $\sigma_n(z;\mathrm{d}\lambda)$ in the nth convergent of the Jacobi continued fraction* (see Ex. 2.7) *admits the integral representation*

$$\sigma_n(z)=\int_{\mathbb{R}}\frac{\pi_n(z)-\pi_n(t)}{z-t}\,\mathrm{d}\lambda(t),\quad n=0,1,2,\ldots. \tag{2.18}$$

Answer. Let the integral on the right of (2.18) be denoted by $s_n=s_n(z)$. Then, trivially, $s_0=0$, and $s_1=\beta_0$ since $\pi_1(z)-\pi_1(t)=z-\alpha_0-(t-\alpha_0)=z-t$. Moreover, since

$$\begin{aligned}\pi_{k+1}(z)&-\pi_{k+1}(t)\\&=(z-\alpha_k)\pi_k(z)-\beta_k\pi_{k-1}(z)-[(t-\alpha_k)\pi_k(t)-\beta_k\pi_{k-1}(t)]\\&=z\pi_k(z)-t\pi_k(t)-\alpha_k[\pi_k(z)-\pi_k(t)]-\beta_k[\pi_{k-1}(z)-\pi_{k-1}(t)]\end{aligned}$$

and

$$z\pi_k(z)-t\pi_k(t)=z[\pi_k(z)-\pi_k(t)]+(z-t)\pi_k(t),$$

we have for $k>0$, dividing by $z-t$ and integrating,

$$s_{k+1}=zs_k+\int_{\mathbb{R}}\pi_k(t)\mathrm{d}\lambda(t)-\alpha_k s_k-\beta_k s_{k-1}.$$

But the integral on the right is zero by orthogonality, so that

$$s_{k+1}=(z-\alpha_k)s_k-\beta_k s_{k-1},\quad k=1,2,3,\ldots.$$

This is precisely the recurrence relation satisfied by $\sigma_n(z)$ (cf. (2.14), and since the initial values $s_0=0$, $s_1=\beta_0$ are also the same, there follows $s_k=\sigma_k(z)$ for $k=0,1,2,\ldots$.

Exercise 2.9. (Ex. 15 of [27, §2].) *Let $\rho_n(z)=\rho_n(z;\mathrm{d}\lambda)$* (cf. (4.13)), *$\rho_{-1}(z)=1$, and $\pi_n(z)=\pi_n(z;\mathrm{d}\lambda)$. Show that*

$$\lim_{n\to\infty}\frac{\rho_n(z)}{\pi_n(z)}=0$$

(cf. [27, Eq. (2.30)]) *implies $\lim_{n\to\infty}\rho_n/y_n=0$, where y_n is any solution of the three-term recurrence relation (satisfied by ρ_n and π_n; see* (2.14) *and* (2.16)) *which is linearly independent of ρ_n.* Thus, $\{\rho_n\}$ is indeed a minimal solution as claimed in Ex. 15 of [27, §2].

Answer. Since

$$\begin{vmatrix}\pi_0 & \pi_{-1}\\ \rho_0 & \rho_{-1}\end{vmatrix}=1\neq 0,$$

the two solutions $\{\pi_n\}$, $\{\rho_n\}$ of the three-term recurrence relation (cf. Theorem 1.43 of [24]) are linearly independent. Therefore, any solution $\{y_n\}$ not a constant multiple of ρ_n must be of the form

$$y_n = c_1\pi_n + c_2\rho_n, \quad c_1 \neq 0.$$

There follows

$$\frac{\rho_n}{y_n} = \frac{\rho_n}{c_1\pi_n + c_2\rho_n} = \frac{1}{c_1(\pi_n/\rho_n) + c_2} \to 0,$$

since $c_1 \neq 0$ and $\pi_n/\rho_n \to \infty$ by our premise.

Exercise 2.10. (Ex. 16 of [27, §2].) *Show that the minimal solutions of a three-term recurrence relation form a one-dimensional manifold.*

Answer. Suppose there are two linearly independent minimal solutions, $f_1(n)$ and $f_2(n)$. The minimality of f_1 implies $f_1(n)/f_2(n) \to 0$ and the one of f_2 implies $f_2(n)/f_1(n) \to 0$, a contradiction. Thus, the assumed linear independence cannot be true.

Exercise 2.11. (Ex. 4 of [27, §4].) *Let* $\pi_k(\cdot;\mathrm{d}\lambda)$, $k = 0,1,2,\ldots$, *be the orthogonal polynomials relative to an absolutely continuous measure* $\mathrm{d}\lambda$. *Show that for each* $N \geq 2$, *the first* N *of them are orthogonal with respect to the discrete inner product*

$$(p,q)_N = \sum_{\nu=0}^{N-1} \lambda_\nu^G p(\tau_\nu^G) q(\tau_\nu^G),$$

where τ_ν^G, λ_ν^G *are the nodes and weights of the* N*-point Gauss formula for* $\mathrm{d}\lambda$. *Moreover,* $\|\pi_k\|_N^2 = \|\pi_k\|_{\mathrm{d}\lambda}^2$ *for* $k \leq N-1$.

Answer. Let $N \geq 2$, $k,\ell \leq N-1$, and put $\pi_k(\cdot;\mathrm{d}\lambda) = \pi_k(\cdot)$. Then, since $\pi_k\pi_\ell \in \mathbb{P}_{2N-2}$, using N-point Gauss quadrature for the measure $\mathrm{d}\lambda$ (having nodes and weights τ_ν^G, λ_ν^G), we have for $k \neq \ell$

$$0 = (\pi_k, \pi_\ell)_{\mathrm{d}\lambda} = \int_{\mathbb{R}} \pi_k(t)\pi_\ell(t)\mathrm{d}\lambda(t) = \sum_{\nu=1}^{N} \lambda_\nu^G \pi_k(\tau_\nu^G)\pi_\ell(\tau_\nu^G) = (\pi_k, \pi_\ell)_N$$

and

$$\|\pi_k\|_{\mathrm{d}\lambda}^2 = \int_{\mathbb{R}} \pi_k^2(t)\mathrm{d}\lambda(t) = \sum_{\nu=1}^{N} \lambda_\nu^G \pi_k^2(\tau_\nu^G) = \|\pi_k\|_N^2,$$

proving both assertions.

2.2 ▪ Symmetry

Exercise 2.12. (Ex. 6 of [27, §2].) *Let* $\mathrm{d}\lambda(t) = w(t)\mathrm{d}t$ *be symmetric on* $[-a,a]$, $a > 0$, *that is,* $w(-t) = w(t)$ *on* $[-a,a]$. *Show that* $\alpha_k(\mathrm{d}\lambda) = 0$ *for all* $k \geq 0$.

Answer. We first show that π_k is an even (odd) polynomial if k is even (odd). For this, let $p_k(t):=(-1)^k\pi_k(-t)$. Clearly, p_k is monic, and

$$\int_{-a}^{a} p_k(t)p_\ell(t)\mathrm{d}\lambda(t)=\int_{-a}^{a} p_k(-\tau)p_\ell(-\tau)w(-\tau)\mathrm{d}\tau$$
$$=(-1)^{k+\ell}\int_{-a}^{a}\pi_k(\tau)\pi_\ell(\tau)w(\tau)\mathrm{d}\tau=0,\quad k\neq\ell.$$

By uniqueness of monic orthogonal polynomials, there follows $p_k(t)\equiv\pi_k(t)$, and hence $\pi_k(-t)=(-1)^k\pi_k(t)$ as claimed. In particular, π_k^2 is even, and therefore

$$(t\pi_k,\pi_k)_{\mathrm{d}\lambda}=\int_{-a}^{a} t\pi_k^2(t)w(t)\mathrm{d}t=0,$$

that is, $\alpha_k=0$ by the first relation in (2.7).

Exercise 2.13.* (Ex. 7* of [27, §2].) *Let* $\mathrm{d}\lambda(t)=w(t)\mathrm{d}t$ *be symmetric in the sense of* Ex. 2.12.

(a) *Show that*

$$\pi_{2k}(t;\mathrm{d}\lambda)=\pi_k^+(t^2),\quad \pi_{2k+1}(t;\mathrm{d}\lambda)=t\pi_k^-(t^2),$$

where $\pi_k^\pm$ *are the monic polynomials orthogonal on* $[0,a^2]$ *with respect to* $\mathrm{d}\lambda^\pm(t)=t^{\mp 1/2}$ $w(t^{1/2})\mathrm{d}t$.

(b) *Let* (cf. Ex. 2.12)

$$\pi_{k+1}(t)=t\pi_k(t)-\beta_k\pi_{k-1}(t),\quad k=0,1,2,\ldots,$$
$$\pi_{-1}(t)=0,\quad \pi_0(t)=1,$$

be the recurrence relation for $\{\pi_k(\cdot;\mathrm{d}\lambda)\}$, *and let* $\alpha_k^\pm$, $\beta_k^\pm$ *be the recurrence coefficients for* $\{\pi_k^\pm\}$. *Show that*

$$\beta_1=\alpha_0^+,$$
$$\left.\begin{aligned}\beta_{2k}&=\beta_k^+/\beta_{2k-1}\\ \beta_{2k+1}&=\alpha_k^+-\beta_{2k}\end{aligned}\right\}\ k=1,2,3,\ldots.$$

(c) *Derive relations similar to those in* (b) *which involve* α_0^+ *and* α_k^-, β_k^-.

(d) *Write a MATLAB program that checks the numerical stability of the nonlinear recursions in* (b) *and* (c) *when* $\{\pi_k\}$ *are the monic Legendre polynomials.*

Answer

(a) The polynomial $\pi_{2k}(t)=\pi_{2k}(t;\mathrm{d}\lambda)$ is an even monic polynomial of degree $2k$; hence a monic polynomial π_k^+ of degree k in t^2. Since

$$\int_{-a}^{a}\pi_{2k}(t)\pi_{2\ell}(t)w(t)\mathrm{d}t=0,\quad k\neq\ell,$$

and $\pi_{2k}(t)=\pi_k^+(t^2)$, we have

$$0=\int_{-a}^{a}\pi_k^+(t^2)\pi_\ell^+(t^2)w(t)\mathrm{d}t=2\int_0^{a}\pi_k^+(t^2)\pi_\ell^+(t^2)w(t)\mathrm{d}t$$
$$=\int_0^{a^2}\pi_k^+(\tau)\pi_\ell^+(\tau)\frac{w(\sqrt{\tau})}{\sqrt{\tau}}\mathrm{d}\tau,\quad k\neq\ell,$$

proving the claim for $\mathrm{d}\lambda^+$. The one for $\mathrm{d}\lambda^-$ is proved similarly.

(b) From

$$\pi_{k+1}^+(t)=(t-\alpha_k^+)\pi_k^+(t)-\beta_k^+\pi_{k-1}^+(t), \quad k=0,1,2,\dots,$$
$$\pi_{-1}^+(t)=0, \ \ \pi_0^+(t)=1,$$

replacing t by t^2 and using the results of (a), we get, when $k=0$,

$$\pi_2(t)=t^2-\alpha_0^+, \tag{2.19}$$

and, for $k=1,2,3,\dots$,

$$\pi_{2k+2}(t)=(t^2-\alpha_k^+)\pi_{2k}(t)-\beta_k^+\pi_{2k-2}(t). \tag{2.20}$$

On the other hand,

$$\pi_2(t)=t\pi_1(t)-\beta_1=t^2-\beta_1, \tag{2.21}$$

and, for $k\geq 1$,

$$\begin{aligned}\pi_{2k+2}(t) &= t\pi_{2k+1}(t)-\beta_{2k+1}\pi_{2k}(t)\\ &= t[t\pi_{2k}(t)-\beta_{2k}\pi_{2k-1}(t)]-\beta_{2k+1}\pi_{2k}(t)\\ &= (t^2-\beta_{2k+1})\pi_{2k}(t)-\beta_{2k}t\pi_{2k-1}(t).\end{aligned}$$

Therefore, since

$$t\pi_{2k-1}(t)=\pi_{2k}(t)+\beta_{2k-1}\pi_{2k-2}(t),$$

we get

$$\pi_{2k+2}(t)=(t^2-\beta_{2k+1})\pi_{2k}(t)-\beta_{2k}[\pi_{2k}(t)+\beta_{2k-1}\pi_{2k-2}(t)],$$

that is,

$$\pi_{2k+2}(t)=(t^2-\beta_{2k+1}-\beta_{2k})\pi_{2k}(t)-\beta_{2k}\beta_{2k-1}\pi_{2k-2}(t). \tag{2.22}$$

Subtracting (2.19) from (2.21) gives

$$\alpha_0^+-\beta_1=0,$$

and subtracting (2.20) from (2.22) gives

$$(\alpha_k^+-\beta_{2k+1}-\beta_{2k})\pi_{2k}(t)+(\beta_k^+-\beta_{2k}\beta_{2k-1})\pi_{2k-2}(t)=0;$$

hence, comparing the coefficients of t^{2k} and t^{2k-2} on the left-hand and right-hand sides, first

$$\alpha_k^+-\beta_{2k+1}-\beta_{2k}=0, \quad k\geq 1,$$

and then

$$\beta_k^+-\beta_{2k}\beta_{2k-1}=0, \quad k\geq 1.$$

Altogether,

$$\left.\begin{aligned}\beta_1 &= \alpha_0^+,\\ \beta_{2k} &= \beta_k^+/\beta_{2k-1}\\ \beta_{2k+1} &= \alpha_k^+-\beta_{2k}\end{aligned}\right\} \quad k=1,2,3,\dots. \tag{2.23}$$

This proves the assertion.

(c) Starting with the three-term recurrence relation for the π_k^-, a procedure similar to the one in (b) yields

$$\left.\begin{aligned} \beta_1 &= \alpha_0^+ \\ \beta_{2k} &= \alpha_{k-1}^- - \beta_{2k-1} \\ \beta_{2k+1} &= \beta_k^- / \beta_{2k} \end{aligned}\right\} \quad k = 1, 2, 3, \ldots. \tag{2.24}$$

(d) For the Legendre measure $d\lambda$ on $[-1, 1]$ the measures $d\lambda^+(t) = t^{-1/2}dt$, $d\lambda^-(t) = t^{1/2}dt$ are Jacobi measures on $[0, 1]$ with Jacobi parameters $\alpha = 0$ and $\beta = -1/2$, $\beta = 1/2$, respectively. The nonlinear recursions in (2.23) and (2.24) are implemented respectively in the MATLAB functions `legendre_plus.m` and `legendre_minus.m`, and the stability test in the script `stab_test.m`:

```
function y=legendre_plus(n)
%LEGENDRE_PLUS Recurrence coefficients for Legendre polynomials.
%   LEGENDRE_PLUS(N) generates the first N recurrence
%   coefficients BETA_K for Legendre polynomials from (2.23).

abj=r_jacobi01(floor(n/2)+1,0,-1/2);
ablp=zeros(n,2); ablp(1,2)=abj(1,2);
ablp(2,2)=abj(1,1);
for k=1:floor(n/2)
  if 2*k+1 <= n
    ablp(2*k+1,2)=abj(k+1,2)/ablp(2*k,2);
  end
  if 2*k+2 <= n
    ablp(2*k+2,2)=abj(k+1,1)-ablp(2*k+1,2);
  end
end
y=ablp;

function y=legendre_minus(n)
%LEGENDRE_MINUS Recurrence coefficients for Legendre polynomials.
%   LEGENDRE_MINUS(N) generates the first N recurrence
%   coefficients BETA_K for Legendre polynomials from (2.24).

abj0=r_jacobi01(1,0,-1/2); a0p=abj0(1,1);
abj=r_jacobi01(floor(n/2),0,1/2);
ablm=zeros(n,2); ablm(1,2)=2;
ablm(2,2)=a0p; ablm(3,2)=abj(1,1)-a0p;
for k=1:floor(n/2)-1
  if 2*k+2 <= n
    ablm(2*k+2,2)=abj(k+1,2)/ablm(2*k+1,2);
  end
  if 2*k+3 <= n
    ablm(2*k+3,2)=abj(k+1,1)-ablm(2*k+2,2);
  end
end
y=ablm;
```

The script comparing the stability properties of (2.23) and (2.24) in the case of Legendre polynomials is `stab_test.m`:

```
%STAB_TEST

disp('    n        errp           errm')
for n=[5 10 25 50 100 250 500]
  y=r_jacobi(n);
  yp=legendre_plus(n);
  ym=legendre_minus(n);
  errp=max(abs((yp(:,2)-y(:,2))./y(:,2)));
  errm=max(abs((ym(:,2)-y(:,2))./y(:,2)));
  fprintf('%6.0f %12.4e %12.4e\n',n,errp,errm)
end
```

The output

```
>> stab_test
    n       errp          errm
    5   2.1588e-016   4.3715e-016
   10   2.1982e-016   6.6408e-016
   25   6.6584e-016   8.8776e-016
   50   8.8809e-016   1.5540e-015
  100   1.5543e-015   1.5540e-015
  250   1.5543e-015   1.7763e-015
  500   1.5543e-015   2.4425e-015
>>
```

shows that both nonlinear recursions are perfectly stable.

2.3 ▪ Orthogonality on two separate (symmetric) intervals

Exercise 2.14.* (Ex. 9* of [27, §2].) *Let* $0<\xi<1$ *and consider orthogonal polynomials* π_k *relative to the weight function*

$$w(t)=\begin{cases} |t|^{\gamma}(1-t^2)^{\alpha}(t^2-\xi^2)^{\beta}, & t\in[-1,-\xi]\cup[\xi,1],\\ 0 & \textit{otherwise}.\end{cases}$$

Here, $\gamma\in\mathbb{R}$ *and* $\alpha>-1$, $\beta>-1$. *Evidently,* w *is a symmetric weight function* (in the sense of Ex. 2.12). *Define* $\pi_k^{\pm}$ *as in* Ex. 2.13*(a).

(a) *Transform the polynomials* $\pi_k^{\pm}$ *orthogonal on* $[\xi^2,1]$ *to orthogonal polynomials* $\mathring{\pi}_k^{\pm}$ *on the interval* $[-1,1]$ *and obtain the respective weight function* $\mathring{w}^{\pm}$ *and recurrence relation.*

(b) *Express* β_{2k} *and* β_{2k+1} *in terms of* $\mathring{\gamma}_r^{\pm}$, *the leading coefficient of the orthonormal polynomial of degree* r *relative to the weight function* $\mathring{w}^{\pm}$ *on* $[-1,1]$. {Hint: Use $\beta_r=\|\pi_r\|^2/\|\pi_{r-1}\|^2$ (cf. Eq. (1.2)) and relate this to the leading coefficients γ_k, $\gamma_k^{\pm}$, and $\mathring{\gamma}_k^{\pm}$, with obvious notations.}

(c) *Prove that*

$$\lim_{k\to\infty}\beta_{2k}=\frac{1}{4}(1-\xi)^2,\quad \lim_{k\to\infty}\beta_{2k+1}=\frac{1}{4}(1+\xi)^2.$$

{Hint: Use the result of (b) in combination with the asymptotic equivalence

$$\overset{\circ}{\gamma}{}_k^{\pm} \sim 2^k \overset{\circ}{\gamma}{}^{\pm}, \quad \overset{\circ}{\gamma}{}^{\pm} = \pi^{-1/2} \exp\left\{-\frac{1}{2\pi}\int_{-1}^{1} \ln \overset{\circ}{w}{}^{\pm}(x)(1-x^2)^{-1/2}\mathrm{d}x\right\}, \quad \text{as } k \to \infty$$

(cf. [61, Eq. (12.7.2)]). You may also want to use

$$\int_0^1 \ln(1-a^2x^2)(1-x^2)^{-1/2}\mathrm{d}x = \pi \ln \frac{1+(1-a^2)^{1/2}}{2}, \quad a^2 < 1$$

(see [40, Eq. 4.295.29]).}

(d) *Prove that*

$$\lim_{k\to\infty} \alpha_k^{\pm} = \frac{1+\xi^2}{2}, \quad \lim_{k\to\infty} \beta_k^{\pm} = \left(\frac{1-\xi^2}{4}\right)^2.$$

(e) The recurrence coefficients $\{\beta_k\}$ must satisfy the two nonlinear recursions of Exs. 2.13*(b),(c). Each of them can be interpreted as a pair of fixed point iterations for the even-indexed and for the odd-indexed subsequence, the fixed points being respectively the limits in (c). *Show that, asymptotically, both fixed points are "attracting" for the recursion in* Ex. 2.13*(b), *and "repelling" for the one in* Ex. 2.13*(c). *Also show that, in the latter, the fixed points become attracting if they are switched. What are the numerical implications of all this?*

(f) *Consider the special case $\gamma = \pm 1$ and $\alpha = \beta = -\frac{1}{2}$. In the case $\gamma = 1$, use MATLAB to run the nonlinear recursion of* Ex. 2.13*(b) *and compare the results with the known answers* (see [17, §5.1])

$$\beta_{2k} = \frac{1}{4}(1-\xi)^2 \frac{1+\eta^{2k-2}}{1+\eta^{2k}}, \quad k = 1,2,3,\ldots, \quad 0 \le t \le 1$$

and

$$\beta_1 = \frac{1}{2}(1+\xi^2), \ \beta_{2k+1} = \frac{1}{4}(1+\xi)^2 \frac{1+\eta^{2k+2}}{1+\eta^{2k}}, \quad k = 1,2,3,\ldots,$$

where $\eta = (1-\xi)/(1+\xi)$. Likewise, in the case $\gamma = -1$, run the nonlinear recursion of Ex. 2.13*(c) *and compare the results with the exact answers*

$$\beta_2 = \tfrac{1}{2}(1-\xi)^2, \ \beta_{2k} = \tfrac{1}{4}(1-\xi)^2, \quad k = 2,3,\ldots,$$

and

$$\beta_1 = \xi, \ \beta_{2k+1} = \tfrac{1}{4}(1+\xi)^2, \quad k = 1,2,3,\ldots.$$

Comment on what you observe.

Answer

(a) From the orthogonality of the polynomials $\pi_k^{\pm}$,

$$\int_{\xi^2}^{1} \pi_k^{\pm}(t)\pi_\ell^{\pm}(t)w(t^{1/2})t^{\mp 1/2}\mathrm{d}t = 0, \quad k \ne \ell,$$

changing variables,

$$t = \frac{1-\xi^2}{2}x + \frac{1+\xi^2}{2}, \quad -1 \le x \le 1, \tag{2.25}$$

and letting

$$\left(\frac{1-\xi^2}{2}\right)^{-k}\pi_k^{\pm}\left(\frac{1-\xi^2}{2}x+\frac{1+\xi^2}{2}\right)=:\mathring{\pi}^{\pm}(x), \tag{2.26}$$

we get the orthogonality

$$\int_{-1}^{1}\mathring{\pi}_k^{\pm}(x)\mathring{\pi}_\ell^{\pm}(x)\left(x+\frac{1+\xi^2}{1-\xi^2}\right)^{(\gamma\mp 1)/2}(1-x)^\alpha(1+x)^\beta\mathrm{d}x=0,\quad k\neq\ell, \tag{2.27}$$

for the polynomials $\mathring{\pi}_k^{\pm}$, with weight function

$$\mathring{w}^{\pm}(x)=\left(x+\frac{1+\xi^2}{1-\xi^2}\right)^{(\gamma\mp 1)/2}(1-x)^\alpha(1+x)^\beta,\quad -1<x<1. \tag{2.28}$$

Moreover, from

$$\pi_{k+1}^{\pm}(t)=(t-\alpha_k^{\pm})\pi_k^{\pm}(t)-\beta_k^{\pm}\pi_{k-1}^{\pm}(t),$$

replacing t by the expression in (2.25) and multiplying through by $\left(\frac{1-\xi^2}{2}\right)^{-(k+1)}$ we get, by virtue of (2.26),

$$\mathring{\pi}_{k+1}^{\pm}(x)=\left(\frac{1-\xi^2}{2}\right)^{-1}\left(\frac{1-\xi^2}{2}x+\frac{1+\xi^2}{2}-\alpha_k^{\pm}\right)\mathring{\pi}_k^{\pm}(x)-\left(\frac{1-\xi^2}{2}\right)^{-2}\beta_k^{\pm}\mathring{\pi}_{k-1}^{\pm}(x),$$

and thus

$$\begin{aligned}\alpha_k^{\pm}&=\frac{1-\xi^2}{2}\mathring{\alpha}_k^{\pm}+\frac{1+\xi^2}{2},\quad k=0,1,2,\ldots,\\ \beta_k^{\pm}&=\left(\frac{1-\xi^2}{2}\right)^2\mathring{\beta}_k^{\pm},\quad k=1,2,\ldots.\end{aligned} \tag{2.29}$$

(b) The leading coefficient of any orthonormal polynomial $\tilde{\pi}_k$ relative to a measure $\mathrm{d}\lambda$ is

$$\gamma_k=\frac{1}{\|\pi_k\|_{\mathrm{d}\lambda}}, \tag{2.30}$$

where π_k is the respective monic orthogonal polynomial. For the measure $\mathrm{d}\lambda(t)=w(t)\mathrm{d}t$ currently under discussion, we have from (2.9) that

$$\beta_{2k}=\frac{\|\pi_{2k}\|_{\mathrm{d}\lambda}^2}{\|\pi_{2k-1}\|_{\mathrm{d}\lambda}^2}=\left(\frac{\gamma_{2k-1}}{\gamma_{2k}}\right)^2,\quad \beta_{2k+1}=\frac{\|\pi_{2k+1}\|_{\mathrm{d}\lambda}^2}{\|\pi_{2k}\|_{\mathrm{d}\lambda}^2}=\left(\frac{\gamma_{2k}}{\gamma_{2k+1}}\right)^2. \tag{2.31}$$

To express γ_{2k} in terms of $\mathring{\gamma}_k^+$, we note that the measure associated with $\mathring{\gamma}_k^+$ is $\mathrm{d}\mathring{\lambda}^+(x)=\mathring{w}^+(x)\mathrm{d}x$ given by (2.28). Therefore,

$$\|\mathring{\pi}_k^+\|_{\mathrm{d}\mathring{\lambda}^+}^2=\int_{-1}^{1}[\mathring{\pi}_k^+(x)]^2\left(x+\frac{1+\xi^2}{1-\xi^2}\right)^{(\gamma-1)/2}(1-x)^\alpha(1+x)^\beta\mathrm{d}x,$$

and hence, by (2.26),

$$\begin{aligned}\|\mathring{\pi}_k^+\|_{\mathrm{d}\mathring{\lambda}^+}^2=\left(\frac{1-\xi^2}{2}\right)^{-2k}&\int_{-1}^{1}\left[\pi_k^+\left(\frac{1-\xi^2}{2}x+\frac{1+\xi^2}{2}\right)\right]^2\\ &\times\left(x+\frac{1+\xi^2}{1-\xi^2}\right)^{(\gamma-1)/2}(1-x)^\alpha(1+x)^\beta\mathrm{d}x.\end{aligned}$$

Changing variables according to (2.25), noting that

$$x+\frac{1+\xi^2}{1-\xi^2}=\frac{2}{1-\xi^2}\,t,\;\; 1-x=\frac{2}{1-\xi^2}(1-t),\;\; 1+x=\frac{2}{1-\xi^2}(t-\xi^2),\;\; \mathrm{d}x=\frac{2}{1-\xi^2}\mathrm{d}t, \tag{2.32}$$

yields

$$\|\mathring{\pi}_k^+\|^2_{\mathrm{d}\mathring{\lambda}^+}=C_k^2\int_{\xi^2}^1[\pi_k^+(t)]^2t^{(\gamma-1)/2}(1-t)^\alpha(t-\xi^2)^\beta\mathrm{d}t,\quad C_k^2=\left(\frac{2}{1-\xi^2}\right)^{2k+(\gamma+1)/2+\alpha+\beta}.$$

Letting $t=\tau^2$, and making use of $\pi_k^+(\tau^2)=\pi_{2k}(\tau)$, then gives

$$\|\mathring{\pi}_k^+\|^2_{\mathrm{d}\mathring{\lambda}^+}=C_k^2\|\pi_{2k}\|^2_{\mathrm{d}\lambda}.$$

Thus, finally (cf. (2.30)),

$$\gamma_{2k}=C_k\mathring{\gamma}_k^+,\quad C_k=\left(\frac{2}{1-\xi^2}\right)^{k+(\gamma+1)/4+(\alpha+\beta)/2}.$$

Similarly, one finds

$$\gamma_{2k+1}=\left(\frac{2}{1-\xi^2}\right)^{1/2}C_k\mathring{\gamma}_k^-.$$

From (2.31), therefore,

$$\begin{aligned}\beta_{2k}&=\left[\frac{\left(\frac{2}{1-\xi^2}\right)^{1/2}C_{k-1}\mathring{\gamma}_{k-1}^-}{C_k\mathring{\gamma}_k^+}\right]^2=\frac{1-\xi^2}{2}\left(\frac{\mathring{\gamma}_{k-1}^-}{\mathring{\gamma}_k^+}\right)^2,\\ \beta_{2k+1}&=\left[\frac{C_k\mathring{\gamma}_k^+}{\left(\frac{2}{1-\xi^2}\right)^{1/2}C_k\mathring{\gamma}_k^-}\right]^2=\frac{1-\xi^2}{2}\left(\frac{\mathring{\gamma}_k^+}{\mathring{\gamma}_k^-}\right)^2.\end{aligned} \tag{2.33}$$

(c) We prove the first limit relation; the second is proved similarly. For convenience, we let

$$a:=\frac{1-\xi^2}{1+\xi^2}. \tag{2.34}$$

Using the asymptotic equivalence for $\mathring{\gamma}_k^\pm$ stated in the hint to this part of the exercise, we have

$$\begin{aligned}\mathring{\gamma}_{k-1}^-&\sim 2^{k-1}\pi^{-1/2}\exp\left\{-\frac{1}{2\pi}\int_{-1}^1\ln\mathring{w}^-(x)(1-x^2)^{-1/2}\mathrm{d}x\right\},\\ \mathring{\gamma}_k^+&\sim 2^k\pi^{-1/2}\exp\left\{-\frac{1}{2\pi}\int_{-1}^1\ln\mathring{w}^+(x)(1-x^2)^{-1/2}\mathrm{d}x\right\}.\end{aligned}$$

Therefore,

$$\frac{\mathring{\gamma}_{k-1}^-}{\mathring{\gamma}_k^+}\sim\frac{1}{2}\exp\left\{\frac{1}{2\pi}\int_{-1}^1\ln\frac{\mathring{w}^+(x)}{\mathring{w}^-(x)}(1-x^2)^{-1/2}\mathrm{d}x\right\}. \tag{2.35}$$

By virtue of (2.28),

$$\frac{\mathring{w}^+(x)}{\mathring{w}^-(x)}=\left(x+\frac{1}{a}\right)^{-1},$$

so that the integral in (2.35) is

$$
\begin{aligned}
-\int_{-1}^{1} \ln\left(x+\frac{1}{a}\right)(1-x^2)^{-1/2}dx &= -\left(\int_{-1}^{0}+\int_{0}^{1}\right)\ln\left(x+\frac{1}{a}\right)(1-x^2)^{-1/2}dx \\
&= -\left[\int_0^1 \ln\left(\frac{1}{a}-t\right)(1-t^2)^{-1/2}dt + \int_0^1 \ln\left(\frac{1}{a}+x\right)(1-x^2)^{-1/2}dx\right] \\
&= -\int_0^1 \ln\left(\frac{1}{a^2}-x^2\right)(1-x^2)^{-1/2}dx = -\int_0^1 \ln\left[\frac{1}{a^2}(1-a^2x^2)\right](1-x^2)^{-1/2}dx \\
&= -\frac{\pi}{2}\ln\frac{1}{a^2} - \int_0^1 \ln(1-a^2x^2)(1-x^2)^{-1/2}dx,
\end{aligned}
$$

that is, using the identity provided in the hint for the integral on the right,

$$\pi\left(\ln a - \ln\frac{1+(1-a^2)^{1/2}}{2}\right) = -\pi\ln\frac{1+(1-a^2)^{1/2}}{2a}.$$

Consequently, from (2.35),

$$
\begin{aligned}
\frac{\mathring{\gamma}_{k-1}^{-}}{\mathring{\gamma}_k^{+}} &\sim \frac{1}{2}\exp\left\{-\frac{1}{2}\ln\frac{1+(1-a^2)^{1/2}}{2a}\right\} \\
&= \frac{1}{2}\exp\left\{\ln\left(\frac{1+(1-a^2)^{1/2}}{2a}\right)^{-1/2}\right\} \\
&= \frac{1}{2}\left(\frac{2a}{1+(1-a^2)^{1/2}}\right)^{1/2}.
\end{aligned}
$$

Substituting here the value of a given in (2.34) yields, by an elementary calculation,

$$\left(\frac{\mathring{\gamma}_{k-1}^{-}}{\mathring{\gamma}_k^{+}}\right)^2 \sim \frac{1}{2}\frac{1-\xi}{1+\xi},$$

and hence, by the first relation in (2.33),

$$\beta_{2k} \sim \frac{1-\xi^2}{2}\cdot\frac{1}{2}\frac{1-\xi}{1+\xi} = \frac{1}{4}(1-\xi)^2, \quad k\to\infty,$$

as was to be shown.

(d) From (2.23), and using the results of (c), one gets

$$\alpha_k^{+} = \beta_{2k}+\beta_{2k+1} \to \frac{1}{4}(1-\xi)^2+\frac{1}{4}(1+\xi)^2 = \frac{1}{2}(1+\xi^2)$$

and

$$\beta_k^{+} = \beta_{2k}\beta_{2k-1} \to \frac{1}{4}(1-\xi)^2\cdot\frac{1}{4}(1+\xi)^2 = \left(\frac{1-\xi^2}{4}\right)^2.$$

Similarly, from (2.24), one obtains the same limits for α_k^{-} and β_k^{-}.

(e) We now compare the two nonlinear recursions (2.23) and (2.24). What they have in common is that both allow the $x_k=\beta_{2k}$ and $y_k=\beta_{2k-1}$ to be computed by nonlinear recursions. Looking at the asymptotic behavior of these recursions as $k \to \infty$, we may replace $\alpha_k^\pm$ and $\beta_k^\pm$ by their limits

$$\alpha_\infty = \alpha_\infty^\pm, \quad \beta_\infty = \beta_\infty^\pm$$

and get from (2.23) the fixed point iterations

(FG) $$x_{k+1} = f(x_k), \quad y_{k+1} = g(y_k),$$

and from (2.24) the fixed point iterations

(GF) $$x_{k+1} = g(x_k), \quad y_{k+1} = f(y_k),$$

where

$$f(x) := \frac{\beta_\infty}{\alpha_\infty - x}, \quad g(y) := \alpha_\infty - \frac{\beta_\infty}{y}.$$

The two iterations in (FG) and (GF) are the same but with switched roles. Both have the same fixed points, namely, the solutions $z_{1,2}$ of the quadratic equation

$$z^2 - \alpha_\infty z + \beta_\infty = 0,$$

which happen to be the limits $z_1 = \lim_{k\to\infty} \beta_{2k}$ and $z_2 = \lim_{k\to\infty} \beta_{2k-1}$.

Now

$$f'(z_1) = g'(z_2) = \left(\frac{1-\xi}{1+\xi}\right)^2 < 1,$$

so that the limit points of the two iterations in (FG) are both attracting. This is not so in (GF), where

$$g'(z_1) = f'(z_2) = \left(\frac{1+\xi}{1-\xi}\right)^2 > 1,$$

so that the expected limit points in (GF) are repelling, thus preventing convergence to these limits and instead causing eventual convergence to the switched limits z_2 and z_1, which are both attracting points for the respective iterations.

(f) In both cases, $\gamma = 1$ and $\gamma = -1$, the polynomials $\mathring{\pi}_k^\pm$ in (2.28), when the upper sign is chosen in the case $\gamma = 1$ and the lower sign in the case $\gamma = -1$, are the monic Chebyshev polynomials (of the first kind), so that in these cases

$$\mathring{\alpha}_k^\pm = 0 \text{ for } k \geq 0, \quad \mathring{\beta}_1^\pm = \tfrac{1}{2}, \quad \mathring{\beta}_k^\pm = \tfrac{1}{4} \text{ for } k > 1. \tag{2.36}$$

There follows, by (2.29),

$$\begin{aligned} \alpha_k^\pm &= \frac{1+\xi^2}{2} \quad (k \geq 0), \\ \beta_1^\pm &= \frac{1}{2}\left(\frac{1-\xi^2}{2}\right)^2, \quad \beta_k^\pm = \frac{1}{4}\left(\frac{1-\xi^2}{2}\right)^2 \quad (k > 1). \end{aligned} \tag{2.37}$$

Consider first the case $\gamma = 1$. Since (2.23) involves only quantities (α_k^+, β_k^+) associated with the polynomials $\mathring{\pi}_k^+$, we can use (2.36) and (2.37) to get

$$\begin{aligned} \beta_1 &= \tfrac{1}{2}(1+\xi^2), \\ \beta_2 &= \tfrac{1}{8}(1-\xi^2)^2/\beta_1, \\ \beta_3 &= \tfrac{1}{2}(1+\xi^2) - \beta_2, \\ &\left.\begin{aligned} \beta_{2k} &= \tfrac{1}{16}(1-\xi^2)^2/\beta_{2k-1} \\ \beta_{2k+1} &= \tfrac{1}{2}(1+\xi^2) - \beta_{2k} \end{aligned}\right\} \quad k = 2, 3, \ldots. \end{aligned} \tag{2.38}$$

This is implemented in the MATLAB script `fpl.m` for the first 100 values of the β's. The script also supplies the maximum relative error. (For the explicit values of β_k see Eqs. (5.6) and (5.7) in [17, §5.1].)

```
%FPL

f1='%10.4f %12.2e\n';
N=100;
beta=zeros(N,1); betaex=zeros(N,1);
disp('        xi         max error')
for xi=[.0001 .001 01 .1:.1:.9 .99 .999 .9999]
  xi2=xi^2;
  eta=(1-xi)/(1+xi);
  betaex(1)=.5*(1+xi2);
  for k=1:floor(N/2)
    betaex(2*k)=.25*(1-xi)^2*(1+eta^(2*k-2))/(1+eta^(2*k));
    if 2*k+1<=N
      betaex(2*k+1)=.25*(1+xi)^2*(1+eta^(2*k+2))/(1+eta^(2*k));
    end
  end
  beta(1)=(1+xi2)/2;
  beta(2)=(1-xi2)^2/(8*beta(1));
  beta(3)=(1+xi2)/2-beta(2);
  for k=2:floor(N/2)
    beta(2*k)=(1-xi2)^2/(16*beta(2*k-1));
    if 2*k+1<=N
      beta(2*k+1)=(1+xi2)/2-beta(2*k);
    end
  end
  err=max(abs((beta-betaex)./betaex));
  fprintf(f1,xi,err)
end
```

The output

```
>>fpl
      xi         max error
    0.0001    2.89e-015
    0.0010    3.89e-015
    0.0100    2.41e-015
    0.1000    4.11e-016
    0.2000    6.94e-016
    0.3000    5.66e-016
    0.4000    6.17e-016
    0.5000    2.22e-016
    0.6000    3.48e-016
    0.7000    4.63e-016
    0.8000    3.47e-016
    0.9000    1.73e-016
    0.9900    1.22e-015
    0.9990    2.90e-014
```

```
    0.9999     5.02e-013
>>
```

shows that the nonlinear recursion (2.38) is perfectly stable for $0 < \xi < 1$, deteriorating only slightly in the immediate neighborhood of $\xi = 1$. (The case $\xi = 1$ itself is meaningless!)

Consider now the case $\gamma = -1$. Here the appropriate recursion is (2.24) since it involves only quantities $(\alpha_{k-1}^-, \beta_k^-)$ associated with the polynomials $\mathring{\pi}_k^-$, with one important exception: α_0^+. To evaluate α_0^+ we note that the polynomials π_k^+ are orthogonal on $[\xi^2, 1]$ relative to the weight function $w(t^{1/2})t^{-1/2}$ (cf. part (a) of this exercise); therefore

$$\alpha_0^+ = \frac{\displaystyle\int_{\xi^2}^1 t w(t^{1/2})t^{-1/2}dt}{\displaystyle\int_{\xi^2}^1 w(t^{1/2})t^{-1/2}dt}. \tag{2.39}$$

Since we are in the case $\gamma = -1$, we have $w(x) = x^{-1}(1-x^2)^{-1/2}(x^2-\xi^2)^{-1/2}$, and thus

$$\begin{aligned} w(t^{1/2})t^{-1/2} &= t^{-1/2}(1-t)^{-1/2}(t-\xi^2)^{-1/2} \cdot t^{-1/2} \\ &= t^{-1}(1-t)^{-1/2}(t-\xi^2)^{-1/2}, \quad \xi^2 < t < 1. \end{aligned}$$

The numerator in (2.39), therefore, is

$$\int_{\xi^2}^1 t \cdot t^{-1}(1-t)^{-1/2}(t-\xi^2)^{-1/2}dt = \int_{\xi^2} (1-t)^{-1/2}(t-\xi^2)^{-1/2}dt.$$

Transforming variables as in (2.25) and noting the relations in (2.32), we get

$$\int_{\xi^2}^1 (1-t)^{-1/2}(t-\xi^2)^{-1/2}dt = \int_{-1}^1 (1-x^2)^{-1/2}dx = \pi.$$

For the denominator we obtain similarly

$$\begin{aligned} \int_{\xi^2}^1 w(t^{1/2})t^{-1/2}dt &= \int_{-1}^1 \frac{1}{\frac{1-\xi^2}{2}x + \frac{1+\xi^2}{2}}(1-x^2)^{-1/2}dx \\ &= \frac{2}{1-\xi^2}\int_{-1}^1 \frac{1}{x+\frac{1+\xi^2}{1-\xi^2}}(1-x^2)^{-1/2}dx \\ &= \frac{2}{1-\xi^2}F\left(-\frac{1+\xi^2}{1-\xi^2}\right), \end{aligned}$$

with $F(z)$ being the Stieltjes transform of the Chebyshev weight function,

$$F(z) = \int_{-1}^1 \frac{1}{t-z}(1-t^2)^{-1/2}dt.$$

Thus, altogether,

$$\alpha_0^+ = \frac{(1-\xi^2)\pi}{2F\left(-\frac{1+\xi^2}{1-\xi^2}\right)}. \tag{2.40}$$

It remains to evaluate the Stieltjes transform F. Letting

$$\frac{1+\xi^2}{1-\xi^2}=\frac{1}{a},$$

we have

$$F\left(-\frac{1}{a}\right)=\int_{-1}^{1}\frac{1}{t+\frac{1}{a}}(1-t^2)^{-1/2}\mathrm{d}t,$$

or, with $t=\cos\theta$,

$$F\left(-\frac{1}{a}\right)=\int_0^{\pi}\frac{\mathrm{d}\theta}{\cos\theta+\frac{1}{a}}=a\int_0^{\pi}\frac{\mathrm{d}\theta}{1+a\cos\theta}.$$

The integral on the right is elementary; it can be rationalized by using the identity $\cos\theta=(1-\tan^2\frac{\theta}{2})/(1+\tan^2\frac{\theta}{2})$ and making the change of variables $x=\tan\frac{\theta}{a}$, i.e., $\theta=2\arctan x$, $\mathrm{d}\theta=2\mathrm{d}x/(1+x^2)$. One finds

$$\int_0^{\pi}\frac{\mathrm{d}\theta}{1+a\cos\theta}=\frac{2}{1-a}\int_0^{\infty}\frac{\mathrm{d}x}{x^2+\frac{1+a}{1-a}}.$$

Letting $x=\sqrt{(1+a)/(1-a)}\,t$ then yields

$$F\left(-\frac{1}{a}\right)=a\frac{2}{1-a}\sqrt{\frac{1+a}{1-a}}\cdot\frac{1-a}{1+a}\int_0^{\infty}\frac{\mathrm{d}t}{t^2+1}=\frac{a}{\sqrt{1-a^2}}\pi.$$

Re-inserting the value of a gives

$$F\left(-\frac{1+\xi^2}{1-\xi^2}\right)=\frac{1-\xi^2}{2\xi}\pi.$$

Finally, by (2.40),

$$\alpha_0^+=\frac{(1-\xi^2)\pi\cdot 2\xi}{2(1-\xi^2)\pi}=\xi. \tag{2.41}$$

The nonlinear recursion (2.24) can now be written, using (2.41) and (2.37), with the minus sign chosen in the superscripts,

$$\begin{aligned}
\beta_1&=\xi,\\
\beta_2&=\tfrac{1}{2}(1+\xi^2)-\beta_1,\\
\beta_3&=\tfrac{1}{8}(1-\xi^2)^2/\beta_2,\\
\left.\begin{aligned}\beta_{2k}&=\tfrac{1}{2}(1+\xi^2)-\beta_{2k-1}\\ \beta_{2k+1}&=\tfrac{1}{16}(1-\xi^2)^2/\beta_{2k}\end{aligned}\right\}&\quad k=2,3,\dots.
\end{aligned} \tag{2.42}$$

(See Eq. (5.9) in [17, §5.1] for the explicit values of the β_k.) This is implemented in the MATLAB script `fmin.m` below.

```
%FMIN

f1='%6.0f %16.12f %12.4e %12.4e\n';
N=28; Nh=floor(N/2);
beta=zeros(N,1); betaex=zeros(N,1); err=zeros(N,1); err0=zeros(N,1);
xi=.9
fprintf('                              xi= %4.4f\n',xi)
disp('    k          beta              err            err0')
xi2=xi^2;
betaex(1)=xi;
betaex(2)=.5*(1-xi)^2;
betaex(3)=.25*(1+xi)^2;
for k=2:Nh
  betaex(2*k)=.25*(1-xi)^2;
  if 2*k+1<=N
    betaex(2*k+1)=.25*(1+xi)^2;
  end
end
beta(1)=xi;
err(1)=abs((beta(1)-betaex(1))/betaex(1));
err0(1)=abs((beta(1)-.25*(1-xi)^2)/(.25*(1-xi)^2));
beta(2)=.5*(1+xi2)-beta(1);
err(2)=abs((beta(2)-betaex(2))/betaex(2));
err0(2)=abs((beta(2)-.25*(1+xi)^2)/(.25*(1+xi)^2));
beta(3)=.125*(1-xi2)^2/beta(2);
err(3)=abs((beta(3)-betaex(3))/betaex(3));
err0(3)=abs((beta(3)-.25*(1-xi)^2)/(.25*(1-xi)^2));
for k=2:Nh
  beta(2*k)=.5*(1+xi2)-beta(2*k-1);
  err(2*k)=abs((beta(2*k)-betaex(2*k))/betaex(2*k));
  err0(2*k)=abs((beta(2*k)-.25*(1+xi)^2)/.25*(1+xi)^2);
  if 2*k+1<=N
    beta(2*k+1)=.0625*(1-xi2)^2/beta(2*k);
    err(2*k+1)=abs((beta(2*k+1)-betaex(2*k+1))/betaex(2*k+1));
    err0(2*k+1)=abs((beta(2*k+1)-.25*(1-xi)^2)/(.25*(1-xi)^2));
  end
end
for k=2:2:N
  fprintf(f1,k,beta(k),err(k),err0(k))
end
x=(1:N/2)'; y=err(2*x);
y0=err0(2*x);
set(0,'DefaultAxesLineWidth',1.5)
set(0,'DefaultAxesFontSize',15)
figure
semilogy(x,y,'-k',x,y0,'ok','LineWidth',1.5)
axis([0 15 10^(-16) 10^5])
xlabel('k','FontSize',15)
fprintf('\n')
for k=1:2:N-1
```

```
  fprintf(f1,k,beta(k),err(k),err0(k))
end
x=(1:N/2)'; y=err(2*x-1); y(1)=eps;
y0=err0(2*x-1);
figure
semilogy(x,y,'-k',x,y0,'ok','LineWidth',1.5)
axis([0 15 10^(-16) 10^5])
xlabel('k','FontSize',15)
```

We run it with $\xi = .9$, which causes the two fixed points to be strongly repelling. Accordingly, both the even- and the odd-indexed β's do not converge to their supposed limit values z_1 and z_2 (cf. (e)), but instead move away from them and converge to the other fixed point. This is shown in Fig. 2.1, where the figure on the left deals with the even-indexed β's and the figure on the right with the odd-indexed ones. In both figures the solid line represents the distance of the respective β's from their supposed limit values, and the circled line the distance from the other fixed point.

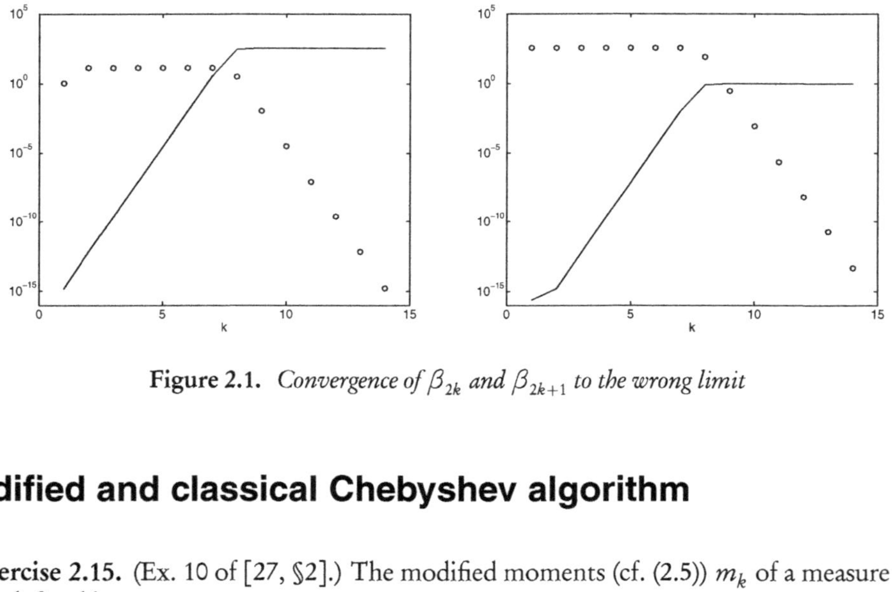

Figure 2.1. *Convergence of β_{2k} and β_{2k+1} to the wrong limit*

2.4 ▪ Modified and classical Chebyshev algorithm

Exercise 2.15. (Ex. 10 of [27, §2].) The modified moments (cf. (2.5)) m_k of a measure $d\lambda$ are defined by

$$m_k = \int_{\mathbb{R}} p_k(t)d\lambda(t), \quad k = 0, 1, 2, \ldots,$$

where $\{p_k\}$ are monic polynomials satisfying a three-term recurrence relation

$$\begin{gathered} p_{k+1}(t) = (t - a_k)p_k(t) - b_k p_{k-1}(t), \quad k = 0, 1, 2, \ldots, \\ p_0(t) = 1, \quad p_{-1}(t) = 0. \end{gathered} \tag{2.43}$$

(When $a_k = b_k = 0$, then $m_k = \mu_k$ are the ordinary moments (2.4).) The mixed moments are defined by

$$\sigma_{k,\ell} = \int_{\mathbb{R}} \pi_k(t; d\lambda)p_\ell(t)d\lambda(t), \quad k, \ell \geq -1.$$

Prove the validity of Algorithm 1 in [27, §2.2]*:*

Algorithm 1 Modified Chebyshev algorithm

initialization:

$$\begin{aligned}&\alpha_0 = a_0 + m_1/m_0, \quad \beta_0 = m_0,\\&\sigma_{-1,\ell} = 0, \quad \ell = 1,2,\dots,2n-2,\\&\sigma_{0,\ell} = m_\ell, \quad \ell = 0,1,\dots,2n-1;\end{aligned}$$

continuation (if $n > 1$):

for $k = 1,2,\dots,n-1$ do

$$\begin{aligned}\sigma_{k\ell} &= \sigma_{k-1,\ell+1} - (\alpha_{k-1} - a_\ell)\sigma_{k-1,\ell} - \beta_{k-1}\sigma_{k-2,\ell}\\&\quad + b_\ell \sigma_{k-1,\ell-1}, \quad \ell = k, k+1, \dots, 2n-k-1,\\\alpha_k &= a_k + \frac{\sigma_{k,k+1}}{\sigma_{kk}} - \frac{\sigma_{k-1,k}}{\sigma_{k-1,\ell-1}}, \quad \beta_k = \frac{\sigma_{kk}}{\sigma_{k-1,k-1}}.\end{aligned}$$

(a) *Verify the initialization part.*

(b) *Combine* $\sigma_{k+1,k-1} = 0$ *with the three-term recurrence relation for both* π_k *and* p_k *to prove the formula for* β_k *in the continuation part.*

(c) *Combine* $\sigma_{k+1,k} = 0$ *with the three-term recurrence relation for both* π_k *and* p_k*, and use the result of* (b)*, to prove the formula for* α_k *in the continuation part.*

Answer

(a) Integrating (2.43) with $k = 0$ gives

$$m_1 = \int_{\mathbb{R}} t\,\mathrm{d}\lambda(t) - a_0 m_0.$$

Since $\int_{\mathbb{R}} t\mathrm{d}\lambda(t) = \alpha_0 \int_{\mathbb{R}} \mathrm{d}\lambda(t) = \alpha_0 m_0$, we get $m_1 = \alpha_0 m_0 - a_0 m_0$, that is, $\alpha_0 = a_0 + m_1/m_0$. Clearly, $\beta_0 = \int_{\mathbb{R}} \mathrm{d}\lambda(t) = m_0$. Finally, $\sigma_{-1,\ell} = 0$ since $\pi_{-1}(t) = 0$, and $\sigma_{0,\ell} = \int_{\mathbb{R}} p_\ell(t)\mathrm{d}\lambda(t) = m_\ell$.

(b) We have

$$\sigma_{k+1,k-1} = \int_{\mathbb{R}} \pi_{k+1}(t) p_{k-1}(t)\mathrm{d}\lambda(t) = 0$$

by orthogonality. Using the three-term recurrence relation for the π_k gives

$$\int_{\mathbb{R}} [(t - \alpha_k)\pi_k(t) - \beta_k \pi_{k-1}(t)] p_{k-1}(t)\mathrm{d}\lambda(t) = 0.$$

Now inserting here the three-term recurrence relation for the p_k, written in the form

$$t p_{k-1}(t) = p_k(t) + a_{k-1} p_{k-1}(t) + b_{k-1} p_{k-2},$$

yields

$$\begin{aligned}&\int_{\mathbb{R}} [p_k(t) + a_{k-1} p_{k-1}(t) + b_{k-1} p_{k-2}(t)] \pi_k(t)\mathrm{d}\lambda(t)\\&\quad = \int_{\mathbb{R}} [\alpha_k \pi_k(t) + \beta_k \pi_{k-1}(t)] p_{k-1}(t)\mathrm{d}\lambda(t).\end{aligned}$$

By orthogonality, the last two terms on the left and the first term on the right are zero, giving

$$\sigma_{kk}=\beta_k\sigma_{k-1,k-1}, \quad \beta_k=\frac{\sigma_{kk}}{\sigma_{k-1,k-1}}.$$

(c) We have by orthogonality

$$\sigma_{k+1,k}=\int_{\mathbb{R}}\pi_{k+1}(t)p_k(t)\mathrm{d}\lambda(t)=0.$$

Thus, from the three-term recurrence relation for the π_k we get

$$\int_{\mathbb{R}}[(t-\alpha_k)\pi_k(t)-\beta_k\pi_{k-1}(t)]p_k(t)\mathrm{d}\lambda(t)=0$$

and, using the same procedure as in (b), but with p_k replacing p_{k-1}, yields

$$\begin{aligned}\int_{\mathbb{R}}&[p_{k+1}(t)+a_kp_k(t)+b_kp_{k-1}(t)]\pi_k(t)\mathrm{d}\lambda(t)\\ &=\int_{\mathbb{R}}[\alpha_k\pi_k(t)+\beta_k\pi_{k-1}(t)]p_k(t)\mathrm{d}\lambda(t),\end{aligned}$$

and thus

$$\sigma_{k,k+1}+a_k\sigma_{kk}=\alpha_k\sigma_{kk}+\beta_k\sigma_{k-1,k}=\alpha_k\sigma_{kk}+\frac{\sigma_{kk}}{\sigma_{k-1,k-1}}\sigma_{k-1,k},$$

that is,

$$\alpha_k=a_k+\frac{\sigma_{k,k+1}}{\sigma_{kk}}-\frac{\sigma_{k-1,k}}{\sigma_{k-1,k-1}}.$$

Exercise 2.16.* (Ex. 11* of [27, §2].) Orthogonal polynomials $\{\pi_k(\cdot;w)\}$ relative to the weight function ("hat function")

$$w(t)=\begin{cases}1+t & \text{if } -1\le t\le 0,\\ 1-t & \text{if } 0\le t\le 1,\\ 0 & \text{otherwise.}\end{cases}$$

(a) *Develop a modified Chebyshev algorithm for generating the first n recurrence coefficients $\beta_k(w)$, $k=0,1,\dots,n-1$* (all $\alpha_k(w)=0$; why?). *Define modified moments with respect to a suitable system of (monic) orthogonal polynomials.*

(b) *What changes in the algorithm are required if one wants $\{\pi_k(\cdot;1-w)\}$, or $\{\pi_k(\cdot;w(1-w))\}$, or $\{\pi_k(\cdot;w^p)\}$ where $p>-1$?*

(c) *Download from* `OPQ` *the routine* `chebyshev.m`*; develop a procedure for computing the modified moments to be used in conjunction with* `chebyshev.m` *to implement* (a) *with a MATLAB function* `r_hatfcn_mcheb.m`*. Illustrate by an example.*

(d) *Devise a 2-component discretization scheme for computing the first n recurrence coefficients $\beta_k(w)$, $k=0,1,2,\dots,n-1$, which uses an n-point discretization of the inner product on each component interval of w, and that will yield exact answers* (in the absence of rounding errors).

(e) *Same as* (b), *but for the discretization scheme.*

(f) *Download from* `OPQ` *the routine* `mcdis.m`; *write a quadrature routine* `qhatf.m` *necessary to implement* (d), *and append a script to the driver of* (c) *that produces results of the discretization procedure for selected values of n. Download whatever additional routines you need. Run the procedure with* `irout` = 1 *and* `irout` ≠ 1 *and observe the respective timings and the maximum discrepancy between the two sets of answers. Verify that the routine "converges" after one iteration if* `idelta` *is properly set. Compare the results with those of* (a).

(g) *Use the routines* `acondG.m` *and* `rcondG.m` *to print the absolute and relative condition numbers of the relevant moment map* G_n (cf. [24, §§2.1.4 and 2.1.5]). *Do any of these correlate well with the numerical results obtained in* (f)*? If not, why?*

Answer

(a),(c) For the polynomials p_k defining modified moments we choose the monic Legendre polynomials. Since p_k for k odd is an odd polynomial and w an even weight function, the corresponding modified moments are zero. Thus we need only the modified moments

$$m_{2k} = \int_{-1}^{1} p_{2k}(t)w(t)\mathrm{d}t, \quad k = 0,1,2,\dots,n-1, \tag{2.44}$$

of even order. For the generation of the required polynomials we derive a three-term recurrence relation as follows. We have

$$\begin{aligned} p_{2k+2}(t) &= t\,p_{2k+1}(t) - b_{2k+1}p_{2k}(t) \\ &= t[t\,p_{2k}(t) - b_{2k}\,p_{2k-1}(t)] - b_{2k+1}p_{2k}(t) \\ &= t^2 p_{2k}(t) - b_{2k}\cdot t\,p_{2k-1}(t) - b_{2k+1}p_{2k}(t), \end{aligned}$$

where the b_k's are the β-recurrence coefficients for the monic Legendre polynomials (all α-coefficients being zero). Now

$$t\,p_{2k-1}(t) = p_{2k}(t) + b_{2k-1}p_{2k-2}(t),$$

giving

$$p_{2k+2}(t) = t^2 p_{2k}(t) - b_{2k}[p_{2k}(t) + b_{2k-1}p_{2k-2}(t)] - b_{2k+1}p_{2k}(t),$$

so that

$$\begin{aligned} p_{2k+2}(t) &= [t^2 - (b_{2k} + b_{2k+1})]p_{2k}(t) - b_{2k}b_{2k-1}p_{2k-2}(t), \\ &\qquad\qquad k = 1,2,\dots,N-2, \qquad (2.45) \\ p_0(t) &= 1, \quad p_2(t) = t^2 - b_1. \end{aligned}$$

This is implemented in the MATLAB function `mope.m` for even-degree polynomials orthogonal relative to a symmetric measure.

```
function y=mope(N,x,ab)
%MOPE Even-degree monic orthogonal polynomial for symmetric weight.
%   Y=MOPE(N,X,AB) generates the Nx1 array Y of the first N
%   even-degree monic orthogonal polynomials P_0(X), P_2(X),...,
%   P_{2N-2}(X). The input variable AB is the (2N-1)x2 array of
%   recurrence coefficients for the monic polynomials. This array is
%   slightly larger than needed here because of later applications.
```

```
y=zeros(N,1); y(1)=1; y(2)=x^2-ab(2,2);
for k=2:N-1
  y(k+1)=(x^2-(ab(2*k-1,2)+ab(2*k,2)))*y(k) ...
    -ab(2*k-1,2)*ab(2*k-2,2)*y(k-1);
end
```

To compute the modified moments (2.44), we write

$$m_{2k}=\int_{-1}^{1} p_{2k}(t)w(t)\mathrm{d}t=\int_{-1}^{0}(1+t)p_{2k}(t)\mathrm{d}t+\int_{0}^{1}(1-t)p_{2k}(t)\mathrm{d}t.$$

In the first integral the change of variables $t=-\tau$ gives

$$\int_{1}^{0}(1-\tau)p_{2k}(-\tau)(-\mathrm{d}\tau)=\int_{0}^{1}(1-t)p_{2k}(t)\mathrm{d}t,$$

so that

$$m_{2k}=2\int_{0}^{1}(1-t)p_{2k}(t)\mathrm{d}t,\quad k=0,1,2,\dots,n-1. \tag{2.46}$$

We compute this by applying the n-point Gauss–Legendre quadrature rule on $[0,1]$ (with nodes τ_ν and weights ω_ν),

$$m_{2k}=2\sum_{\nu=1}^{n}\omega_\nu(1-\tau_\nu)p_{2k}(\tau_\nu),\quad k=0,1,2,\dots,n-1, \tag{2.47}$$

which produces exact answers (modulo rounding errors). To do this efficiently, we precompute the $n\times n$ matrix

$$P=\begin{bmatrix} p_0(\tau_1) & p_0(\tau_2) & \cdots & p_0(\tau_n)\\ p_2(\tau_1) & p_2(\tau_2) & \cdots & p_2(\tau_n)\\ \vdots & \vdots & & \vdots\\ p_{2n-2}(\tau_1) & p_{2n-2}(\tau_2) & \cdots & p_{2n-2}(\tau_n)\end{bmatrix},$$

and use the rows of P to compute the modified moments in (2.47).

We are now ready to invoke the MATLAB routine chebyshev.m, implementing the modified Chebyshev algorithm, to obtain the desired β-recurrence coefficients for the weight function w. This is done in the routine r_hatfcn_mcheb.m.

```
function ab=r_hatfcn_mcheb(N)
%R_HATFCN_MCHEB Recurrence coefficients for the hat weight
%function.
%   AB=R_HATFCN_MCHEB(N) generates the Nx2 array AB of the
%   first N recurrence coefficients for the weight function
%   W(T)=1+T if -1 <= T <= 0, 1-T if 0 <= T <= 1, and 0
%   otherwise, using the modified Chebyshev algorithm.

P=zeros(N);
abl=r_jacobi(2*N-1); abj=r_jacobi01(N);
```

```
xwj=gauss(N,abj); xj=xwj(:,1); wj=xwj(:,2);
for j=1:N
  P(:,j)=mope(N,xj(j),abl);
end
mom=zeros(1,2*N);
for k=1:N
  mom(2*k-1)=2*sum(wj.*(1-xj).*P(k,:)');
end
ab=chebyshev(N,mom,abl)
```

Example. $I=\int_{-1}^{1} e^x w(x)\mathrm{d}x = 2(\cosh 1 - 1) = 1.08616126963049$.

We compare Gauss quadrature relative to the weight function w with Gauss–Legendre quadrature on $[0,1]$ applied to

$$I=\int_0^1 (1-x)(e^x+e^{-x})\mathrm{d}x.$$

The script `order.m` below determines the respective order n of Gauss quadrature required to get results accurate to machine precision. It is found that the former requires $n=8$, while the latter requires $n=11$.

```
%ORDER

N=20; I=2*(cosh(1)-1);
abw=r_hatfcn_mcheb(N); abj=r_jacobi01(N);
Ig=0; n=0;
while abs(Ig-I)>eps
  n=n+1;
  if n>N
    error('divergence')
  end
  xw=gauss(n,abw); x=xw(:,1); w=xw(:,2);
  Ig=sum(w.*exp(x));
end
n1=n
Ig=0; n=0;
while abs(Ig-I)>eps
  n=n+1;
  if n>N
    error('divergence')
  end
  xw=gauss(n,abj); x=xw(:,1); w=xw(:,2);
  Ig=sum(w.*(1-x).*(exp(x)+exp(-x)));
end
n2=n
```

Thus, Gauss–Legendre not only requires two more quadrature points, but also involves twice as many function evaluations as Gauss quadrature of the same order relative to the weight function w. Trying to apply Gauss–Legendre quadrature directly on the integral as given in the example is hopeless because of the discontinuity of the first derivative of

w at $x = 0$. Indeed, 500-point Gauss–Legendre yields only five correct digits after the decimal point!

(b) The three alternative weight functions suggested are all even, and therefore require only one change: the definition of the modified moments (2.46) and their evaluation. They are

$$m_{2k} = \begin{cases} 2\int_0^1 t\, p_{2k}(t)\mathrm{d}t & \text{for } 1-w, \\ 2\int_0^1 t(1-t)p_{2k}(t)\mathrm{d}t & \text{for } w(1-w), \\ 2\int_0^1 (1-t)^p\, p_{2k}(t)\mathrm{d}t & \text{for } w^p. \end{cases}$$

Since $k \le n-1$, the first integral can again be evaluated by n-point Gauss–Legendre, the second by $(n+1)$-point Gauss–Legendre or n-point Gauss–Jacobi with Jacobi parameters $\alpha = \beta = 1$, and the last integral by n-point Gauss–Jacobi with $\alpha = p$, $\beta = 0$, all relative to the interval $[0,1]$.

(d),(f) Since

$$(p,q)_w = \int_0^1 p(-t)q(-t)(1-t)\mathrm{d}t + \int_0^1 p(t)q(t)(1-t)\mathrm{d}t, \tag{2.48}$$

we can discretize both integrals on the right by Gauss–Jacobi quadrature on $[0,1]$ with Jacobi parameters $\alpha = 1$, $\beta = 0$, and then apply a 2-component discretization. If we are interested in the first n recurrence coefficients for the weight function w, taking the n-point Gauss–Jacobi quadrature rule will yield exact results. This is implemented in the routine `r_hatfcn_dis.m`,

```
function ab=r_hatfcn_dis(N)
%R_HATFCN_DIS Recurrence coefficients for the hat weight
%function.
%   AB=R_HATFCN_DIS(N) generates the Nx2 array AB of the
%   first N recurrence coefficients for the weight function
%   W(T)=1+T if -1 <= T <=0, 1-T if 0 <= T <= 1, and 0
%   otherwise, using a 2-component discretization procedure.

global mc mp iq idelta irout AB
mc=2; mp=0; iq=1; idelta=2;
irout=1;
%irout=0;
Mmax=N+10; eps0=.5e-14;
[ab,Mcap,kount]=mcdis(N,eps0,@qhatf,Mmax);
[Mcap kount]
```

which uses the `OPQ` routine `mcdis.m` in conjunction with the quadrature routine `qhatf.m` shown below:

```
function xw=qhatf(N,i)
% QHATF Quadrature routine for discretization.
%   QHATF(N,I) generates the N-point Gaussian quadrature
%   rule for discretizing the hat function inner product
%   in the Ith component interval, I=1 or I=2.

  ab=r_jacobi01(N,1,0);
```

```
    xw=gauss(N,ab);
    if i==2
      xw(:,1)=-xw(:,1);
    end
```

The script compare_chebdis.m and its output (the first line of each ans being [Mcap kount] from the output of the routine mcdis.m and the second line [N err], where err is the maximum relative discrepancy between the moment-based and discretization-based methods)

```
%COMPARE_CHEBDIS

f0='%10.0f %12.4e\n';
for N=[10 20 50 100]
  abm=r_hatfcn_mcheb(N);
  abd=r_hatfcn_dis(N);
  err=max(abs((abm(:,2)-abd(:,2))./abd(:,2)));
  fprintf(f0,N,err)
end

>> compare_chebdis

ans =
        11       1
        10  2.1320e-015
ans =
        21       1
        20  6.1224e-015
ans =
        51       1
        50  1.2671e-014
ans =
       101       1
       100  1.8987e-014
>>
```

confirm that the discretization procedure converges after just one iteration and shows perfect agreement between the results obtained by the moment-based and the discretization-based methods, regardless of whether stieltjes.m or lanczos.m is used in the routine r_hatfcn_dis.m.

Running the routine r_hatfcn_dis.m a thousand times for $n = 10, 20, 50, 100$, once with stieltjes.m and once with lanczos.m, reveals that the former is slightly faster (56.8 sec) than the latter (60.2 sec).

(e) The changes required are just in the choice of measures in (2.48), which are, resp., $t\,dt$, $t(1-t)dt$, and $(1-t)^p dt$ and can be discretized similarly as in (d) and (f).

(g) The script compare_cond.m,

```
%COMPARE_COND

f0='%10.0f %10.4f %10.4f\n';
mom=zeros(1,200);
ab=r_hatfcn_mcheb(100);
```

```
ab0=r_jacobi(200);
disp('        n      acondG      rcondG')
for n=[10 20 50 100]
  ac=acondG(n,ab,ab0);
  rc=rcondG(n,0,ab,ab0,mom);
  fprintf(f0,n,ac,rc)
end
```

and its output

```
>> compare_cond
        n      acondG      rcondG
       10     15.5013     15.0935
       20     40.3829     39.3481
       50    151.8205    148.2050
      100    422.2388    412.5020
>>
```

show that the two condition numbers correlate well among themselves and with the results in (f).

Exercise 2.17. Half-range Hermite polynomials.

These are orthogonal on $[0,\infty]$ with respect to the Hermite weight function $w(t)=e^{-t^2}$.

(a) *Determine the moments* μ_k *of* w.

(b) *Write a symbolic routine* `smom_hrh.m` *that evaluates the first* $2N$ *moments in* `dig`*-digit arithmetic.*

(c) *Write a symbolic routine* `dig_hrh.m` *that determines the number* `dig` *of digits required for the classical Chebyshev algorithm to obtain the first* N *recurrence coefficients* α_k, β_k, $k=0,1,2,\dots,N-1$, *to an accuracy of* `nofdig` *digits.* (Use the `SOPQ` routine `schebyshev.m` to carry out the Chebyshev algorithm in variable-precision arithmetic.) *Along with* `dig`*, output also the* $N\times 2$ *array* `ab` *of recurrence coefficients to* `nofdig=32` *digits.*

(d) *Use the routine of* (c) *to find out how many digits are required to obtain the* $N\times 2$ *array* `ab` *accurate to 32 digits when* $N=200$.

(e) *Write a symbolic routine* `sr_hrh.m` *generating the first* N *recurrence coefficients in* `dig`*-digit arithmetic and placing them into the array* `ab` *to* `nofdig` *digits.*

Answer

(a) Using the change of variables $t^2=\tau$, one gets

$$\mu_k=\int_0^\infty t^k e^{-t^2}\mathrm{d}t=\frac{1}{2}\int_0^\infty \tau^{(k-1)/2}e^{-\tau}\mathrm{d}\tau=\frac{1}{2}\Gamma\left(\frac{k+1}{2}\right),\quad k=0,1,2,\dots.$$

(b) The requested routine is `smom_hrh.m` (cf. Appendix B.3)

```
function mom=smom_hrh(dig,N)
%SMOM_HRH Moments of the half-range Hermite weight function.

digits(dig); mom=vpa(zeros(2*N,1));
for k=1:2*N
  mom(k)=gamma(vpa(k/2))/2;
end
```

(c) The requested routine is (cf. Appendix B.1)

```
function [ab,dig]=dig_hrh(N,dig0,dd,nofdig)
%DIG_HRH Number of digits needed for half-range Hermite
%polynomials.

i=dig0-dd;
maxerr=1;
while maxerr>.5*10^(-nofdig)
  i=i+dd; dig=i
  mom=smom_hrh(dig,N);
  if i==dig0
    ab0=schebyshev(dig,N,mom');
  else
    ab1=schebyshev(dig,N,mom');
    serr=abs((ab1-ab0)./ab1);
    err=double(serr);
    maxerr=max(max(err));
    ab0=ab1;
  end
end
ab=vpa(ab1,nofdig);
```

(d) Experimenting with the routine in (c) led to an estimate of 250 digits. When run with $N = 200$, dig0 = 250, dd = 5, nofdig = 32, the routine produced the output below.

```
>> [ab,dig]=dig_hrh(200,250,5,32)

ab =

[ 0.56418958354775628694807945156077, 0.88622692545275801364908374167057]
[ 0.98842539284680028548706335878879, 0.18169011381620932846223247325497]
[  1.2859676193639399602827887260072, 0.34132512895943919856417178056476]
[  1.5247208440801153035130022763796, 0.50496215298800163193575115541755]

                  ⋮                                    ⋮

[  11.445516966681474635755649824606,  32.666737529228585667959844490388]
[  11.474603523305770966075411553897,  32.833403836177910164695827234664]
[  11.503616535347326051312859572029,  33.000070146760877097090615081598]
[  11.532556557869880465476641160337,  33.166736460922705263951009870619]

dig =
  255
>>
```

confirming that 250-digit arithmetic is indeed sufficient to obtain ab to an accuracy of 32 digits.

The complete array ab is made available in the file ab_hrhermite.txt (cf. Appendix A.5).

(e) The routine is (cf. Appendix B.1)

```
function ab=sr_hrh(dig,nofdig,N)
%SR_HRH Recurrence coefficients for the half-range Hermite
%weight function.

digits(dig);
mom=smom_hrh(dig,N);
ab=schebyshev(dig,N,mom');
ab=vpa(ab,nofdig);
```

Exercise 2.18. *Freud polynomials* are orthogonal on the whole real line $\mathbb{R}$ relative to the weight function

$$w(t)=|t|^{\mu}e^{-|t|^{\nu}}, \quad \mu>-1, \ \nu>0.$$

Subrange Freud polynomials are orthogonal with respect to the same weight function, but on strict subintervals of $\mathbb{R}$. The object of this and the next five exercises is to compute these polynomials and to study their zeros when considered as functions of the parameter ν, in particular the potentially monotone behavior of the zeros on appropriate intervals of the variable ν. (This supplies computational details for the work in [34].)

Lower one-sided subrange Freud polynomials π_n.

Here, the interval of orthogonality is $[0,c]$, $0<c<\infty$.

(a) *Determine the moments*

$$\mu_k=\int_0^c t^k w(t)\mathrm{d}t, \quad k=0,1,2,\ldots,$$

of the weight function w *and write a symbolic MATLAB routine* `smom_l1sf.m` *to generate the first* $2N$ *of them in* `dig`*-digit arithmetic.*

(b) *Write a symbolic routine* `dig_l1sf.m` *that, for given values of* N, c, μ, ν, *determines* (as in Ex. 2.17(c)) *the number of digits required for the classical Chebyshev algorithm to obtain the first* N *recurrence coefficients* α_k,β_k, $k=0,1,2,\ldots,N-1$, *to an accuracy of* `nofdig` *digits.*

(c) *Experiment with the routine in* (b) *to determine the number of digits required when the first* `N=15` *or* `N=30` *recurrence coefficients are wanted to 16-digit accuracy* (`nofdig=16`), *for* $\mu=-1/2,0,1/2$ *and selected values of* c *and* ν *in the respective ranges* $1\le c\le 6$, $.1\le\nu\le 10$.

(d) *Write a symbolic routine* `sr_l1sf.m` *that, for given* c, μ, *and* ν, *generates the first* N *recurrence coefficients in* `dig`*-digit arithmetic and produces the* $N\times 2$ *array* `ab` *of recurrence coefficients to* `nofdig` *digits.*

(e) *Use the routine of* (d) *with* `nofdig=16` *in a variable-precision script* `zeros_l1sf.m` *that computes all zeros of the lower one-sided subrange Freud polynomial* π_N *for consecutive values of* ν *in some given interval and assigns to a variable* `mon` *the values*

`mon=1`	*if all these zeros increase monotonically,*
`mon=-1`	*if all these zeros decrease monotonically,*
`mon=0`	*if none of the above.*

Have the script also monitor the absolute value of the increments or decrements of the zeros for two successive values of ν *and determine the minimum and maximum of them.* {Hint: Make use of the information gathered in (c).}

(f) *Run the script* zeros_l1sf.m *of* (e) *with* $N = 15$ *and* $N = 30$, *using* $c = 1$, $\mu = 0$, *and* nofdig = 16. *Provide plots of the zeros.*

(g) *Same as* (f), *but with* $c = 2$ *and* $N = 1, 7, 15, 30$. *Comment on the results.*

(h) *Same as* (g), *but with* $c = 6$, $N = 15$, *and* $N = 30$.

Answer

(a) The moments are

$$\mu_k = \int_0^c t^k w(t)\mathrm{d}t = \int_0^c t^{k+\mu} e^{-t^\nu}\mathrm{d}t, \quad k = 0, 1, 2, \ldots .$$

Changing variables, $t^\nu = \tau$, yields

$$\mu_k = \frac{1}{\nu}\int_0^{c^\nu} \tau^{(k+\mu+1)/\nu-1} e^{-\tau}\mathrm{d}\tau = \frac{1}{\nu}\gamma\left(\frac{k+\mu+1}{\nu}, c^\nu\right),$$

where $\gamma(a,x) = \int_0^x t^{a-1}e^{-t}\mathrm{d}t$ is the (lower) incomplete gamma function, computed here by the routine sgam_lowinc.m of Ex. 5.3(b). The routine for the moments thus is (cf. Appendix B)

```
function mom=smom_l1sf(dig,N,c,mu,nu)
%SMOM_L1SF Moments of the lower 1-sided Freud weight function.

if c<=0|mu<=-1|nu<=0, error('c or mu or nu out of range'); end
digits(dig); mom=vpa(zeros(2*N,1));
c=vpa(c); mu=vpa(mu); nu=vpa(nu);
for k=1:2*N
  a=(k+mu)/nu; x=c^nu;
  mom(k)=sgam_lowinc(dig,a,x)/nu;
end
```

(b) The routine, dig_l1sf.m, is the same as dig_hrh.m of Ex. 2.17(c), but with more input variables and calling the appropriate moment-generating routine smom_l1sf.m.

(c) When $N = 15$, the number of digits is found to be between 27 and 34, and between 46 and 56 when $N = 30$.

(d) The requested routine is (cf. Appendix B.1)

```
function ab=sr_l1sf(dig,nofdig,N,c,mu,nu)
%SR_L1SF Recurrence coefficients for the lower 1-sided Freud
%weight function.

digits(dig);
mom=smom_l1sf(dig,N,c,mu,nu);
ab=schebyshev(dig,N,mom');
ab=vpa(ab,nofdig);
```

(e) In accordance with the results in (c) we use dig = 40 when $N = 15$, and dig = 60 when $N = 30$. The script zeros_l1sf.m is

```
%ZEROS_L1SF

c=1; mu=0;
N=15; dig=40;
%N=30; dig=60;
nofdig=16;
inu=0;
dmin=100; dmax=0;
for nu=.1:.1:10
  inu=inu+1;
  ab=sr_l1sf(dig,nofdig,N,c,mu,nu);
  xw=gauss(N,double(ab));
  Z(inu,1:N+1)=[nu xw(:,1)'];
  if inu>1
    d=(Z(inu,2:N+1)-Z(inu-1,2:N+1))';
    dmi=min(abs(d)); dma=max(abs(d));
    if dmi<dmin, dmin=dmi; end
    if dma>dmax, dmax=dma; end
    if all(d<0)
      mon=-1;
    elseif all(d>0)
      mon=1;
    else
      mon=0;
    end
    [nu mon dmi dma]
%     if mon~=-1
%     if mon~=0
%     if  mon~=1
%       [nu mon dmi dma]
%       [nu-.1 nu]
%       error('quit')
%     end
  end
end
[dmin dmax]
nu=(.1:.1:10)';
for k=1:N
%  semilogy(nu,Z(:,k+1));set(gca,'FontSize',14)
   plot(nu,Z(:,k+1));set(gca,'FontSize',14)
%  axis([0 2 -.1 1.1])
  axis([0 10 -.1 1.1])
  xlabel('nu')
  hold on
end
```

(f) The plots produced by `zeros_l1sf.m` for $N=15$ and $N=30$ are shown in Fig. 2.2. In both cases, collective monotone behavior of the zeros was never observed, except for the first two ν-values. However, the change of the zeros from one ν-value to the next is quite minute, as small as 1.935×10^{-7} in absolute value, and never larger than 1.629×10^{-3}

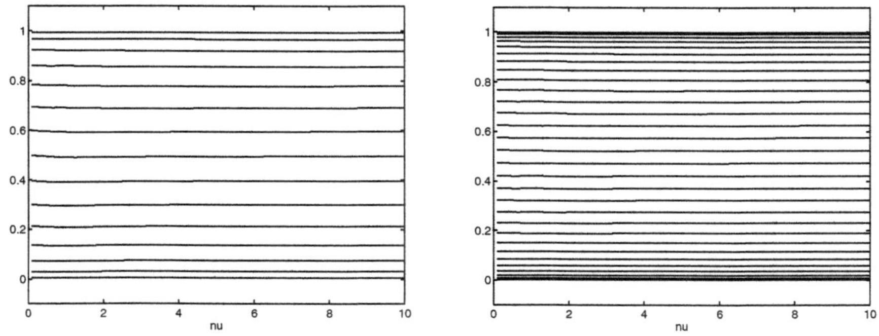

Figure 2.2. *The zeros of* π_N *when* $c = 1$ *and* $\mu = 0$ *for* $N = 15$ *(on the left) and* $N = 30$ *(on the right)*

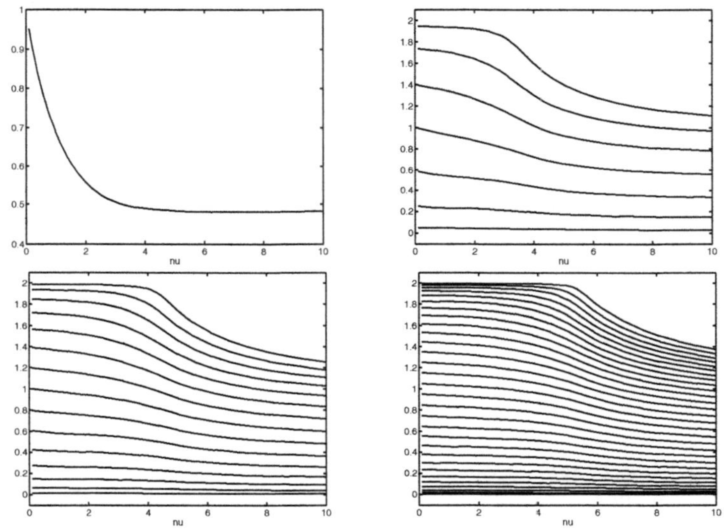

Figure 2.3. *The zeros of* π_N *when* $c = 2$, $\mu = 0$, *for* $n = 1, 7, 15, 30$ *(from top left to bottom right)*

when $N = 15$, and similarly for $N = 30$. The zeros are practically constant functions of ν.

(g) The plots produced are shown in Fig. 2.3. The case $N = 1$ is somewhat special: the zero decreases to a minimum value and thereafter increases very slowly. For $N > 1$, the appearance of the graphs resembles that of a waterfall; a gentle one when c is relatively small and a more precipitous one for larger c; see Fig. 2.4, where $c = 6$. It may appear that all zeros are collectively decreasing. However, there are exceptional intervals, for example $[.5, 1.4]$, $[.4, 1.4]$, $[.3, 1.4]$ for, resp., $N = 7, 15, 30$, and more intervals when ν is much larger than 10, in which the variable `mon` (cf. (e)) becomes 0. But all these exceptions occur in the flat parts of the graphs and are quite minute, not really visible.

(h) The plots are shown in Fig. 2.4.

Exercise 2.19. Upper one-sided subrange Freud polynomials π_n.

These are orthogonal on $[c, \infty]$ with respect to the weight function $w(t) = t^\mu e^{-t^\nu}$.

If $c \geq e^{-1/\nu_0}$, $\nu_0 > 0$, it has been shown in [34, §2.1, Theorem 1(b)] that all zeros of π_n decrease monotonically for all ν on $(0, \nu_0)$, regardless of the values of n and μ. One of the

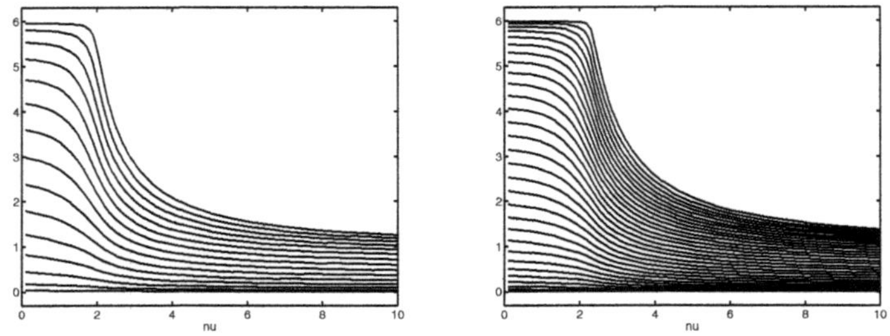

Figure 2.4. *The zeros of π_N when $c=6$ for $N=15$ (on the left) and $N=30$ (on the right)*

objectives of this exercise (see (d)) is to determine empirically the best possible interval $(0,\nu_0^*)$, $\nu_0^*>\nu_0$, for which the assertion of the theorem still holds.

(a) *Write a symbolic MATLAB routine* `smom_u1sf.m` *that generates the first* $2N$ *moments of* w.

(b) *Write a routine* `dig_u1sf.m` *analogous to the routine* `dig_l1sf.m` *of* Ex. 2.18(b), *and experiment with it for values of* N, c, μ *in the ranges* $1\le N\le 10$, $0<c\le 1$, $-1/2\le \mu\le 1$ (needed in (d)) *and for* $c=1$, $\mu=0$, $N=15$ (needed in (e)).

(c) *Write a routine* `sr_u1sf.m` *generating in* `dig`*-digit arithmetic the* $N\times 2$ *array* `ab` *of recurrence coefficients to* `nofdig` *digits. Substitute it for the routine* `sr_l1sf.m` *in the script* `zeros_l1sf.m` *to produce the script* `zeros_u1sf.m`.

(d) Given $\nu_0>0$ and $c=e^{-1/\nu_0}$, we know from the theorem cited at the beginning of this exercise that $(0,\nu_0)$ is a ν-interval in which all zeros of π_n decrease monotonically. The routine `zeros_u1sf.m` of (c) has a provision (if activated) to exit whenever this property does not hold (i.e., `mon` $\neq$ `-1`) and to let us know the respective value of ν. *Use this feature of the routine to determine* $\nu_0^*>\nu_0$ *to five significant digits such that* $(0,\nu_0^*)$ *is the (nearly) largest interval of this form in which all zeros of* π_n *decrease monotonically.*

(e) *Use the routine* `zeros_u1sf.m` *to produce plots of the zeros when* $c=1$, $\mu=0$, *and* $N=15$, *one for the range* $0<\nu<2$, *and another for the range* $2\le\nu\le 10$.

Answer

(a) The moments, similarly as in Ex. 2.18(a), are

$$\mu_k=\int_c^\infty t^{k+\mu}e^{-t^\nu}\mathrm{d}t=\frac{1}{\nu}\Gamma\left(\frac{k+\mu+1}{\nu},c^\nu\right),$$

where $\Gamma(a,x)=\int_x^\infty t^{a-1}e^{-t}\mathrm{d}t$ is the (upper) incomplete gamma function, computed here by the routine `sgam_upinc.m` of Ex. 5.3(c). The routine for the moments is (cf. Appendix B.3)

```
function mom=smom_u1sf(dig,N,c,mu,nu)
%SMOM_U1SF Moments of upper 1-sided Freud weight functions.

if c<=0|mu<=-1|nu<=0, error('c or mu or nu out of range'); end
digits(dig); mom=vpa(zeros(2*N,1));
c=vpa(c); mu=vpa(mu); nu=vpa(nu);
for k=1:2*N
  a=(k+mu)/nu; x=c^nu;
  mom(k)=sgam_upinc(dig,a,x)/nu;
end
```

(b) The routine `dig_u1sf.m` is the same as `dig_l1sf.m`, except for the command generating the appropriate moment. The script

```
%RUN_DIG_U1SF

f0='%8.2f %8.1f %6.0f\n';
mu=0; N=10; dig0=12; dd=1;
disp('     c          nu         dig')
for c=.25:.25:1
  for nu=[.1 1 5 10 50 100]
    [ab,dig]=dig_u1sf(N,c,mu,nu,dig0,dd,16);
    fprintf(f0,c,nu,dig)
  end
  fprintf('\n')
end
```

can be used to run experiments. Table 2.1 shows some typical results when $\mu = 0$ and $N = 10$. They are practically the same for the other values of μ, but of course smaller, in general, when $N < 10$. In the case $c = 1$, $\mu = 0$, one gets `dig = 46` when $N = 15$.

(c) The routine is (cf. Appendix B.1)

```
function ab=sr_u1sf(dig,nofdig,N,c,mu,nu)
%SR_U1SF Recurrence coefficients for upper 1-sided Freud
%weight functions.

digits(dig);
mom=smom_u1sf(dig,N,c,mu,nu);
ab=schebyshev(dig,N,mom');
ab=vpa(ab,nofdig);
```

Table 2.1. *Number of digits required for 16-digit accuracy when* $\mu = 0, N = 10$

ν	c	dig	c	dig	c	dig	c	dig
.1	.25	13	.50	13	.75	13	1.00	13
1.0		18		18		18		18
5.0		22		25		27		31
10.0		23		27		31		37
50.0		25		28		34		49
100.0		25		29		36		55

(d) A little experimentation suggests that the intervals of interest expand to the right as μ and/or n is increased. The least favorable case, therefore, is $n = 1$ and μ very close to -1, say $\mu = -.99999$. Thus we put `N = 1, mu = -.99999` in the routine `zeros_u1sf.m`, and `dig = 16` (obtained by the routine in (b)). With the exit provision activated, and part of the routine slightly edited, we then run it consecutively for $\nu = \nu_0, \nu_0 + .1, \nu + .2, \ldots$. If the routine exits with $\nu = \nu_1$, we have an interval $[\nu_1 - .2, \nu_1]$ which must contain ν_0^*. Now the process can be repeated, starting with $\nu_1 - .2$ and using the increment .01. This will produce a refined interval of length .02 containing ν_0^*. Keep doing this with ever smaller increments .001, .0001, ... until ν_0^* is captured in a sufficiently small interval. This is best done manually in the MATLAB working window, manipulating the script `zeros_u1sf.m` appropriately.

The results are summarized in Table 2.2. It can be seen that the true intervals $(0,\nu_0^*)$ are considerably larger than the intervals $(0,\nu_0)$ predicted by the cited theorem. Also note that the last two entries in the table are very close to the case of half-range Freud polynomials and indicate that in this case, too, there are instances in which the zeros do not all decrease. Recall for half-range Freud polynomials, however, that Table 2.2 relates to a worst-case scenario. Other, more favorable cases for half-range Freud polynomials will be dealt with in the next exercise.

Table 2.2. *Worst-case intervals $(0,\nu_0^*)$ of monotone decrease of all zeros ($n=1$ and $\mu\approx -1$)*

ν_0	$c=e^{-1/\nu_0}$	ν_0^*	ν_0	$c=e^{-1/\nu_0}$	ν_0^*
6	.8464...	39.336	.6	.1887...	5.1103
4	.7788...	26.594	.5	.1353...	4.5158
2	.6065...	13.877	.4	.0820...	3.9393
1	.3678...	7.5703	.3	.0356...	3.3930
.9	.3291...	6.9481	.2	6.738×10^{-3}	2.8985
.8	.2865...	6.3295	.1	4.540×10^{-5}	2.4861
.7	.2396...	5.7162			

(e) The plots of the zeros are shown in Fig. 2.5.

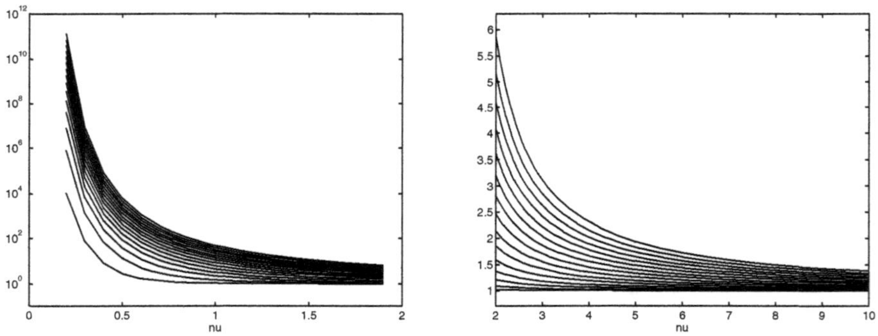

Figure 2.5. *The zeros of the upper one-sided subrange Freud polynomial π_N of degree $N=15$ for $0<\nu<2$ (left) and $2\le\nu\le 10$ (right) when $c=1$ ($\nu_0=\infty$) and $\mu=0$*

Exercise 2.20. Half-range Freud polynomials π_n.

These are orthogonal on $[0,\infty]$ with respect to the weight function $w(t)=t^\mu e^{-t^\nu}$.

(a) *Write a symbolic MATLAB routine* `smom_hrf.m` *that generates the first $2N$ moments of w.*

(b) *Write a routine* `dig_hrf.m` *analogous to the routine* `dig_u1sf.m` *of* Ex. 2.19(b), *and experiment with it for values of N, μ, ν in the ranges $1\le N\le 10$, $-1/2\le\mu\le 1$, $0<\nu\le 100$* (needed in (d)) *and also for $\mu=0$ and $N=15$, $\nu=10$* (needed in (e)).

(c) *Write a routine* `sr_hrf.m` *generating in* `dig`*-digit arithmetic the $N\times 2$ array* `ab` *of recurrence coefficients to* `nofdig` *digits. Substitute it for the routine* `sr_u1sf.m` *in the script* `zeros_u1sf.m` *to produce the script* `zeros_hrf.m`.

(d) *Use the script* `zeros_hrf.m` *of* (c) *in the same manner as in* Ex. 2.19(d) *to determine the (optimal) intervals $(0,\nu^*)$ in which all zeros decrease, for $1\le n\le 6$, $\mu=\mu_-=-.9999999,-1/2,0,1/2,1$. Comment on the results.*

(e) *Amend the script* `zeros_hrf.m` *(as in* `zeros_l1sf.m`*) to plot the zeros for $\mu=0$ and $N=15$, once on the interval $0<\nu<2$ and once on the interval $2\le\nu\le 10$.*

Answer

(a) The moments are

$$\mu_k = \int_0^\infty t^{k+\mu} e^{-t^\nu} dt = \frac{1}{\nu}\Gamma\left(\frac{k+\mu+1}{\nu}\right), \quad k = 0, 1, 2, \ldots,$$

and are generated by the routine (cf. Appendix B.3)

```
function mom=smom_hrf(dig,N,mu,nu)
%SMOM_HRF Moments of half-range Freud weight functions.

if mu<=-1|nu<=0, error('mu and/or nu out of range'); end
digits(dig); mom=vpa(zeros(2*N,1));
mu=vpa(mu); nu=vpa(nu);
for k=1:2*N
  mom(k)=gamma((k+mu)/nu)/nu;
end
```

(b) Obtain the routine `dig_hrf.m` from the routine `dig_u1sf.m` by removing c from the list of input variables and changing the command generating the moments to `mom=smom_hrf(dig,N,mu,nu)`. Running it with the script `run_dig_hrf.m`, an obvious analogue of `run_dig_u1sf.m`, tells us that the number `dig` of digits required for the values of N and μ stated in the exercise is never larger than 23 in the first case and 32 in the second.

(c) The routine is (cf. Appendix B.1)

```
function ab=sr_hrf(dig,nofdig,N,mu,nu)
%SR_HRF Recurrence coefficients for the half-range Freud
%weight function.

digits(dig);
mom=smom_hrf(dig,N,mu,nu);
ab=schebyshev(dig,N,mom');
ab=vpa(ab,nofdig)
```

(d) Since there are no theoretical results (no variable ν_0), we proceed as in Ex. 2.19(d), but with the variable ν in place of ν_0. That is, we determine, by consecutively increasing ν, the largest interval $(0, \nu^*)$ that does not trigger an exit when running the script `zeros_hrf.m`. In accordance with the findings in (b), we take `dig` = 24 in the script. The results found are displayed in Table 2.3. It can be seen, as already mentioned in the answer to Ex. 2.19(d), that the interval $(0, \nu^*)$ expands as either n or μ, or both, are increased. In particular, all zeros of the half-range Freud polynomial π_n decrease monotonically for all ν in the interval $(0, 10]$ when $n \geq 2$, and for all ν in the interval $(0, 100]$ when $n \geq 6$, regardless of the value of $\mu > -1$.

(e) The plots of the zeros are shown in Fig. 2.6.

Exercise 2.21. Lower symmetric subrange Freud polynomials π_n.

These are orthogonal on $[-c, c]$ with respect to the weight function $w(t) = |t|^\mu e^{-|t|^\nu}$.

(a) *Write a symbolic MATLAB routine* `smom_lsf.m` *that generates the first* $2N$ *moments of* w.

(b) *Write a routine* `dig_lsf.m` *analogous to the routine* `dig_l1sf.m` *of* Ex. 2.18(b) *and use it with* `nofdig` =16 *for* $N = 15$ *and* $N = 30$, *and for selected values of* c, μ, ν *in the ranges*

Table 2.3. *The intervals* $(0,\nu^*)$ *of monotonic decrease of all zeros of half-range Freud polynomials* π_n *in dependence on* n *and* μ

$\mu=\mu_-$		$\mu=-1/2$		$\mu=0$		$\mu=1/2$		$\mu=1$	
n	ν^*	n	ν^*	n	ν^*	n	ν^*	n	ν^*
1	2.1662	1	4.5574	1	6.8949	1	9.2204	1	11.541
2	11.541	2	15.371	2	19.204	2	23.039	2	26.874
3	26.874	3	32.233	3	37.593	3	42.955	3	48.318
4	48.318	4	55.207	4	62.098	4	68.990	4	75.882
5	75.882	5	84.303	5	92.725	5	101.15		
6	109.57								

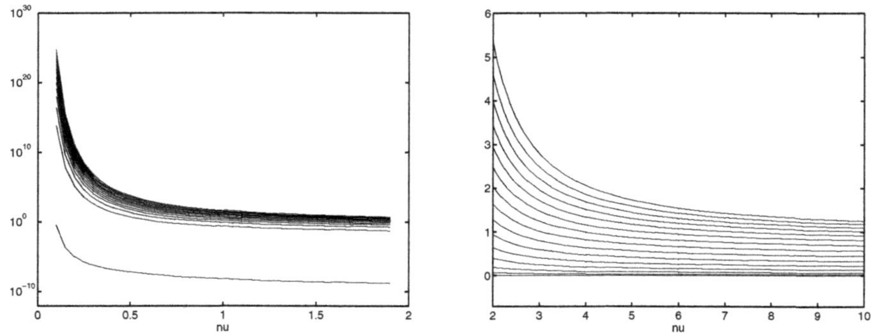

Figure 2.6. *The zeros of the half-range Freud polynomial* π_N *of degree* $N=15$ *when* $\mu=0$, *for* $0<\nu<2$ *(left) and* $2\le\nu\le 10$ *(right)*

$1\le c\le 6$, $-1/2\le\mu\le 1/2$, *and* $1\le\nu\le 10$, *to determine how many digits are required at most.*

(c) *Write a routine* `sr_lsf.m` *generating in* `dig`*-digit arithmetic the* $N\times 2$ *array* `ab` *of recurrence coefficients to* `nofdig` *digits. Use it in a script* `zeros_lsf.m` *that is analogous to the previous three scripts* `zeros_`$\cdots$, *but applicable to symmetric weight functions, and hence computes only the positive zeros.*

(d) *Run the script* `zeros_lsf.m` *of* (c) *with* $N=15$ *and* $N=30$, *using* $c=1$, $\mu=0$, *and* `nofdig = 16`. *Provide plots of the positive zeros.*

(e) *Same as* (d), *but with* $c=2$ *and* $N=2,7,15,30$. *Comment on the results.*

(f) *Same as* (e), *but with* $c=6$, $N=15$, *and* $N=30$.

Answer

(a) The moment μ_k is zero if k is odd, and otherwise

$$\mu_k=2\int_0^c t^{k+\mu}e^{-t^\nu}\mathrm{d}t=\frac{2}{\nu}\gamma\left(\frac{k+\mu+1}{\nu},c^\nu\right),\quad k\,(\text{even})=0,2,4,\ldots$$

(cf. Ex. 2.18(a)). The routine generating them, using `sgam_lowinc.m` of Ex. 5.3(b), is (cf. Appendix B.3)

```
function mom=smom_lsf(dig,N,c,mu,nu)
%SMOM_LSF Moments of lower symmetric Freud weight functions.

if c<=0|mu<=-1|nu<=0, error('c or mu or nu out of range'); end
```

```
digits(dig); mom=vpa(zeros(2*N,1));
c=vpa(c); mu=vpa(mu); nu=vpa(nu);
for k=1:2:2*N
  a=(k+mu)/nu; x=c^nu;
  mom(k)=2*sgam_lowinc(dig,a,x)/nu;
end
```

(b) The routine `dig_lsf.m` differs from `dig_llsf.m` only by one command: the one that generates the required moments. Using the script `run_dig_lsf.m`, an analog of `run_dig_hrf.m`, one finds that at most 27 digits are required for $N = 15$, and at most 35 for $N=30$.

(c) The routine is (cf. Appendix B.1)

```
function ab=sr_lsf(dig,nofdig,N,c,mu,nu)
%SR_LSF Recurrence coefficients for lower symmetric Freud
%weight functions.

digits(dig);
mom=smom_lsf(dig,N,c,mu,nu);
ab=schebyshev(dig,N,mom');
ab=vpa(ab,nofdig);
```

(d) The plots are shown in Fig. 2.7. As in the asymmetric case (cf. Fig. 2.2), all zeros are practically constant as functions of ν.

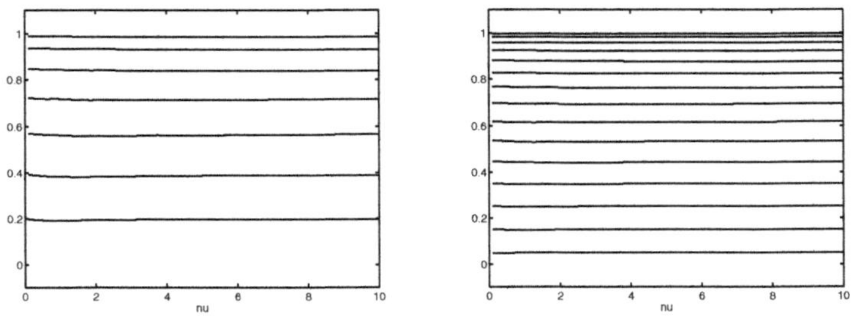

Figure 2.7. *The positive zeros of the lower symmetric subrange Freud polynomial* π_N *of degree* $N = 15$ *(left) and of degree* $N = 30$ *(right) when* $c = 1$ *and* $\mu = 0$

(e) The plots are shown in Fig. 2.8.

(f) The plots are shown in Fig. 2.9.

Altogether, the behavior of the zeros is very similar to the one in the asymmetric case (cf. Ex. 2.18).

Exercise 2.22. Upper symmetric subrange Freud polynomials π_n.

These are orthogonal on $[-\infty,-c] \cup [c,\infty]$ with respect to the weight function $w(t) = |t|^\mu e^{-|t|^\nu}$.

If $c \geq e^{-1/(2\nu_0)}$, $\nu_0 > 0$, it has been shown in [34, Theorem 2(b)] that all positive zeros of π_n decrease monotonically for all ν on $(0, 2\nu_0)$, regardless of the values of n and μ. The objective of this exercise (see (c)) is to determine empirically the best possible interval $(0, \nu_0^*)$, $\nu_0^* > 2\nu_0$, for which the assertion of the theorem still holds.

(a) *Write a symbolic MATLAB routine* `smom_usf.m` *that generates the first* $2N$ *moments of* w.

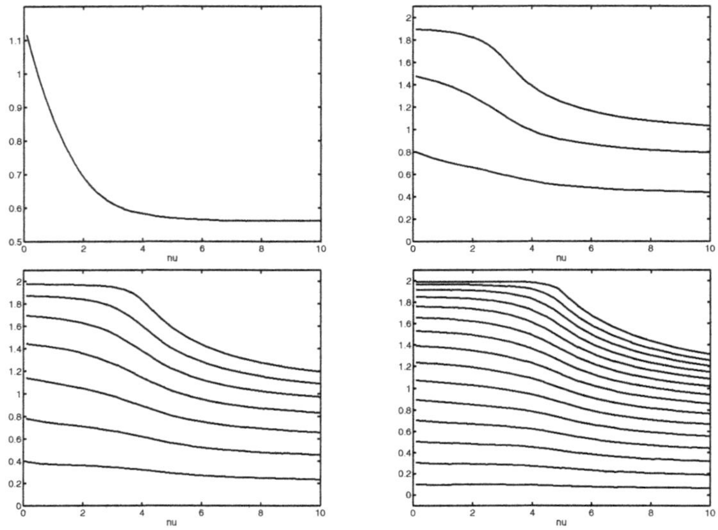

Figure 2.8. *The positive zeros of the lower symmetric subrange Freud polynomial π_N when $c = 2$, $\mu = 0$, for $n = 2, 7, 15, 30$ (from top left to bottom right)*

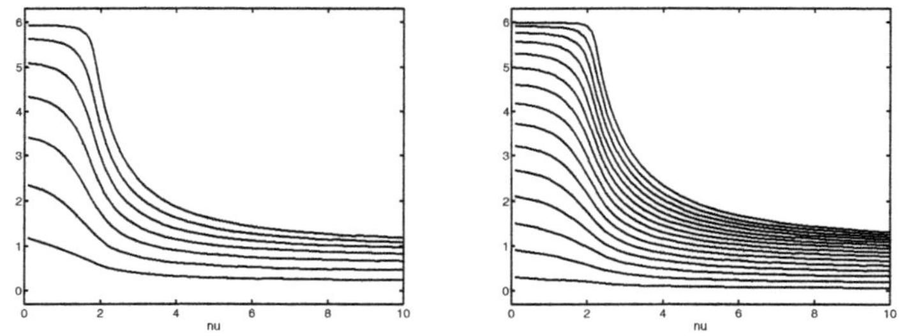

Figure 2.9. *The positive zeros of the lower symmetric subrange Freud polynomial π_N of degree $N = 15$ (left) and of degree $N = 30$ (right) when $c = 6$ and $\mu = 0$*

(b) *Write a routine* `sr_usf.m` *generating in* `dig`*-digit arithmetic the $N \times 2$ array* `ab` *of recurrence coefficients to* `nofdig` *digits. Substitute it for the routine* `sr_lsf.m` *in the script* `zeros_lsf.m` *to produce the script* `zeros_usf.m`.

(c) Given $\nu_0 > 0$ and $c = e^{-1/(2\nu_0)}$, we know from the theorem cited at the beginning of this exercise that $(0, 2\nu_0)$ is a ν-interval in which all zeros of π_n decrease monotonically. The routine `zeros_usf.m` of (b) has a provision (if activated) to exit whenever this property does not hold (i.e., `mon` $\neq$ `-1`) and to let us know the respective value of ν. *Use this feature of the routine to determine $\nu_0^* > 2\nu_0$ to five significant digits such that $(0, \nu_0^*)$ is the (nearly) largest interval of this form in which all zeros of π_n decrease monotonically.*

Answer

(a) The moments are (cf. Ex. 2.19(a))

$$\mu_k = 2\int_c^\infty t^{k+\mu} e^{-t^\nu} \mathrm{d}t = \frac{2}{\nu}\Gamma\left(\frac{k+\mu+1}{\nu}, c^\nu\right), \quad k\,(\text{even}) = 0, 2, 4, \ldots,$$

while $\mu_k = 0$ if k is odd. The routine generating them, using `sgam_upinc.m` of Ex. 5.3(c), is (cf. Appendix B.3)

```
function mom=smom_usf(dig,N,c,mu,nu)
%SMOM_USF Moments of upper symmetric Freud weight functions.

if c<=0|mu<=-1|nu<=0, error('c or mu or nu out of range'); end
digits(dig); mom=vpa(zeros(2*N,1));
c=vpa(c); mu=vpa(mu); nu=vpa(nu);
for k=1:2:2*N
  a=(k+mu)/nu; x=c^nu;
  mom(k)=2*sgam_upinc(dig,a,x)/nu;
end
```

(b) The routine is (cf. Appendix B.1)

```
function ab=sr_usf(dig,nofdig,N,c,mu,nu)
%SR_USF Recurrence coefficients for upper symmetric Freud
%weight functions.

digits(dig);
mom=smom_usf(dig,N,c,mu,nu);
ab=schebyshev(dig,N,mom');
ab=vpa(ab,nofdig);
```

(c) Running the script `zeros_usf.m` similarly as in Ex. 2.19(d), but with $N=2$, yields the results in Table 2.4. Note that ν_0^* is significantly larger than $2\nu_0$ predicted by the cited theorem.

Table 2.4. *Worst-case intervals* $(0,\nu_0^*)$ *of monotonic decrease of all positive zeros* ($n=2$ *and* $\mu\approx -1$)

ν_0	$c=e^{-1/(2\nu_0)}$	ν_0^*	ν_0	$c=e^{-1/(2\nu_0)}$	ν_0^*
6	.9200...	78.671	.6	.4345...	10.221
4	.8824...	53.189	.5	.3678...	9.0315
2	.7788...	27.753	.4	.2865...	7.8786
1	.6065...	15.141	.3	.1888...	6.7860
.9	.5737...	13.896	.2	8.208×10^{-2}	5.7970
.8	.5352...	12.659	.1	6.737×10^{-3}	4.9722
.7	.4895...	11.432			

Exercise 2.23. Freud polynomials π_n.

These are orthogonal on $[-\infty,\infty]$ with respect to the weight function $w(t)=|t|^\mu e^{-|t|^\nu}$.

(a) *Write a symbolic MATLAB routine* `smom_freud.m` *that generates the first* $2N$ *moments of* w.

(b) *Produce the routine* `dig_freud.m` *by making appropriate modifications to the routine* `dig_lsf.m`.

(c) *Write a routine* `sr_freud.m` *generating in* `dig`*-digit arithmetic the* $N\times 2$ *array* `ab` *of recurrence coefficients to* `nofdig` *digits. Substitute it for the routine* `sr_lsf.m` *in the script* `zeros_lsf.m` *to produce the script* `zeros_freud.m`.

(d) *Use the script* `zeros_freud.m` *of* (c) *in the same manner as in* Ex. 2.20(d) *to determine the (optimal) intervals* $(0,\nu^*)$ *in which all zeros decrease, for* $2\le n\le 9$, $\mu=\mu_-=-.9999999,-1/2,0,1/2,1$. *Comment on the results.*

(e) *Plot the positive zeros for* $\mu=0$ *and* $N=15$, *once on the interval* $0<\nu<2$, *and once on the interval* $2\le\nu\le 10$.

Answer

(a) The moments are (cf. Ex. 2.20(a))

$$\mu_k = 2\int_0^\infty t^{k+\mu} e^{-t^\nu} \mathrm{d}t = \frac{2}{\nu}\Gamma\left(\frac{k+\mu+1}{\nu}\right), \quad k\,(\text{even}) = 0,2,4,\ldots,$$

while $\mu_k = 0$ for k odd. They are generated by the routine (cf. Appendix B)

```
function mom=smom_freud(dig,N,mu,nu)
%SMOM_FREUD Moments of the Freud weight function.

if double(mu)<=-1|double(nu)<=0, ...
error('mu and/or nu out of range'); end
digits(dig); mom=vpa(zeros(2*N,1));
mu=vpa(mu); nu=vpa(nu);
for k=1:2:2*N-1
  mom(k)=2*gamma((k+mu)/nu)/nu;
end
```

(b) The modifications consist in removing the variable c in the routine `dig_lsf.m` and changing the command defining the moments to `mom=smom_freud(dig,N,mu,nu)`.

(c) The routine is (cf. Appendix B.1)

```
function ab=sr_freud(dig,nofdig,N,mu,nu)
%SR_FREUD Recurrence coefficients for Freud polynomials.

digits(dig);
mom=smom_freud(dig,N,mu,nu);
ab=schebyshev(dig,N,mom');
ab=vpa(ab,nofdig);
```

(d) The results are shown in Table 2.5.

Table 2.5. *The intervals* $(0,\nu^*)$ *of monotonic decrease of all positive zeros of Freud polynomials* π_n *in dependence on* n *and* μ

$\mu=\mu_-$		$\mu=-1/2$		$\mu=0$		$\mu=1/2$		$\mu=1$	
n	ν_0^*	n	ν_0^*	n	ν_0^*	n	ν_0^*	n	ν_0^*
2	4.3325	2	6.7519	2	9.1147	2	11.457	2	13.790
3	13.790	3	16.117	3	18.441	3	20.762	3	23.082
4	23.082	4	26.910	4	30.741	4	34.574	4	38.408
5	38.408	5	42.242	5	46.077	5	49.913	5	53.749
6	53.749	6	59.106	6	64.465	6	69.825	6	75.186
7	75.186	7	80.548	7	85.910	7	91.273	7	96.636
8	96.636	8	103.52						
9	124.20								

Note that the intervals $(0,\nu^*)$ expand as n and/or μ increase. The results also suggest that all positive zeros of Freud polynomials decrease monotonically on the ν-interval $(0,10]$ if $n\geq 3$, and on the interval $(0,100]$ if $n\geq 9$, regardless of the value of μ.

(e) The plots are shown in Fig. 2.10.

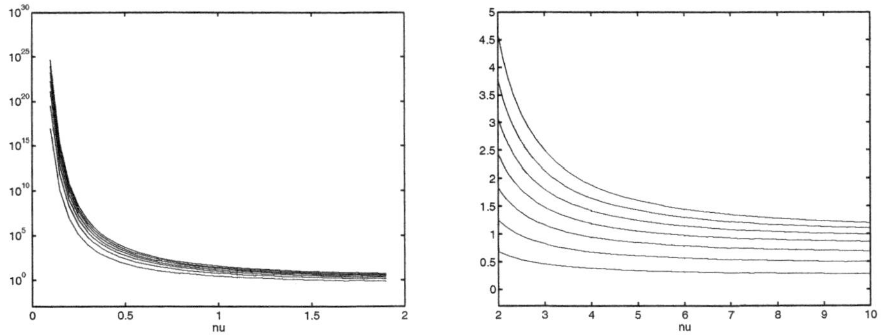

Figure 2.10. *The positive zeros of the Freud polynomial π_N of degree $N=15$ when $\mu=0$, for $0<\nu<2$ (left) and $2\le\nu\le 10$ (right)*

Exercise 2.24. A conjecture of G. Freud.

With β_k, $k=1,2,3,\dots$, denoting the β-recurrence coefficients for the monic Freud polynomials (cf. Ex. 2.18), G. Freud [12] conjectured that, regardless of the value of $\mu>-1$,

$$\lim_{k\to\infty} k^{-1/\nu}\sqrt{\beta_k}=\left(\frac{2\Gamma(\nu)}{\Gamma^2(\nu/2)}\right)^{-1/\nu},\quad \nu>0. \tag{2.49}$$

(The original has a slightly different, but equivalent, expression for the right-hand side of (2.49).) Freud proved the relation (2.49) for $\nu=2,4$, and 6, and D. S. Lubinsky et al. [48] proved it for arbitrary $\nu>0$ and $\mu>-1$. Moreover, when ν is an even positive integer and $\mu=0$, Máté et al. [50] showed the existence of an asymptotic expansion

$$k^{-\nu}\sqrt{\beta_k}=\sum_{j=0}^{\infty}c_{2j}k^{-2j},\quad k\to\infty, \tag{2.50}$$

where c_0, of course, is equal to the right-hand side of (2.49). In particular,

$$\frac{k^{-\nu}\sqrt{\beta_k}-c_0}{c_0}\sim\frac{c_2}{c_0}k^{-2},\quad k\to\infty,\quad \nu=2,4,6,\dots,\ \mu=0. \tag{2.51}$$

Let $\gamma=c_2/c_0$.

(a) *Plot the right-hand side of* (2.49) *as a function of* ν. *Determine the asymptotic behavior of this function as* $\nu\to\infty$ *and as* $\nu\to 0$.

(b) *Use the routine* `sr_freud.m` *of* Ex. 2.23(c) *to compute* γ *for* $\nu=4:2:10$. *Explore the behavior of the left-hand side of* (2.51) *as* $k\to\infty$, *when* $\mu\neq 0$.

(c) *Investigate the validity of* (2.51) *for other values of* ν *in the range* $0<\nu\le 10$.

(d) *Is there a "Freud conjecture" for half-range Freud polynomials?*

Answer

(a) The plots are readily produced, using the `semilogy` command for $\nu\in(0,1)$ and the `plot` command for $\nu\in[1,10]$. For example,

```
nu=(.1:.01:1)';
y=(2*gamma(nu)./(gamma(nu/2).^2)).^(-1./nu);
semilogy(nu,y);set(gca,'FontSize',14)
```

They are shown in Fig. 2.11, on the left for $0<\nu<1$, on the right for $1\le\nu\le 10$.

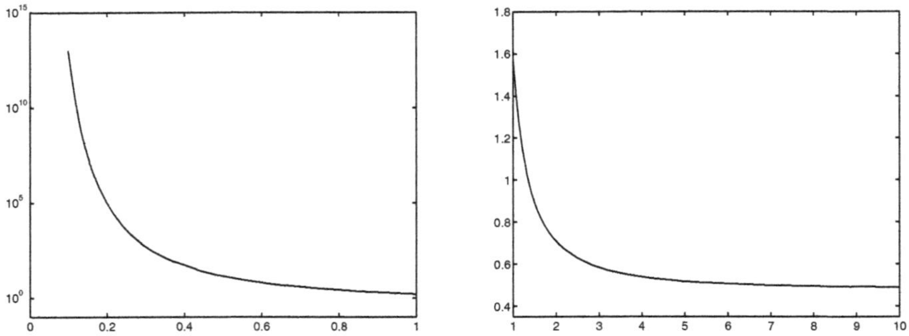

Figure 2.11. *The function* $y=(2\Gamma(\nu)/(\Gamma(\nu/2)^2))^{-1/\nu}$

For the behavior as $\nu\to\infty$, use $\Gamma(x)\sim\sqrt{2\pi}e^{-x}x^{x-1/2}$ (cf. [53, Eq. 5.11.3]) to obtain

$$\frac{2\Gamma(\nu)}{\Gamma^2(\nu/2)}\sim\sqrt{\frac{\nu}{\pi}}\,2^{\nu-1/2},\qquad \nu\to\infty,$$

so that $c_0\sim 1/2$ when $\nu\to\infty$.

For the behavior as $\nu\to 0$, use $\Gamma(x)=\Gamma(x+1)/x$ to obtain

$$\frac{2\Gamma(\nu)}{\Gamma^2(\nu/2)}=\frac{2\Gamma(\nu+1)(\nu/2)^2}{\nu\Gamma^2(\nu/2+1)}\sim\frac{\nu}{2},\qquad \nu\to 0,$$

hence

$$c_0\sim\left(\frac{2}{\nu}\right)^{1/\nu},\qquad \nu\to 0.$$

(b) The case $\nu=2$ is trivial: we are then dealing with Hermite polynomials, for which it is known that $\beta_k=k/2$ (see, for example, [24, Table 1.1]), so that $k^{-1/2}\sqrt{\beta_k}$ is constant equal to $1/\sqrt{2}$, the value on the right of (2.49) when $\nu=2$. Thus, $c_{2j}=0$ for $j>0$ in (2.50), in particular $\gamma=0$.

To determine γ when $\nu\neq 2$, we compute the sequence

$$\gamma_k=k^2\,\frac{k^{-\nu}\sqrt{\beta_k}-c_0}{c_0},\qquad k=1,2,3,\ldots,99, \tag{2.52}$$

in an attempt to investigate its limit behavior. Using the routine `sr_freud.m`, as suggested in the exercise, requires us to know how many digits to carry, for given ν, to obtain β_k, $1\le k\le 99$, to an accuracy of `nofdig` digits. Since we expect some degree of cancellation to occur when computing the difference on the right of (2.52), we take `nofdig` $=28$, to be on the safe side. The routine `dig_freud.m` of Ex. 2.23(b) then tells us that, when $\mu=0$, the number of digits needed, in units of four, is as shown in Table 2.6. The same numbers work also for $\mu=-1/2,1/2$, and 1.

Table 2.6. *The number of digits required to compute* β_k, $1\le k\le 99$, *to* 28 *digits when* $\mu=0$

ν	.1	.5	1.0	3.0	5.0	10.0
dig	28	48	56	72	80	88

The computation of $\{\gamma_k\}_{k=1}^{99}$ can now proceed. It is carried out by the script `freud_conj.m`, which also does some diagnostics and provides plots.

```
%FREUD_CONJ

f0='%8.0f %12.4e %20.12e %12.4e\n';
mu=0; N=100; dig=90; nofdig=28;
nu=1/2
ab=sr_freud(dig,nofdig,N,mu,nu);
digits(nofdig); nu=vpa(nu);
err=vpa(zeros(N-1,1)); cc=vpa(zeros(N-1,1));
co=zeros(N/2,1); ce=zeros(N/2,1);
do=zeros((N-2)/2,1); de=zeros((N-2)/2,1);
dc=zeros(N-2,1);
ainf=(2*gamma(nu)/(gamma(nu/2)^2))^(-1/nu);
for k=2:N
  vk=vpa(k);
  ak=(vk-1)^(-1/nu)*sqrt(ab(k,2));
  err=(ak-ainf)/ainf; c=(vk-1)^2*err;
%  err=(ak-ainf)/ainf; c=(vk-1)*err;
  am=max(abs(double(ak)),abs(double(ainf))); d=abs(double(ak-ainf));
  can=am/d;
  erra=abs(double(err));
  error(k-1)=erra;
  cc(k-1)=c;
  if k>=80
    fprintf(f0,k-1,erra,double(c),can)
  end
end
for k=1:N/2
  co(k)=double(cc(2*k-1));
  if k<N/2
    ce(k)=double(cc(2*k));
  end
end
coe=[co ce]
for k=2:N-1
  dc(k-1)=double(cc(k)-cc(k-1));
end
dc
dmin=min(abs(dc)); dmax=max(abs(dc));
dminmax=[dmin dmax]
for k=3:2:99
  do((k-1)/2)=double(cc(k)-cc(k-2));
  if k<=97
  de((k-1)/2)=double(cc(k+1)-cc(k-1));
  end
end
doe=[do de]
k=(1:99)';
%semilogy(k,error);set(gca,'FontSize',14)
%axis([0 100 10^(-5) 1])
plot(k,cc);set(gca,'FontSize',14)
xlabel('k')
```

For the values $\nu = 4:2:10$ it yields the limits shown in Table 2.7.

Table 2.7. *The coefficient γ for ν even*

ν	γ
4	.020833
6	.027776
8	.031248
10	.033330

A typical plot (for $\nu = 4$) is shown in Fig. 2.12.

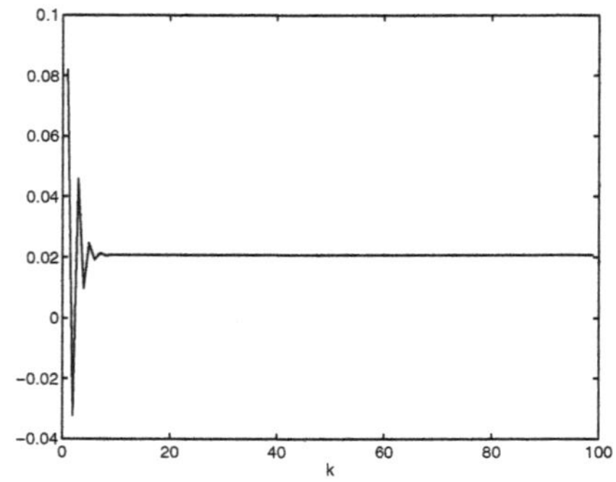

Figure 2.12. *The sequence $k^2(k^{-\nu}\sqrt{\beta_k} - c_0)/c_0$, $k = 1:99$, when $\nu = 4$ and $\mu = 0$*

When $\mu \neq 0$, experimentation with the script `freud_conj.m` reveals, first of all, that $\gamma_k = O(k^{-1})$, not $O(k^{-2})$ as in the case $\mu = 0$, and more precisely,

$$\frac{k^{-\nu}\sqrt{\beta_k} - c_0}{c_0} \sim \begin{cases} \gamma_o k^{-1} & \text{when } k\,(\text{odd}) \to \infty, \\ \gamma_e k^{-1} & \text{when } k\,(\text{even}) \to \infty, \end{cases} \qquad \mu \neq 0, \tag{2.53}$$

for some constants γ_o and γ_e. Moreover, the left-hand side of (2.53) multiplied by k approaches γ_o and γ_e monotonically increasing, resp., decreasing. Numerical data are shown in Table 2.8, the top number in each entry referring to k odd, the bottom number to k even.

Table 2.8. *The coefficients γ_o and γ_e*

$\mu\backslash\nu$	4	6	8	10
−1/2	−.250	−.250	−.250	−.250
	.125	.167	.188	.201
1/2	.250	.250	.250	.250
	−.125	−.166	−.187	−.199
1	.498	.498	.498	.498
	−.249	−.332	−.373	−.398

(c) The script `freud_conj.m` allows us to experiment with the sequence γ_k in (2.52) for values of ν other than the even integers ≥ 4, when $\mu = 0$. Overall, the conclusion is that the γ_k with odd or even indices k seem to converge separately to different values, which may be very close to each other (in some cases, like $\nu = .1$ or $\nu = .2$, even the

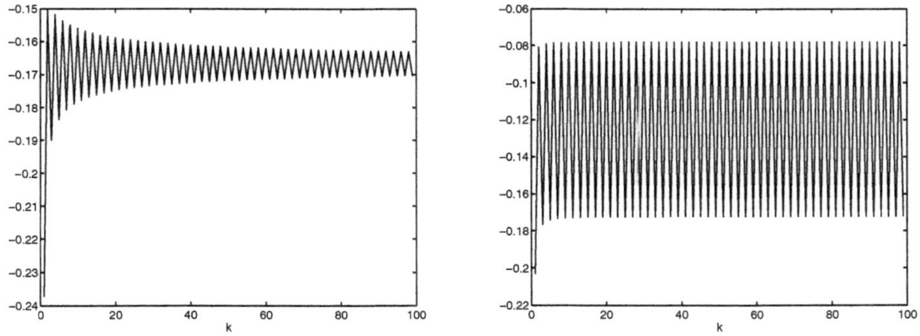

Figure 2.13. *The sequence* γ_k *for* $\nu = .4$ *(left) and* $\nu = .5$ *(right)*

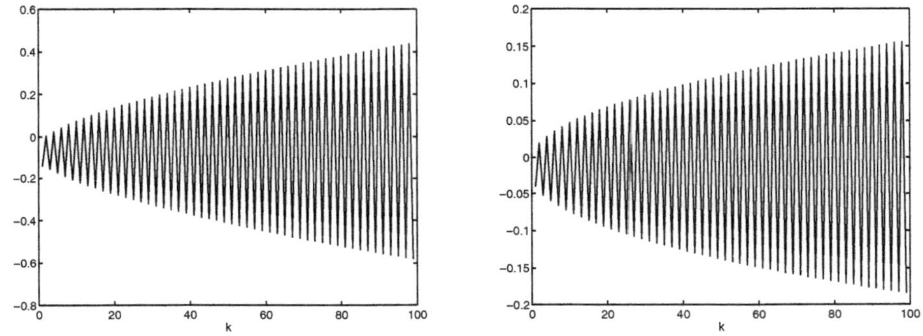

Figure 2.14. *The sequence* γ_k *for* $\nu = .75$ *(left) and* $\nu = 1.5$ *(right)*

same). The manner in which they converge, however, differs considerably from region to region. When $0 < \nu \leq 1/2$, convergence is typically monotone, for example, monotone decreasing for k even, and monotone increasing for k odd, as shown in Fig. 2.13 on the left for $\nu = .4$. The plot on the right is for $\nu = 1/2$, where convergence for both, k even and k odd, is monotone increasing, though at a very small rate (in increments of about 10^{-6} per step). For the range $1/2 < \nu < 2$, the type of convergence (if at all) is illustrated in Fig. 2.14 for $\nu = .75$ on the left and $\nu = 1.5$ on the right. As ν approaches $\nu = 2$, the envelopes at the top and bottom flatten out and at the same time approach each other, in the limit becoming the positive real line, as they should. Between $\nu = 2$ and $\nu = 4$, the behavior is again as in the left of Fig. 2.13. At $\nu = 4$, convergence is monotone increasing for $k \geq 11$, but at a very small rate (about 10^{-8} per step); see Fig. 2.15. In the case of $\nu = 5, 7, 9$, the behavior of the sequence γ_k is very much like the one observed when $\nu = 4, 6, 8$, or 10, and therefore (2.51) may well be valid also for odd integers $\nu \geq 5$, perhaps even for $\nu = 3$ (but not for $\nu = 1$). Possible numerical values are as shown in Table 2.9.

Table 2.9. *The coefficient* γ *for* ν *odd*

ν	γ
5	.024999
7	.029760
9	.032405

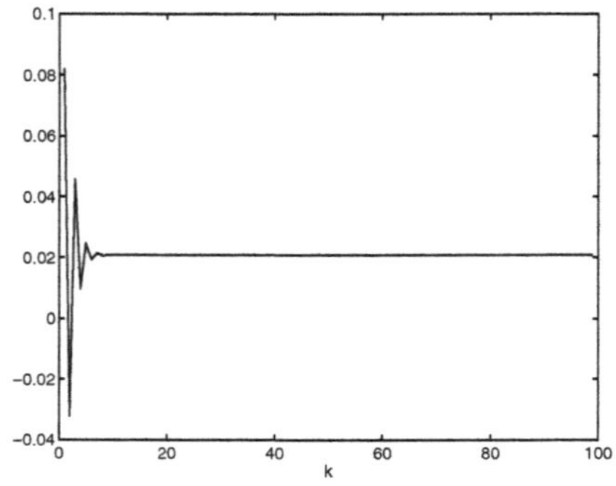

Figure 2.15. *The sequence γ_k for $\nu = 4$*

(d) The answer is a qualified yes. From numerical experiments with the script `hrf_conj.m`, using the function `sr_hrf.m` from Ex. 2.20(c),

```
%HRF_CONJ

f0='%8.0f %29.12e\n';
mu=0; N=100; dig=160; nofdig=28;
nu=10
ab=sr_hrf(dig,nofdig,N,mu,nu);
digits(nofdig); nu=vpa(nu);
b=zeros(N-1,1);
for k=2:N
  vk=vpa(k);
  ak=(vk-1)^(-1/nu)*sqrt(ab(k,2));
  b(k-1)=double(ak);
  fprintf(f0,k-1,double(ak))
end
k=(1:N-1)';
plot(k,b);set(gca,'FontSize',14)
xlabel('k')
```

it appears that the limit of $k^{-1/\nu}\sqrt{\beta_k}$ as $k \to \infty$ does indeed exist. But it depends not only on ν, but also on μ. Numerical values for $\mu = 0$ are shown in Table 2.10. The trivial case here is $\nu = 1$, when the half-range Freud polynomials are the generalized Laguerre polynomials for which $\sqrt{\beta_k} = \sqrt{k(k+\mu)}$ (see, for example, [24, Table 1.1]), and thus $k^{-1}\sqrt{\beta_k} \sim 1$ as $k \to \infty$.

Table 2.10. *The Freud conjecture for half-range Freud polynomials when $\mu = 0$*

ν	$\lim k^{-1/\nu}\sqrt{\beta_k}$	ν	$\lim k^{-1/\nu}\sqrt{\beta_k}$
.1	8.8689×10^{12}	5	.275508
.5	9.8695	6	.266811
1	1.000000	7	.261337
2	.408249	8	.257661
3	.321832	9	.255074
4	.290717	10	.253190

Exercise 2.25. *Consider the half-range Freud weight function* $w(t)=e^{-t^3}$ *on* $[0,\infty]$ *and its close relative*

$$w^*(t)=e^{-t^3/3} \quad \text{on } [0,\infty], \tag{2.54}$$

and let the respective recurrence coefficients be $[\alpha_k,\beta_k]$ *and* $[\alpha_k^*,\beta_k^*]$.

(a) *Show that*

$$\alpha_k^* = 3^{1/3}\alpha_k, \quad k=0,1,2,\ldots;$$
$$\beta_0^* = 3^{1/3}\beta_0, \quad \beta_k^* = 3^{2/3}\beta_k, \quad k=1,2,\ldots.$$

(b) *Compute the first* $N=15$ *coefficients* $[\alpha_k^*,\beta_k^*]$, $k=0,1,2,\ldots,N-1$, *to 32 digits, using* (a) *and the routine* `sr_hrf.m` *of* Ex. 2.20(c). The same recurrence coefficients have been computed to 16 digits in [16], using a different method (discretization), and have been presumed accurate to all 16 digits given there, in contrast to numerical results published in [47]. This part of the exercise is to show that the presumption is indeed justified.

(c) *The same as* (b), *but for* $N=200$.

Answer

(a) Let the respective monic orthogonal polynomials be π_k and π_k^*. Then from

$$\int_0^\infty \pi_k(t)\pi_\ell(t)e^{-t^3}\mathrm{d}t = 3^{-1/3}\int_0^\infty \pi_k(3^{-1/3}\tau)\pi_\ell(3^{-1/3}\tau)e^{-\tau^3/3}\mathrm{d}\tau = 0, \quad k\neq \ell,$$

it follows that

$$\pi_k^*(\tau)=3^{k/3}\pi_k(3^{-1/3}\tau), \quad k=0,1,2,\ldots.$$

Therefore, changing variables $\tau=3^{1/3}t$,

$$\begin{aligned}\alpha_k^* &= \frac{\int_0^\infty \tau[\pi_k^*(\tau)]^2 e^{-\tau^3/3}\mathrm{d}\tau}{\int_0^\infty [\pi_k^*(\tau)]^2 e^{-\tau^3/3}\mathrm{d}\tau} = \frac{\int_0^\infty 3^{1/3}t[\pi_k^*(3^{1/3}t)]^2 e^{-t^3}3^{1/3}\mathrm{d}t}{\int_0^\infty [\pi_k^*(3^{1/3}t)]^2 e^{-t^3}3^{1/3}\mathrm{d}t} \\ &= \frac{3^{(2k+2)/3}\int_0^\infty t[\pi_k(t)]^2 e^{-t^3}\mathrm{d}t}{3^{(2k+1)/3}\int_0^\infty [\pi_k(t)]^2 e^{-t^3}\mathrm{d}t} = 3^{1/3}\alpha_k, \quad k=0,1,2,\ldots,\end{aligned}$$

and similarly

$$\beta_0^* = 3^{1/3}\beta_0, \quad \beta_k^* = 3^{2/3}\beta_k, \quad k=1,2,\ldots.$$

(b) Running the routine `dig_hrf.m` of Ex. 2.20(b) with $N=15$, $\mu=0$, $\nu=3$, `dig0` $=40$, and `dd` $=1$ tells us that for `nofdig` $=32$ one needs 42-digit arithmetic. The script `run_sr_hrf.m` (not shown here), run with this number of digits and incorporating the modifications in (a), then yields for the recurrence coefficients associated with the weight function (2.54),

```
>> run_sr_hrf
ab =
[ 0.72901113294722698141863626470394,  1.28789931685406908720068316002 88]
[  1.04221982567474409113630561280 72, 0.24500097941742094234842574778401]
[  1.25373064220196481298720143752 58,  0.35307351727990706406124026858 49]
[   1.4061820889340038794916026341 03, 0.45380654475472013456007795220385]
[  1.53047170886982661661215729476 39, 0.54670915163293606036098161020572]
[  1.63711468769310102193644787180 41, 0.63279143126565640324575431661673]
[  1.73132652800093136169757882536 98, 0.71359155024155923743765162022757]
[  1.81621572840939896325122845569 16, 0.79017160081817896753892393209209]
[  1.89380331629450606032360803522 83,  0.86327669550039949557728853867 11]
[  1.96548682633123741762042573101 36, 0.93345297398370756848555274696548]
[  2.03227833945823938936323097838 11,  1.00111430192640155665127326087 21]
[   2.0949374105669606086639947880 02,  1.06658307630640518414496520518 17]
[   2.1540505128026897945531915916 87,   1.1301163055170059928542543210 71]
[  2.21008111612034238210057667079 15,   1.1919228879829774855536020259 29]
[  2.26340263870694175539303616311 96,  1.25217543914882990583139850411 36]
>>
```

These agree with those in Table 4.3 of [16] to all 16 decimal digits given there and reproduce the 15-point Gauss quadrature formula of [16, Table 2.2] accurately to all 16 digits.

(c) When $N = 200$, `nofdig` $= 32$, the routine `dig_hrf.m` yields `dig` $= 269$ for the number of digits needed. The script `run_sr_hrf.m`, run with the appropriate changes, then produces the first 200 recurrence coefficients for (2.54) to 32 digits. They are made available in the file `ab_hrf3mod.txt`. For an application to computing complex inhomogeneous Airy functions, see Ex. 5.5.

Exercise 2.26.* (Ex. 12* of [27, §2], revised.) The original exercise is replaced here with the following simplified, and at the same time more general, exercise.

Orthogonal polynomials $\{\pi_k(\cdot\,; w)\}$ relative to the weight function ("exponential integral")

$$w(t) = E_\nu(t),\ \ E_\nu(t) = \int_1^\infty \frac{e^{-tu}}{u^\nu}\,du \ \ \textit{on}\ (0,\infty), \quad \nu > 0.$$

For $\nu = 1$ and 2, these are of interest in problems of radiative transfer; see [3, Ch. II, §23].

(a) *Apply the classical Chebyshev algorithm in high-precision arithmetic (symbolic MATLAB) to generate the first n recurrence coefficients $\alpha_k(w)$, $\beta_k(w)$, $k = 0, 1, \ldots, n-1$. In the case $\nu = 1$ determine numerically the underlying degree of ill-conditioning and check your answers for $n = 20$ against* [5, Table 3]. *Also check against the two- to four-point Gaussian quadrature formulae given in* [55] *for $\nu = 1:5$.*

(b) *Develop a modified Chebyshev algorithm using for p_k an appropriate monic polynomial of degree k. Is it competitive with the classical Chebyshev algorithm?*

(c) *Adapt your work in* (a) *to deal with*

$$w(t) = E_\nu(t) \ \ \textit{on}\ [0, c], \quad 0 < c < \infty, \ \ \nu = m \in \mathbb{N},$$

and check your answers against the 3- and 4-point quadrature formulae given in [45] *for $m = 1$ and $m = 2$.*

Answer

(a) We need to compute the moments of $w(t)=E_\nu(t)$; they are

$$\begin{aligned}\mu_k &= \int_0^\infty t^k E_\nu(t)\mathrm{d}t = \int_0^\infty t^k \mathrm{d}t \int_1^\infty \frac{e^{-tu}}{u^\nu}\mathrm{d}u \\ &= \int_1^\infty \frac{\mathrm{d}u}{u^\nu}\int_0^\infty t^k e^{-ut}\mathrm{d}t = \int_1^\infty \frac{\mathrm{d}u}{u^\nu}\int_0^\infty \left(\frac{\tau}{u}\right)^k e^{-\tau}\frac{\mathrm{d}\tau}{u} \\ &= k!\int_1^\infty \frac{\mathrm{d}u}{u^{\nu+k+1}},\end{aligned}$$

that is,

$$\mu_k = \frac{k!}{\nu+k}, \quad k=0,1,2,\ldots, \tag{2.55}$$

or, evaluated recursively,

$$\begin{aligned}\mu_k &= k\frac{\nu+k-1}{\nu+k}\mu_{k-1}, \quad k=1,2,\ldots, \\ \mu_0 &= \frac{1}{\nu}.\end{aligned} \tag{2.56}$$

This is implemented in the symbolic MATLAB routine smom_Enu.m (cf. Appendix B.3),

```
function mom=smom_Enu(dig,N,nu)
%SMOM_ENU Moments of the exponential integral weight function.

digits(dig);
mom=vpa(zeros(2*N,1)); snu=vpa(nu);
mom(1)=1/snu;
for k=1:2*N-1
  mom(k+1)=(k*(snu+k-1)/(snu+k))*mom(k);
end
```

The double-precision analogue is mom_Enu.m.

The Chebyshev algorithm using these moments is implemented (in symbolic MATLAB) in the routine sr_Enu.m (cf. Appendix B),

```
function ab=sr_Enu(dig,N,nu)
%SR_ENU Recurrence coefficients for the exponential integral
%weight function.

digits(dig);
mom=smom_Enu(dig,N,nu);
ab=schebyshev(dig,N,mom');
%ab
```

using the SOPQ routine schebyshev.m (cf. Appendix B.1). The analogue in MATLAB double precision is r_Enu.m.

It is important to know, however, how much precision is needed to achieve results of given accuracy. For this, it is necessary to estimate the underlying condition numbers. This can be done by comparing dig0-digit results with dig1-digit results, dig1 $\gg$ dig0. If the absolute value of the relative errors are divided by $.5\times 10^{-\mathrm{dig0}}$, this gives fairly realistic estimates of the condition numbers involved. The relevant routine is condEnu.m,

```
function cond=condEnu(N,nu,dig0,dig1)
  %CONDENU Condition number estimates for the exponential integral
  %weight function.
  %   COND=CONDENU(N,NU,DIG0,DIG1) computes the Nx1 array COND of
  %   condition number estimates for obtaining the first n, n<=N,
  %   recurrence coefficients for the weight function W(T)=int_1^Inf
  %   EXP(-Tu)du/u^NU on [0,Inf], NU>0, from its moments by comparing
  %   DIG0-digit results with DIG1-digit results, DIG1>>DIG0.

  cond=zeros(N,1);
  ab0=sr_Enu(dig0,N,nu);
  ab1=sr_Enu(dig1,N,nu);
  err=double(abs((ab0-ab1)./ab1));
  eps0=.5*10^(-dig0); c=err/eps0;
  for n=1:N
    cond(n)=max(c(n,1),c(n,2));
  end
```

When $N = 20$ and $\nu = 1$, for example, the condition numbers produced by the script `run_condEnu.m` (not shown) are displayed in Table 2.11.

Table 2.11. *Condition number estimates for obtaining the recurrence coefficients for the weight function $E_1(t)$ on $[0, \infty]$ from its moments*

n	cond	n	cond	n	cond	n	cond
2	2.8371e-008	7	2.1587e-004	12	4.4659e+000	17	1.2645e+006
3	5.5819e-008	8	2.9996e-003	13	1.1480e+002	18	8.0359e+006
4	6.4849e-007	9	2.0068e-002	14	1.6056e+003	19	3.3973e+007
5	7.1052e-006	10	1.0529e-001	15	1.7399e+004	20	9.7683e+007
6	3.1876e-005	11	3.9288e-001	16	1.6018e+005		

The script also plots the condition numbers for $N = 40$, $\nu = m = 1:5$, in a logarithmic scale as shown in Fig. 2.16. It can be seen that they are relatively insensitive to the particular values of m. For $n = 40$, one must expect a condition number of the order 10^{28}.

Suppose now that we want the first 40 recurrence coefficients accurate to 32 decimal digits. According to Fig. 2.16, we would need a working precision of about $28 + 32 = 60$

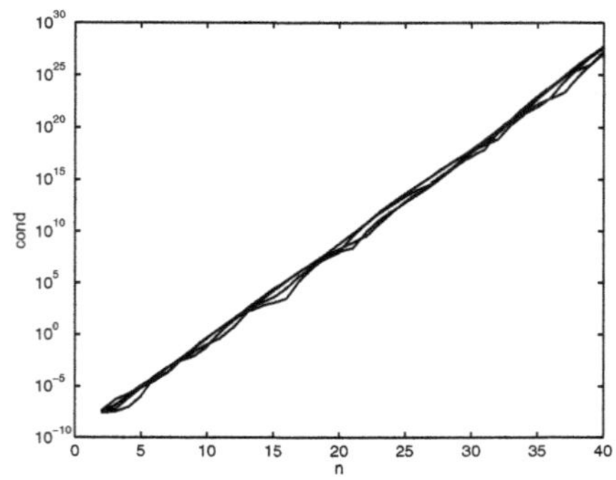

Figure 2.16. *Condition numbers for $E_m(t)$, $m = 1:5$ on $[0, \infty]$*

digits. Indeed, with `dig=60`, the maximum relative discrepancy in the recurrence coefficients, as determined by the script `run_sr_Enu.m`,

```
%RUN_SR_ENU

f0='%8.0f %12.4e\n';
N=40;
disp('       m       err')
for m=1:5
  ab1=sr_Enu(60,N,m);
  ab2=sr_Enu(68,N,m);
  err=max(max(double(abs((ab1-ab2)./ab2))));
  fprintf(f0,m,err)
end
%ab1
%xw=sgauss(60,10,ab1(1:10,:))
%xw=sgauss(60,20,ab1(1:20,:))
```

are as shown in Table 2.12, which should be sufficient to get 32 correct significant digits.

Table 2.12. *Accuracy check for* $E_m(t)$, $m = 1:5$

m	1	2	3	4	5
maxerr	4.31×10^{-34}	4.37×10^{-33}	6.27×10^{-34}	1.03×10^{-33}	8.34×10^{-34}

When $m = 1$, the result for `ab1`, rounded to 32 significant digits, is made available in the file `ab_expint.txt`. The beginning and end thereof is shown below:

```
ab1 =
    [  0.5,                                 1.0                               ]
    [  2.3,                                 0.41666666666666666666666666666667]
    [  4.2346938775510204081632653061224,   2.6133333333333333333333333333333]
    [  6.1991728899520653289533363162975,   6.7549479383590170762182423990004]
    [  8.1760287637692264292699317909546,  12.870010159219961694306927947982]
    [ 10.159422455949061212251865164385,   20.968994696423940837568730033392]

                       ⋮                                  ⋮

    [ 72.062054026674477715769790223624, 1262.4424585985283766786742605580]
    [ 74.061220720875801808371126074796, 1334.4740333609488353760830067157]
    [ 76.060420084426113329606309812572, 1408.5051703790570651814974233587]
    [ 78.059650037853873322599060693536, 1484.5358873226525872793900026879]
```

It took 87 seconds to produce (on an Intel Pentium D 830 3GHz dual core processor with 1GB DDR2 non-ECC SDRAM 2DIM memory). The first 12 lines in the file, incidentally, confirm the 12-digit results in [5, Table 3] and also, if used in `sgauss.m`, those in Table 4 except for two end-figure errors of one unit.

In order to check the 7-digit results in [55], ordinary MATLAB double precision is more than enough. This is done by the script `Reiz.m`,

```
%REIZ
```

```
N=4;
for m=1:5
  fprintf('    m=%1.0f\n',m)
  ab=r_Enu(N,m);
  for n=2:N
    xw=gauss(n,ab)
    fprintf('\n')
  end
end
```

There is agreement with the results in the cited reference except for occasional end-figure discrepancies of one unit.

(b) Let p_k be a monic polynomial of degree k. The corresponding modified moments are

$$\begin{aligned} m_k &= \int_0^\infty p_k(t)E_\nu(t)\mathrm{d}t = \int_0^\infty p_k(t)\mathrm{d}t \int_1^\infty \frac{e^{-tu}}{u^\nu}\mathrm{d}u \\ &= \int_1^\infty \frac{\mathrm{d}u}{u^\nu}\int_0^\infty p_k(t)e^{-tu}\mathrm{d}t = \int_1^\infty \frac{\mathrm{d}u}{u^{\nu+1}}\int_0^\infty p_k\left(\frac{\tau}{u}\right)e^{-\tau}\mathrm{d}\tau. \end{aligned} \tag{2.57}$$

Since $k \le 2n-1$ and $p_k(\tau/u)$ is a polynomial of degree $\le 2n-1$ in the variable τ, the last integral in (2.57) can be exactly evaluated by n-point Gauss–Laguerre quadrature,

$$\int_0^\infty p_k\left(\frac{\tau}{u}\right)e^{-\tau}\mathrm{d}t = \sum_{r=1}^n \lambda_r^L p_k\left(\frac{\tau_r^L}{u}\right),$$

where τ_r^L, λ_r^L are the Gauss–Laguerre nodes and weights. Substituting this in the last integral of (2.57) and changing variables, $u = 1/t$, yields

$$m_k = -\int_1^0 t^{\nu+1}\sum_{r=1}^n \lambda_r^L p_k(t\tau_r^L)t^{-2}\mathrm{d}t = \int_0^1 \left[\sum_{r=1}^n \lambda_r^L p_k(t\tau_r^L)\right]t^{\nu-1}\mathrm{d}t.$$

In the last integral, the sum in brackets, being a polynomial (in t) of degree $\le 2n-1$, can be exactly evaluated by n-point Gauss–Jacobi quadrature on $[0,1]$ with Jacobi parameters $\alpha = 0$, $\beta = \nu - 1$, so that, with τ_s^J, λ_s^J denoting its nodes and weights,

$$m_k = \sum_{r=1}^n\sum_{s=1}^n \lambda_s^J\lambda_r^L p_k(\tau_s^J\tau_r^L), \quad 0 \le k \le 2n-1. \tag{2.58}$$

We are still at liberty to choose p_k. It should be a monic orthogonal polynomial resembling the one we seek, in particular having a measure of orthogonality supported on $[0,\infty]$ which behaves exponentially decreasing at ∞ and is mildly singular at the origin. A choice coming close to fulfilling these requirements is the Laguerre polynomial, perhaps the generalized one with parameter $\alpha < 0$ and $|\alpha|$ small. Thus we will try the monic generalized Laguerre polynomial, $p_k = \hat{L}_k^{(\alpha)}$, for which the modified moments (2.58) are produced by the routine `mmom_Enu.m` below, the symbolic analogue being `smmom_Enu.m`.

```
function mom=mmom_Enu(N,nu,a)
%MMOM_ENU Modified moments of the exponential integral weight
%function.
%   MOM=MMOM_ENU(N,NU,A) generates the 2Nx1 array MOM of the
```

```
%   first 2N modified moments of the weight function
%   W(T)=int_1^Inf EXP(-Tu)du/u^NU on [0,Inf], NU>0, relative
%   to monic generalized Laguerre polynomials with parameter A.

mom=zeros(2*N,1);
abj=r_jacobi01(N,0,nu-1);
abl=r_laguerre(2*N-1,a);
xwj=gauss(N,abj);
xwl=gauss(N,abl);
P=zeros(N,N,2*N);
for r=1:N
  for s=1:N
    x=xwj(s,1)*xwl(r,1);
    P(r,s,1)=sum(xwj(:,2))*sum(xwl(:,2));
    p0=0; p=1;
    for j=1:2*N-1
      pm1=p0; p0=p;
      p=(x-abl(j,1))*p0-abl(j,2)*pm1;
      P(r,s,j+1)=p;
    end
  end
end
for k=1:2*N
  sig=0;
  for r=1:N
    for s=1:N
      w=xwj(s,2)*xwl(r,2);
      sig=sig+w*P(r,s,k);
    end
  end
  mom(k)=sig;
end
```

The numerical condition may be explored similarly as in (a) by the routine condEnumod.m,

```
function cond=condEnumod(N,nu,a,dig0,dig1)
%CONDENUMOD Condition number estimates for the exponential
%integral weight function.
%   COND=CONDENUMOD(N,NU,A,DIG0,DIG1) computes the Nx1 array COND
%   of condition number estimates for obtaining the first n, n<=N,
%   recurrence coefficients for the weight function W(T)=int_1^Inf
%   EXP(-Tu)du/u^NU on [0,Inf], NU>0, from its modified moments
%   relative to monic generalized Laguerre polynomials with
%   parameter A by comparing DIG0-digit results with DIG1-digit
%   results, DIG1>>DIG0.

cond=zeros(N,1);
ab0=sr_Enumod(dig0,N,nu,a);
ab1=sr_Enumod(dig1,N,nu,a);
err=double(abs((ab0-ab1)./ab1));
eps0=.5*10^(-dig0); c=err/eps0;
```

```
for n=1:N
  cond(n)=max(c(n,1),c(n,2));
end
```

using `sr_Enumod.m`, the obvious analogue of `sr_Enu.m`. It is considerably slower than its companion routine, `condEnu.m`, since the routine `sr_Enumod.m` takes about 20 times as much time as `sr_Enu.m`, the culprit being the modified moments (2.58). What is surprising is that the results (produced by the script `run_condEnumod.m` with $\nu = 1$, and shown in Table 2.13), when $\mathtt{a} = 0$ (but also for, say, $\mathtt{a} = 1/10$), are worse than those in (a). This indeed is the first time we have encountered an instance in which the modified Chebyshev algorithm, even with a reasonable choice of modified moments, performs worse than the classical Chebyshev algorithm using ordinary moments. There is no incentive, therefore, to pursue this approach any further.

Table 2.13. *Condition number estimates for obtaining the recurrence coefficients for the weight function $E_1(t)$ on $[0,\infty]$ from its modified moments*

n	cond	n	cond	n	cond	n	cond
2	1.6918e−004	7	3.8705e−002	12	2.1391e+002	17	1.6812e+007
3	4.1860e−004	8	4.0332e−001	13	1.8733e+003	18	1.9693e+008
4	2.3911e−004	9	1.8476e+000	14	7.2312e+003	19	1.9399e+009
5	4.9357e−004	10	5.2069e+000	15	1.3721e+004	20	2.0018e+010
6	1.9166e−003	11	4.1212e+000	16	8.8257e+005		

(c) The discussion in (b) suggests that we should stick with the classical Chebyshev algorithm (in sufficiently high precision) and not try to use modified moments, which in the case of a finite interval $[0,c]$ would not be so easy to compute anyway. So we take ordinary moments,

$$\mu_k = \int_0^c t^k E_m(t)\mathrm{d}t. \tag{2.59}$$

By elementary manipulations, similarly as in (a), one finds

$$\mu_k = \int_1^\infty t^{-(\nu+k+1)}\gamma(k+1,ct)\mathrm{d}t, \tag{2.60}$$

where $\gamma(a,x)$ is the (lower) incomplete gamma function

$$\gamma(a,x) = \int_0^x t^{a-1}e^{-t}\mathrm{d}t.$$

Integration by parts in (2.60), integrating the first factor and differentiating the second, yields

$$\mu_k = \frac{1}{\nu+k}[\gamma(k+1,c)+c^{\nu+k}\Gamma(1-\nu,c)], \quad k=0,1,2,\ldots, \tag{2.61}$$

where

$$\Gamma(a,x) = \int_x^\infty t^{a-1}e^{-t}\mathrm{d}t$$

is the (upper) incomplete gamma function. For large c, we have $\gamma(k+1,c) \sim k!$ and $\Gamma(1-\nu,c) \sim c^{-\nu}e^{-c}$ (cf. [1, Eq. 6.5.32]), so that, for fixed k,

$$\mu_k \sim \frac{1}{\nu+k}(k!+c^k e^{-c}), \quad c\to\infty,$$

recovering (2.55) in the limit as $c\to\infty$.

In order to compute μ_k from (2.61) in variable-precision arithmetic, one needs symbolic routines for incomplete gamma functions. These are provided in Ex. 5.3: the routine sgam_lowinc.m for the lower incomplete gamma function in part (b) of that exercise, and in part (c) the routine sgam_upinc.m for the upper incomplete gamma function. The resulting symbolic routine for μ_k, $k=0,1,\ldots,2n-1$, therefore is (cf. Appendix B.3)

```
function mom=smom_Enufin(dig,N,nu,c)
%SMOM_ENUFIN Moments of the finite exponential integral weight
%function.

digits(dig);
c=vpa(c); nu=vpa(nu);
mom=vpa(zeros(2*N,1));
for k=1:2*N
  mom(k)=(sgam_lowinc(k,c)+c^(nu+k-1)*sgam_upinc(1-nu,c)/(nu+k-1);
end
```

The routine sr_Enufin.m (cf. Appendix B.1) then implements the Chebyshev algorithm,

```
function ab=sr_Enufin(dig,N,nu,c)
%SR_ENUFIN Recurrence coefficients for the finite exponential
%integral weight function.

digits(dig);
mom=smom_Enufin(dig,N,nu,c);
ab=schebyshev(dig,N,mom');
%ab
```

The double-precision analogue is r_Enufin.m.

There is, however, again the question of conditioning. We can estimate condition numbers similarly as in (a) by a routine named condEnufin.m, which is an obvious analogue of the routines condEnu.m and condEnumod.m. If run (with the script run_condEnufin.m) for $N=40$, $m=1:5$, and $c=[.1\ .5\ 1\ 2\ 4\ 8\ 16\ 32\ 64\ 128]$, it produced the plots shown in Fig. 2.17 for $m=1$ (on the left) and $m=5$ (on the right).

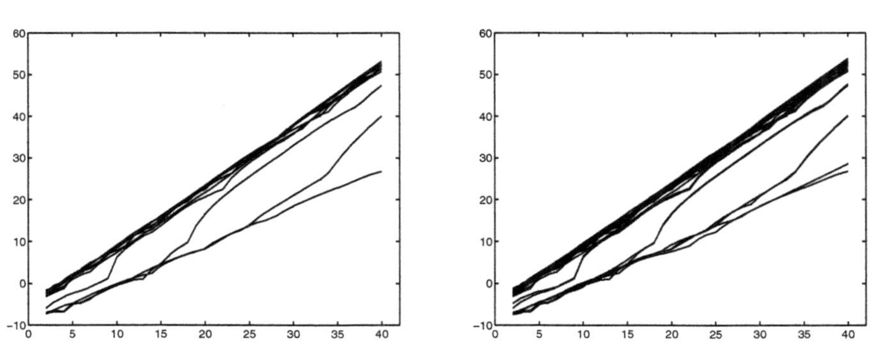

Figure 2.17. *Condition numbers for $E_m(t)$, $m=1$ and $m=5$, on $[0,c]$, $c=[.1\ .5\ 1\ 2\ 4\ 8\ 16\ 32\ 64\ 128]$*

As in the case $c=\infty$ (cf. Fig. 2.16), it is evident that the condition numbers are relatively insensitive to the choice of m. For moderately large values of c (say, $0<c\le 16$), and $n=40$, the condition number is of the order 10^{53}, suggesting that, to obtain results

to 32 correct digits, the number d of digits to be used should be about 85. As c becomes larger, the condition numbers slowly approach the ones for $c = \infty$, being of the order 10^{46} when $c = 32$, of the order 10^{39} when $c = 64$, and of the order 10^{27} when $c = 128$.

Actually, the script `run_sr_Enufin.m`,

```
%RUN_SR_ENUFIN

N=40; m=1; c=.1;
ab1=sr_Enufin(82,N,m,c);
ab2=sr_Enufin(90,N,m,c);
err=max(double(abs((ab1-ab2)./ab2)))
```

can be used to determine by trial and error the exact number of digits d needed. In the first call to the routine `sr_Enufin.m`, the number of digits (the first input argument) is guessed (82 in the above display for $m = 1$, $c = .1$) and then adjusted, either upward or downward, in accordance with the maximum error produced. In the second call to `sr_Enufin.m`, the number of digits is kept fixed (at 90). The results agree very closely with the estimates given above: for 32-digit accuracy and $N = 40$, $c = [16\ 32\ 64\ 128]$, the required number d of digits is shown in Table 2.14 for $m = 1:5$. Note also that the results are practically independent of m in the range $1 \le m \le 5$.

Table 2.14. *Number d of digits needed to get results to 32 correct decimal digits*

m	c	d	m	c	d	m	c	d	m	c	d	m	c	d
1	16	83	2	16	84	3	16	84	4	16	84	5	16	84
	32	79		32	79		32	79		32	79		32	79
	64	72		64	72		64	72		64	72		64	72
	128	61		128	61		128	60		128	60		128	60

We now use the script `Kegel.m`, calling on the double-precision routine `r_Enufin.m` and the `OPQ` routine `gauss.m`,

```
%KEGEL

f0='%8.0f %5.1f %12.8f %12.8f %12.8f\n';
f1='%27.8f %12.8f %12.8f\n';
f2='%8.0f %5.1f %12.8f %12.8f %12.8f %12.8f\n';
f3='%27.8f %12.8f %12.8f %12.8f\n';
disp('       m     c                        nodes/weights')
for m=1:2
  for c=[.1:.1:1 1.2:.2:4 4.5:.5:8 9:15]
    ab=r_Enufin(4,m,c);
    xw=gauss(3,ab); x=xw(:,1); w=xw(:,2);
    fprintf(f0,m,c,x(1),x(2),x(3))
    fprintf(f1,w(1),w(2),w(3))
    xw=gauss(4,ab); x=xw(:,1); w=xw(:,2);
    fprintf(f2,m,c,x(1),x(2),x(3),x(4))
    fprintf(f3,w(1),w(2),w(3),w(4))
    fprintf('\n')
  end
end
```

to reproduce Tables 1–4 in [45]. Spot-checking the results supports the author's claim of accuracies up to one or two units in the last decimal digit.

Exercise 2.27.* *Consider the measure*

$$\mathrm{d}\lambda(t) = w_p(t;\alpha)\mathrm{d}t, \quad w_p(t;\alpha) = t^\alpha[\ln(1/t)]^{p-1} \text{ on } [0,1], \tag{2.62}$$

where $\alpha > -1$ is nonintegral and $p = 2$ or 3. When $\alpha = -1/2$ this measure arises in the summation of certain slowly convergent series (see Ex. 5.24).

(a) *Let $p = 2$. Evaluate the modified moments of* (2.62),

$$m_n(\alpha) = \int_0^1 P_n^*(t)w_2(t;\alpha)\mathrm{d}t, \quad n = 0,1,2,\ldots, \tag{2.63}$$

relative to the shifted Legendre polynomials $P_n^*(t) = P_n(2t-1)$. {Hint: Use the known formula [9, p. 316, Eq. (15)]

$$\int_{-1}^1 (1+t)^\alpha P_n(t)\mathrm{d}t = \frac{2^{\alpha+1}\Gamma^2(\alpha+1)}{\Gamma(\alpha+n+2)\Gamma(\alpha+1-n)}, \quad \alpha > -1,\ n \geq 0,$$

and the one obtained by differentiation with respect to α.}

(b) *Do the same as in* (a), *but for $p = 3$.* {Hint: Differentiate $w_p(t;\alpha)$ with respect to α.}

(c) *For $p = 2$ and $p = 3$ use the modified Chebyshev algorithm with the respective modified moments* (relative to monic shifted Legendre polynomials $(k!^2/(2k)!)P_k^*(t)$) *to generate the first N recurrence coefficients α_k, β_k for $w_p(\cdot;\alpha)$.* In the case $p = 2$, this has been implemented in the `OPQ` routine `r_jaclog.m` (cf. Appendix A.1) and the symbolic `SOPQ` routine `sr_jaclog.m` (cf. Appendix B.1). *Write analogous routines* `r_jaclogsq.m` *and* `sr_jaclogsq.m` *for the case $p = 3$. Find out the number* `dig` *of digits needed to get 30-digit accuracy in the case $N = 100$, $\alpha = -1/2$, both for $p = 2$ and $p = 3$.*

Run both symbolic routines in the appropriate `dig`*-digit arithmetic, with $\alpha = -1/2$, to check Tables* 1 *and* 2 *of* [18]. *Comment on your results.*

Answer (cf. also [14] and [24, Example 2.27]).

(a) We first transform to the interval $[-1,1]$,

$$\begin{aligned} m_n(\alpha) &= \int_0^1 P_n(2t-1)w_2(t;\alpha)\mathrm{d}t = \frac{1}{2}\int_{-1}^1 P_n(\tau)w_2\left(\frac{1+\tau}{2};\alpha\right)\mathrm{d}\tau \\ &= \frac{1}{2}\int_{-1}^1 P_n(t)\left(\frac{1+t}{2}\right)^\alpha \ln\left(\frac{2}{1+t}\right)\mathrm{d}t \\ &= \frac{1}{2^{\alpha+1}}\int_{-1}^1 P_n(t)(1+t)^\alpha[\ln 2 - \ln(1+t)]\mathrm{d}t. \end{aligned}$$

Using the hint, we get

$$m_n(\alpha) = \frac{\Gamma^2(\alpha+1)}{\Gamma(\alpha+n+2)\Gamma(\alpha+1-n)}\ln 2 - \frac{1}{2^{\alpha+1}}\int_{-1}^1 P_n(t)(1+t)^\alpha \ln(1+t)\mathrm{d}t.$$

To deal with the integral on the right, we differentiate the identity in the Hint with respect to α,

$$\int_{-1}^{1} P_n(t)(1+t)^\alpha \ln(1+t)\mathrm{d}t = \frac{\mathrm{d}}{\mathrm{d}\alpha}\left(\frac{2^{\alpha+1}\Gamma^2(\alpha+1)}{\Gamma(\alpha+n+2)\Gamma(\alpha+1-n)}\right)$$

$$= \frac{1}{\Gamma^2(\alpha+n+2)\Gamma^2(\alpha+1-n)}\{\Gamma(\alpha+n+2)\Gamma(\alpha+1-n)[2^{\alpha+1}\ln 2\Gamma^2(\alpha+1)$$

$$+2^{\alpha+1}\cdot 2\Gamma(\alpha+1)\Gamma'(\alpha+1)] - 2^{\alpha+1}\Gamma^2(\alpha+1)[\Gamma'(\alpha+n+2)\Gamma(\alpha+1-n)$$

$$+\Gamma(\alpha+n+2)\Gamma'(\alpha+1-n)]\}$$

$$= 2^{\alpha+1}\frac{[\Gamma^2(\alpha+1)\ln 2 + 2\Gamma(\alpha+1)\Gamma'(\alpha+1)] - \Gamma^2(\alpha+1)[\psi(\alpha+n+2)+\psi(\alpha+1-n)]}{\Gamma(\alpha+n+2)\Gamma(\alpha+1-n)},$$

where $\psi(x)=\Gamma'(x)/\Gamma(x)$. Thus, simplifying,

$$\int_{-1}^{1} P_n(t)(1+t)^\alpha \ln(1+t)\mathrm{d}t$$
$$= \frac{2^{\alpha+1}\Gamma^2(\alpha+1)}{\Gamma(\alpha+n+2)\Gamma(\alpha+1-n)}[\ln 2 + 2\psi(\alpha+1) \tag{2.64}$$
$$-\psi(\alpha+n+2)-\psi(\alpha+1-n)].$$

Using the well-known identities $\Gamma(x+1)=x\Gamma(x)$, $\psi(x+1)=\psi(x)+1/x$, both applied from left to right and from right to left, we have

$$\Gamma(\alpha+n+2) = (\alpha+n+1)(\alpha+n)\cdots(\alpha+1)\Gamma(\alpha+1)$$
$$= (\alpha+1)\Gamma(\alpha+1)\prod_{k=1}^{n}(\alpha+1+k),$$
$$\Gamma(\alpha+1-n) = \frac{1}{\alpha+1-n}\Gamma(\alpha+2-n) = \frac{1}{\alpha+1-n}\,\frac{1}{\alpha+2-n}\Gamma(\alpha+3-n)$$
$$= \cdots = \frac{1}{(\alpha+1-n)(\alpha+2-n)\cdots\alpha}\Gamma(\alpha+1) = \Gamma(\alpha+1)\frac{1}{\prod_{k=1}^{n}(\alpha+1-k)},$$

$$\psi(\alpha+n+2) = \psi(\alpha+n+1) + \frac{1}{\alpha+n+1} = \psi(\alpha+n) + \frac{1}{\alpha+n+1} + \frac{1}{\alpha+n}$$
$$= \cdots = \psi(\alpha+1) + \frac{1}{\alpha+1} + \sum_{k=1}^{n}\frac{1}{\alpha+1+k},$$
$$\psi(\alpha+1-n) = \psi(\alpha+2-n) - \frac{1}{\alpha+1-n} = \psi(\alpha+3-n) - \frac{1}{\alpha+1-n} - \frac{1}{\alpha+2-n}$$
$$= \cdots = \psi(\alpha+1) - \sum_{k=1}^{n}\frac{1}{\alpha+1-k}.$$

Altogether, therefore,

$$m_n(\alpha) = \prod_{k=1}^{n}\frac{\alpha+1-k}{\alpha+1+k}\left\{\frac{1}{(\alpha+1)^2} + \frac{1}{\alpha+1}\sum_{k=1}^{n}\left(\frac{1}{k+\alpha+1} + \frac{1}{k-\alpha-1}\right)\right\}. \tag{2.65}$$

(For integer values of $\alpha \geq 0$, see [14, Eq. (1)].) This is implemented in the routines `mm_log.m` (cf. Appendix A.5) and `smm_log.m` (cf. Appendix B.3) with $n = 2N$.

(b) (Cf. [18, §5].) To get the result for $p = 3$, we follow the hint and differentiate w_p with respect to α,

$$\frac{\mathrm{d}}{\mathrm{d}\alpha} w_p(t;\alpha) = -t^\alpha[\ln(1/t)]^p = -w_{p+1}(t;\alpha).$$

Thus, taking the negative derivative of (2.63), that is, of (2.65), yields the modified moments in the case $p = 3$. To simplify the computation, we let

$$\begin{aligned}\Pi &= \sum_{k=1}^{n} \frac{\alpha+1-k}{\alpha+1+k},\\ \Sigma_1 &= \sum_{k=1}^{n} \frac{k}{(k+\alpha+1)(k-\alpha-1)},\\ \Sigma_2 &= \sum_{k=1}^{n} \frac{k}{(k+\alpha+1)^2(k-\alpha-1)^2}\end{aligned}$$

and note that

$$\sum_{k=1}^{n} \left(\frac{1}{k+\alpha+1} + \frac{1}{k-\alpha-1} \right) = 2\Sigma_1,$$

so that, by (2.65),

$$m_n(\alpha)|_{p=2} = \Pi \left\{ \frac{1}{(\alpha+1)^2} + \frac{2}{\alpha+1}\Sigma_1 \right\}. \tag{2.66}$$

Since

$$\frac{\mathrm{d}}{\mathrm{d}\alpha}\left(\frac{\alpha+1-k}{\alpha+1+k} \right) = \frac{2k}{(\alpha+1+k)^2},$$

we have

$$\begin{aligned}\frac{\mathrm{d}}{\mathrm{d}\alpha}\Pi &= \sum_{\ell=1}^{n} \left(\prod_{k\neq\ell} \frac{\alpha+1-k}{\alpha+1+k} \right) \frac{2\ell}{(\alpha+1+\ell)^2}\\ &= \left(\sum_{\ell=1}^{n} \frac{2\ell}{(\alpha+1+\ell)^2} \frac{\alpha+1+\ell}{\alpha+1-\ell} \right)\Pi\\ &= -2\Sigma_1\Pi.\end{aligned}$$

Furthermore,

$$\frac{\mathrm{d}}{\mathrm{d}\alpha}\Sigma_1 = \sum_{k=1}^{n} \frac{2(\alpha+1)k}{(k+\alpha+1)^2(k-\alpha-1)^2} = 2(\alpha+1)\Sigma_2.$$

Therefore, the negative derivative of (2.66) is

$$\begin{aligned}m_n(\alpha)|_{p=3} &= -\left(\frac{\mathrm{d}}{\mathrm{d}\alpha}\Pi \right) \left\{ \frac{1}{(\alpha+1)^2} + \frac{2}{\alpha+1}\Sigma_1 \right\}\\ &\quad + \Pi \left\{ \frac{2}{(\alpha+1)^3} + \frac{2}{(\alpha+1)^2}\Sigma_1 - \frac{2}{\alpha+1}\frac{\mathrm{d}}{\mathrm{d}\alpha}\Sigma_1 \right\}\\ &= 2\Sigma_1\Pi \left\{ \frac{1}{(\alpha+1)^2} + \frac{2}{\alpha+1}\Sigma_1 \right\} + \Pi \left\{ \frac{2}{(\alpha+1)^3} + \frac{2}{(\alpha+1)^2}\Sigma_1 - 4\Sigma_2 \right\},\end{aligned}$$

that is,

$$m_n(\alpha)|_{p=3} = \frac{2}{(\alpha+1)^3}\left\{1+2(\alpha+1)\Sigma_1+2(\alpha+1)^2\Sigma_1^2-2(\alpha+1)^3\Sigma_2\right\}\Pi. \tag{2.67}$$

(c) The two symbolic routines implementing the cases $p = 2$ and $p = 3$ are shown below along with the relevant modified moment routines (cf. Appendix B).

```
function ab=sr_jaclog(dig,N,a)
%SR_JACLOG  Recurrence coefficients for an algebraic/logarithmic
%weight function.

if nargin<3, a=0; end;
if((N<=0)|(a<=-1)) error('parameter(s) out of range'), end
abj=sr_jacobi01(dig,2*N);
mom=smm_log(dig,N,a);
ab=schebyshev(dig,N,mom,abj);

function mm=smm_log(dig,N,a)
%SMM_LOG Modified moments of an algebraic/logarithmic weight
%function.

if a<=-1, error('a out of range in mm_log'), end
digits(dig);
c=vpa(1);
for n=1:2*N
  if(a-floor(a)==0 & a<n-1)
    p=vpa(n-a-1:n+a);
    mm(n)=vpa((-1)^(n-1-a)/prod(p));
    mm(n)=vpa((gamma(a+1))^2*mm(n));
  else
    if n==1
      mm(1)=vpa(1/(a+1)^2);
    else
      k=1:n-1;
      s=vpa(1./(a+1+k)-1./(a+1-k));
      p=vpa((a+1-k)./(a+1+k));
      mm(n)=vpa((1/(a+1)+sum(s))*prod(p)/(a+1));
    end
  end
  mm(n)=c*mm(n); c=vpa(n*c/(2*(2*n-1)));
end

function ab=sr_jaclogsq(dig,N,a)
%SR_JACLOGSQ Recurrence coefficients for an algebraic/logarithmic
%weight function.

if nargin<2, a=0; end;
if((N<=0)|(a<=-1)) error('parameter(s) out of range'), end
abj=sr_jacobi01(dig,2*N);
mom=smm_logsq(dig,N,a);
ab=schebyshev(dig,N,mom,abj);
```

```
function mm=smm_logsq(dig,N,a)
%SMM_LOGSQ Modified moments of an algebraic/logarithmic
%weight function.

if a<=-1, error('a out of range in mm_log'), end
digits(dig);
c=vpa(1);
for n=1:2*N
  if n==1
    mm(1)=vpa(2/(a+1)^3);
  else
    k=1:n-1;
    s1=vpa(k./((k+a+1).*(k-a-1))); sig1=sum(s1);
    s2=vpa(k./((k+a+1).^2.*(k-a-1).^2)); sig2=sum(s2);
    p0=vpa((a+1-k)./(a+1+k)); p=prod(p0);
    mm(n)=vpa((2/(a+1)^3)*(1+2*(a+1)*sig1+2*(a+1)^2*sig1^2 ...
      -2*(a+1)^3*sig2)*p);
  end
  mm(n)=c*mm(n); c=vpa(n*c/(2*(2*n-1)));
end
```

The routine relevant to determining the number of digits needed for a given accuracy requirement, and delivering the array ab of recurrence coefficients to that accuracy, is dig_jaclog.m,

```
function [ab,dig]=dig_jaclog(N,a,dig0,dd,nofdig)
%DIG_JACLOG Number of digits needed for polynomials orthogonal
%with respect to an algebraic/logarithmic weight function.
%   [AB,DIG]=DIG_JACLOG(N,A,DIG0,DD,NOFDIG) determines the number
%   DIG of digits needed to obtain the first N recurrence
%   coefficients for the weight function W(T)=T^A*LOG(1/T) on
%   [0,1], A>-1, to an accuracy of NOFDIG decimal digits using
%   the modified Chebyshev algorithm. It successively increments
%   an initial estimate DIG0 by DD units until the required
%   accuracy is achieved. The recurrence coefficients to NOFDIG
%   digits are placed into the Nx2 array AB.

i=dig0-dd;
maxerr=1;
while maxerr>.5*10^(-nofdig)
  i=i+dd; dig=i
  mom=smm_log(dig,N,a);
  if i==dig0
    abj=sr_jacobi01(dig,2*N);
    ab0=schebyshev(dig,N,mom,abj);
  else
    abj=sr_jacobi01(dig,2*N);
    ab1=schebyshev(dig,N,mom,abj);
    serr=abs((ab1-ab0)./ab1);
    err=double(serr);
    maxerr=max(max(err));
```

```
      ab0=ab1;
    end
  end
  ab=vpa(ab1,nofdig);
```

Run with `N` $= 100$, `a` $= -1/2$, `dig0` $= 30$, `dd` $= 1$, `nofdig` $= 30$, to check the 25-digit Table 1 of [18], yields `dig` $= 31$, indicating that the modified Chebyshev algorithm in this case is quite stable. Comparison with Table 1 revealed agreement for all n up to 78, apart from occasional end-figure discrepancies in the last digit, or last two digits, both in the α_k and β_k. At $n = 79$, the discrepancy involves five terminal figures for α_n, and two for β_n. Thereafter, for all $n \geq 80$, the last 7 digits in the α_k, and the last 7–8 digits in the β_k are found to be incorrect in Table 1. This was verified by rerunning the routine with `dig=36` and confirming complete agreement to all 30 digits with the `dig=31` results.

Table 1 has been computed in Fortran quadruple precision arithmetic. Apparently, something went slightly wrong when $79 \leq n \leq 99$. We do not believe that the cause was ill-conditioning, since that would result in gradually worsening discrepancies, not in a constant degree of discrepancy arising abruptly. At any rate, we reproduce below an excerpt of the results obtained with `sr_jaclog.m` in 36-digit computation, rounded to 25 digits, in particular the last twenty lines of the table. The complete table, ready for the MATLAB `loadvpa` command, is available in the file `ab_jaclog.txt`.

```
>> ab=sr_jaclog(36,100,-1/2)

ab =

[ 0.1111111111111111111111111,                           4.0]
[ 0.4661483641075477810171688, 0.02765432098765432098765432]
[ 0.4880690581976426561739654, 0.05534292684170711183265476]
[ 0.4938743419208057331274822, 0.05940526298488865183067045]

                   ⋮                                  ⋮

[ 0.4999872490550774735344419,  0.0624959115439857486505043]
[ 0.4999875568874173722473441, 0.06249601257600626686934612]
[ 0.4999878536789730303318159,  0.0624961099159577925909712]
[ 0.4999881399521823295228062, 0.06249620374138053092395969]
[ 0.4999884161989171589408439, 0.06249629421927693285244579]
[ 0.4999886828826058172041719, 0.06249638150685309048657687]
[ 0.4999889404401853721131961, 0.06249646575220000340037342]
[ 0.4999891892838993771738144, 0.06249654709492022390839268]
[ 0.4999894298029547913878182, 0.06249662566670482827358916]
[ 0.4999896623650505697222936, 0.06249670159186516232276505]
[ 0.4999898873177891631924994, 0.06249677498782336708641053]
[  0.499990104989981070863317, 0.06249684596556529520655234]
[ 0.4999903156928516087286043, 0.06249691463005907698036343]
[ 0.4999905197211581865961978, 0.06249698108064228078379871]
[  0.499990717354225599513551, 0.06249704541138033176196802]
[  0.499990908856906140990331, 0.06249710771139860070742267]
[ 0.4999910944804707149856202, 0.06249716806519035065204342]
[ 0.4999912744634365576057699,  0.0624972265529025254004647]
```

```
[ 0.4999914490323366727962659, 0.0624972832506011833342799 7]
```

In the case $p = 3$, the routine analogous to `dig_jaclog.m` is `dig_jaclogsq.m` and differs only in the way the modified moments are computed, which is now by the command `mom=smm_logsq`. Run with `N`$=100$, `a`$=-1/2$, `dig0`$=30$, `dd`$=1$, `nofdig`$=30$, to check the 20-digit Table 2 of [18], it yields the same number `dig`$=30$ of digits needed. Comparison with Table 2 revealed a gradual deterioration of accuracy, affecting 3–4 terminal digits when $n = 10$, and up to 7 terminal digits when $n = 99$. The validity of the results obtained in 30-digit computation has been confirmed by comparing them with results obtained with `dig`$=36$. Since the inaccuracies here show up for much smaller values of n than in the case $p = 2$, there may be still other factors involved here that cause them.

Below we show an excerpt of the results of the 36-digit calculation, using `sr_jaclogsq`, believed to be correct after rounding them to 25 decimal digits. The complete table is posted in the file `ab_jaclog_sq.txt`.

```
>>      ab=sr_jaclogsq(46,100,-1/2)

     ab =

     [ 0.03703703703703703703703704,                            16.0]
     [  0.3581128866978306091699794, 0.006628257887517146776406036]
     [  0.4429359676434602031139694,  0.04115455101736139541496346]
     [  0.4692533332263028409621199,  0.05174178200393609100891121]

                   ⋮                               ⋮

     [  0.4999313975746292078320907,  0.06248105702754002195785549]
     [  0.4999330601632353388309418,   0.0624815208186779222924071]
     [  0.4999346629904377648475519,  0.06248196779505153301539905]
     [    0.49993620888761787578562,  0.06248239875946842920659297]
     [  0.4999377005203966921145379,  0.06248281446740126534129196]
     [  0.4999391404001471638329172,  0.06248321563029731407310331]
     [  0.4999405308945842245309177,  0.06248360291862117804480463]
     [  0.4999418742375161072660727,  0.06248397696465499604586768]
     [  0.4999431725378319634435093,  0.06248433836507796927391713]
     [   0.499944427787793433913241,  0.06248468768334486960247844]
     [  0.4999456418706910638466354,  0.06248502545188122753273391]
     [  0.4999468165679206037468075,  0.06248535217411117191434126]
     [  0.4999479535655288575297782,  0.06248566832633234254259627]
     [  0.4999490544602740837744401,  0.06248597435945090974787222]
     [  0.4999501207652416409945435,  0.06248627070058851296672718]
     [  0.4999511539150518101777379,  0.06248655775457179845894799]
     [  0.4999521552706933324232727,  0.06248683590531427233582785]
     [  0.4999531261240130508790767,  0.06248710551709926858871101]
     [   0.499954067701889313102467,   0.0624873669357720268419659]
     [  0.4999549811701143982382435,  0.06248762048984816635622313]

>>
```

Exercise 2.28. (a) Bose–Einstein measure,

$$\mathrm{d}\lambda(t)=\frac{t}{e^t-1}\,\mathrm{d}t \quad \text{on } [0,\infty].$$

Prove that the moments are

$$\mu_k=\int_0^\infty t^k\,\mathrm{d}\lambda(t)=\Gamma(k+2)\zeta(k+2), \quad k=0,1,2,\ldots,$$

where ζ *is the Riemann zeta function.* {Hint: Use the first of the two integral representations (cf. [53, 25.5.1, 25.5.2])

$$\zeta(s)=\frac{1}{\Gamma(s)}\int_0^\infty \frac{t^{s-1}}{e^t-1}\,\mathrm{d}t, \quad \zeta(s)=\frac{1}{\Gamma(s+1)}\int_0^\infty \frac{e^t t^s}{(e^t-1)^2}\,\mathrm{d}t$$

of the zeta function.}

(b) Squared Bose–Einstein measure,

$$\mathrm{d}\lambda(t)=\left(\frac{t}{e^t-1}\right)^2\mathrm{d}t \quad \text{on } [0,\infty].$$

Derive an expression for the moments similar to the one in (a). {Hint: Use both equations in the hint to (a).}

(c) Square root Bose–Einstein measure,

$$\mathrm{d}\lambda(t)=\frac{t^{1/2}}{e^t-1}\,\mathrm{d}t \quad \text{on } [0,\infty].$$

Obtain the moments of this measure.

(d) *Apply the classical Chebyshev algorithm in sufficiently high-precision arithmetic to generate the first eighty recurrence coefficients* α_k, β_k *for the measures in* (a)–(c) *to an accuracy of 32 decimal digits.*

Answer

(a) Using the first equation of the hint, with $s=k+2$, we get immediately

$$\mu_k=\int_0^\infty \frac{t^{k+1}}{e^t-1}\,\mathrm{d}t=\Gamma(k+2)\zeta(k+2).$$

(b) Using the first equation of the hint to (a), we have

$$\begin{aligned}\Gamma(s)\zeta(s)&=\int_0^\infty \frac{t^{s-1}}{e^t-1}\,\mathrm{d}t=\int_0^\infty \frac{e^t-1}{(e^t-1)^2}\,t^{s-1}\mathrm{d}t\\ &=\int_0^\infty \frac{e^t t^{s-1}}{(e^t-1)^2}\,\mathrm{d}t-\int_0^\infty \frac{t^{s-1}}{(e^t-1)^2}\,\mathrm{d}t.\end{aligned}$$

The second equation of the Hint, applied to the first integral on the right, gives

$$\Gamma(s)\zeta(s)=\Gamma(s)\zeta(s-1)-\int_0^\infty t^{s-3}\left(\frac{t}{e^t-1}\right)^2\mathrm{d}t.$$

Therefore, putting $s=k+3$,

$$\mu_k = \int_0^\infty t^k \left(\frac{t}{e^t-1}\right)^2 \mathrm{d}t = \Gamma(k+3)[\zeta(k+2)-\zeta(k+3)]. \tag{2.68}$$

(c) The first equation of the hint to (a), with $s=k+3/2$, gives immediately

$$\mu_k = \int_0^\infty \frac{t^{k+1/2}}{e^t-1}\,\mathrm{d}t = \Gamma(k+3/2)\zeta(k+3/2).$$

(d) The routine producing the required array `ab` is (cf. Appendix B)

```
function ab=sr_BE(i)
%SR_BE Recurrence coefficients for Bose-Einstein weight functions.

syms mom ab
  if i==1
  digits(99); dig=99;
  for k=1:160
    mom(k)=gamma(vpa(k+1,dig))*zeta(vpa(k+1,dig));
  end
elseif i==2
  digits(135); dig=135;
  for k=1:160
    mom(k)=gamma(vpa(k+2,dig))*(zeta(vpa(k+1,dig)) ...
    -zeta(vpa(k+2,dig)));
  end
elseif i==3
  digits(98); dig=98;
  for k=1:160
    mom(k)=gamma(vpa(k+1/2))*zeta(vpa(k+1/2));
  end
else
  error('i not in range')
end
ab=schebyshev(dig,80,mom);
ab=vpa(ab,32);
```

By trial and error, it was found that the precisions needed for `i=1`, `i=2`, `i=3` are, resp., `dig=99`, `dig=135`, `dig=98`. The reason for the much higher precision required in the case `i=2` is the increasing amount of cancellation occurring in computing the difference of the two zeta functions on the right of (2.68) as k becomes larger. For $k=160$, for example, the two values agree in their first 48 digits! Thus, one has to contend not only with ill-conditioning, but also with cancellation.

The results are stored respectively in the files `ab_BE.txt`, `ab_BEsq.txt`, and `ab_BEsqrtm1.txt` in a form ready to be downloaded by the command `loadvpa`. The beginning and end of the files are shown below.

ab_BE

```
  1.4615259388028769974520734626154     1.6449340668482264364724151666646
  3.7041914443293395135022621125385     1.8117836906421124892896543719250
  5.7709986569280917921598820137766     5.6762887062602777876808564814393
  7.8060807007294852726114066513060    11.5674440865786914884452962566860
            ⋮                                   ⋮
141.9561808043103601709409176186     4967.1568419344897468989870652100
143.95648747502634844387227684702    5109.1349706159827937428127960198
145.95678779456994551973442548458    5253.1132518327021260990843104864
147.95708197921272766960771205244    5399.0916824365442210242640827596
149.95737023505308849045191341288    5547.0702593862379179933159356302
151.95765275862341222384363629656    5697.0489797423362371857223680167
153.9579297374535308339361307722     5849.0278406625060958047474184248
155.95820135059413696741381157812    6003.0068393970945456872134807664
157.95846776910347351840637832948    6158.9859732849519451262935889684
159.9587291565003067548633890785     6316.9652397494940967987340151247
```

ab_BEsq

```
 0.8110623843271969462056716818663     0.8857543273772643021453540102691
 2.0824895633609331136782973236937     0.5721932339461376964598358346409
 3.1804727324923422760495136783942     1.7599621950108990929573889024867
 4.2304646872325677288996593014101     3.4523849192031173545509573057624
            ⋮                                   ⋮
71.4381624270072792506871995371750   1258.0503434492590369200437876531
72.43859349999275149649642295961     1293.7849664501379931795733549223
73.43901567742910871152807595242     1330.0196954587692130959174672011
74.43942926122920376050700531017     1366.7545283128283422056237817806
75.4398345391476630486895355084      1403.9894629225033524706342675126
76.44023178562350226116972429559     1441.7244972671359564379848633959
77.44062126256223438406491923073     1479.9596293920603718573856258321
78.44100322006254009377019990203     1518.6948574056254496129287941750
79.44137789709208653617505049875     1557.9301794763873341903788502876
80.44174552211664818384389746170     1597.6655938304608733743528270010
```

ab_BEsqrtm1

```
  0.77026867019278179736191584141818    2.3151573733941170004258194691180
  3.1875985567615246793664140678620     1.0240846879834073034233867314096
  5.2637469230456075097135902570342     4.4142904264700503516592160133692
  7.3017185139793212284724011418749     9.8113025695540054901760247552196
            ⋮                                   ⋮
141.45612698083632750973701354341     4931.9074013163092693359639863947
143.45643476107765202062294125243     5073.3855285433894052545193622675
```

```
145.45673615244529077211667744401   5216.8638082990570280145570665983
```

145.45673615244529077211667744401	5216.8638082990570280145570665983
147.45703137298296479311842042137	5362.3422374364977361851365555591
149.45732063045542362519096033735	5509.8208129156237971183669329098
151.45760412296324184459291225245	5659.2995317980747701647122116466
153.45788203951325814744968006095	5810.7783912425152240401086836018
155.45815456054839058441500336572	5964.2573885002082527804624670548
157.45842185844020353565276223721	6119.7365209108452753560242221322
159.45868409794728243625732737003	6277.2157858986142150565720418051

The first 40 values of the recurrence coefficients in the files `ab_BE.txt` and `ab_BEsq.txt` are in agreement to all 25 digits with results given in Tables 1 and 2 of [35] and obtained in a different way by discretization procedures.

Exercise 2.29. (a) Fermi measure,

$$d\lambda(t) = \frac{1}{e^t+1}\,dt \quad \text{on } [0,\infty].$$

Prove that the moments are

$$\mu_k = \int_0^\infty t^k d\lambda(t) = \begin{cases} \ln 2 \;\; \textit{if } k=0, \\ (1-2^{-k})\Gamma(k+1)\zeta(k+1) \;\; \textit{if } k=1,2,\dots, \end{cases}$$

where ζ is the Riemann zeta function. {Hint: Use the first of the two integral representations (cf. [53, 25.5.3, 25.5.4])

$$\zeta(s) = \frac{1}{(1-2^{1-s})\Gamma(s)} \int_0^\infty \frac{t^{s-1}}{e^t+1}\,dt, \quad \zeta(s) = \frac{1}{(1-2^{1-s})\Gamma(s+1)} \int_0^\infty \frac{e^t t^s}{(e^t+1)^2}\,dt$$

of the zeta function.}

(b) Squared Fermi measure,

$$d\lambda(t) = \left(\frac{1}{e^t+1}\right)^2 dt \quad \text{on } [0,\infty].$$

Derive an expression for the moments similar to the one in (a). {Hint: Use both equations in the hint to (a).}

(c) Square root Fermi measure,

$$d\lambda(t) = \frac{t^{-1/2}}{e^t+1}\,dt \quad \text{on } [0,\infty].$$

Obtain the moments of this measure.

(d) *Apply the classical Chebyshev algorithm in sufficiently high-precision arithmetic to generate the first eighty recurrence coefficients α_k, β_k for the measures in* (a)–(c) *to an accuracy of 32 decimal digits.*

Answer

(a) Putting $s=k+1$ in the first equation of the hint immediately yields

$$\mu_k = \int_0^\infty \frac{t^k}{e^t+1}\,dt = (1-2^{-k})\Gamma(k+1)\zeta(k+1)$$

if $k>0$. If $k=0$, one finds $\mu_0=\ln 2$, either directly by doing the integral, or by taking the limit $s\to 1$ in [53, 25.5.3], using [53, 25.2.3].

(b) Similarly as in Ex. 2.28(b), using the two equations in the hint to (a), one finds, if $k>1$,

$$\mu_k=\int_0^\infty \frac{t^k}{(e^t+1)^2}\,\mathrm{d}t=\Gamma(k+1)[(1-2^{-k})\zeta(k+1)-(1-2^{1-k})\zeta(k)]. \tag{2.69}$$

For $k=0$ and $k=1$, one must take the limit $s\to 1$ as in (a), in combination with [53, 25.6.1], that is, $\zeta(0)=-1/2$, $\zeta(2)=\pi^2/6$, to get

$$\mu_0=\ln 2-\tfrac{1}{2},\quad \mu_1=\tfrac{1}{12}\pi^2-\ln 2.$$

(c) Putting $s=k+1/2$ in the first equation of the hint to (a) gives

$$\mu_k=\int_0^\infty \frac{k^{-1/2}}{e^t+1}\,\mathrm{d}t=(1-2^{-k+1/2})\Gamma(k+1/2)\zeta(k+1/2).$$

(d) The routine producing the required array ab is (cf. Appendix B)

```
function ab=sr_fermi(i)
%SR_FERMI Recurrence coefficients for Fermi weight functions.

syms mom ab
if i==1
  digits(98); dig=98;
  mom(1)=log(vpa(2,dig));
  for k=2:160
    mom(k)=(1-2^(-vpa(k-1,dig)))*gamma(vpa(k,dig))*zeta(vpa(k,dig));
  end
elseif i==2
  digits(135); dig=135;
  mom(1)=log(vpa(2,dig))-1/2;
  mom(2)=vpa(pi,dig)^2/12-log(vpa(2,dig));
  for k=3:160
    mom(k)=gamma(vpa(k,dig))*((1-2^(vpa(1-k,dig)))*zeta(vpa(k,dig)) ...
    -(1-2^(vpa(2-k,dig)))*zeta(vpa(k-1,dig)));
  end
elseif i==3
  digits(98); dig=98;
  for k=1:160
    mom(k)=(1-2^(vpa(-k+3/2,dig)))*gamma(vpa(k-1/2))*zeta(vpa(k-1/2));
  end
else
  error('i not in range')
end
ab=schebyshev(dig,80,mom);
ab=vpa(ab,32);
```

By trial and error, it was found that the precisions needed for i=1, i=2, i=3 are, resp., dig=98, dig=135, dig=98. The reason for the much higher precision required in the case `i=2` is exactly the same as noted in the answer to Ex. 2.28(d), that is, severe cancellation in the right-hand side of (2.69).

The results are stored, resp., in the files `ab_Fermi.txt`, `ab_Fermisq.txt`, and `ab_Fermisqrtm1.txt` in a form ready to be downloaded by the command `loadvpa`. The beginning and end of the files are shown below.

```
                          ab_Fermi

1.1865691104156254528217229759472  0.69314718055994530941723212145818
3.0963542153967773858680971566504  1.1933560457895086591789457590149
5.0722272795356037485262795403069  4.1918064245490424515900825633069
7.0601226279075802377733621702027  9.2153673854346419765449324411212

          ⋮                                   ⋮

141.01288221829984014386044678506  4900.9076788158989445111637683806
143.01279141944159846131340464377  5041.9140711896172546827402146282
145.01270251373554568624500581896  5184.9204191946523502493551461142
147.01261543629965818156737042222  5329.9267237417639421546177974237
149.01253012532278538913427357573  5476.9329857109888953542761510896
151.01244652188027717268051588269  5625.9392059530721381975155015281
153.01236456976296214603227490368  5776.9453852908130533537196822259
155.01228421531834929699421093817  5929.9515245203333697981639082391
157.01220540730303370916084382646  6084.9576244122720754330267216559
159.01212809674538399640765920105  6241.9636857129124172634485823247

                          ab_Fermisq

0.66954046385384382323872273614275  0.19314718055994530941723212145818
1.6646861330098296800110754089394   0.37052787108516848568066219171069
2.6194318410157584766041915851217   1.1549078536526013653658237950709
3.5938653669740722997562785471112   2.4167828719375690277239980531822

          ⋮                                   ⋮

70.51827923521568229063714186328    1225.6431145617100256939333842219
71.518149542862634990940718965088   1260.8976250945840050340991320733
72.518022572256043146509524330086   1296.6521044638962997638483317592
73.517898229513545416382318931959   1332.9065533060853129017107746679
74.517776425224689660749576602718   1369.660972236241431579656649676
75.517657074180738504552672983499   1406.9153618490951180969133511301
76.51754009512416381439153690895    1444.6697227199470377042328892772
77.517425410516156655909983165441   1482.924055405544320323955777184
78.517312946320641296750859179901   1521.6783604449067159265667590682
79.517202631803426270855169822448   1560.9326383601060974295670964362

                          ab_Fermisqrtm1

0.63245886971850936236610463079063  1.0721549299401913395308974465756
2.6184924841473600282012219919112   0.67521709631759430156397124373479
4.5795649636410436101717029410862   3.1675835602583982388211775940841
6.5640307792994767460871911441287   7.6961592305189008195692774297764
```

```
            ⋮                                  ⋮
140.51291476556475167220251847091   4865.9040949969279004484752553223
```

⋮	⋮
140.51291476556475167220251847091	4865.9040949969279004484752553223
142.51282327845082342769917698953	5006.4105131709799453614527728606
144.51273370839054226291634129346	5148.9168864296144891219026473084
146.51264598935624570779579580135	5293.4232157025973737105124800433
148.51256005846077224225904611056	5439.9295018880602088796180714504
150.51247585576798651057426204309	5588.435745853986675137462816069
152.51239332411709148675628802445	5738.9419484396102500536944805069
154.51231240895955748115050867312	5891.4481104567297240260150997861
156.51223305820761098769959646902	6045.9542326909483380020065918485
158.51215522209332724680490439783	6202.4603159028418929514921341101

The first 40 values of the recurrence coefficients in the files `ab_Fermi.txt` and `ab_Fermisqrtm1.txt` are in agreement to all 25 digits with results given in Tables 3 and 4 of [35] and obtained in a different way by discretization procedures.

Exercise 2.30. *Consider the measure*

$$\mathrm{d}\lambda(t) = (1 + \sin(1/t))\mathrm{d}t \quad \textit{on } [0,1].$$

(a) *Determine the moments*

$$\mu_k = \int_0^1 t^k \mathrm{d}\lambda(t), \quad k = 0,1,2,\ldots.$$

{Hint: Derive a recurrence relation for the "core moments" $\mu_k^0 = \int_0^1 t^k \sin(1/t)\mathrm{d}t$ involving the sine and cosine integrals Si, Ci.}

(b) *Apply the Chebyshev algorithm in sufficiently high-precision arithmetic to obtain the first $N = 80$ recurrence coefficients for the orthogonal polynomials $\{\pi_k(\cdot;\mathrm{d}\lambda)\}_{k=0}^{N-1}$ to an accuracy of* 16 *significant decimal digits. How much precision is needed in Chebyshev's algorithm? Store the results in the file* `ab_sin0.txt`.

(c) *Use the file* `ab_sin0.txt` *to generate the appropriate Gaussian quadrature rules for computing*

$$\text{(i) } I_1 = \int_0^1 t(1-t)\sin(1/t)\mathrm{d}t, \quad \text{(ii) } I_2 = \int_0^1 \frac{\sin(1/t)}{x-t}\mathrm{d}t, \quad x > 1.$$

to an accuracy of 12 *decimal digits. Display the results and the respective relative errors, for $x = [1.01, 1.05, 1.1, 1.5, 2, 5, 10]$ in the case of* (ii), *along with the number n of Gauss quadrature nodes required.*

Answer

(a) We have

$$\mu_k = \frac{1}{k+1} + \mu_k^0. \tag{2.70}$$

Changing variables, $t \mapsto 1/t$, in the integral for μ_k^0 gives

$$\mu_k^0 = \int_1^\infty t^{-(k+2)} \sin t \, \mathrm{d}t.$$

Two integrations by parts, integrating the first factor and differentiating the second, yields

$$\mu_{k+1}^0 = \frac{1}{k+2}\left[\frac{1}{k+1}(\cos 1 - \mu_{k-1}^0) + \sin 1\right], \quad k = 0, 1, 2, \ldots.$$

Here, when $k = 0$, we need (cf. [1, Eqs. 5.2.1 and 5.2.25])

$$\mu_{-1}^0 = \int_1^\infty \frac{\sin t}{t}\, dt = \int_0^\infty \frac{\sin t}{t}\, dt - \int_0^1 \frac{\sin t}{t}\, dt = \frac{\pi}{2} - \mathrm{Si}(1),$$

and when $k = 1$, using integration by parts and [1, Eq. 5.2.27],

$$\mu_0^0 = \int_1^\infty \frac{\sin t}{t^2}\, dt = \sin 1 - \mathrm{Ci}(1).$$

Thus, the first N core moments can be obtained recursively by

$$\begin{aligned} \mu_{k+1}^0 &= \frac{1}{k+2}\left[\frac{1}{k+1}(\cos 1 - \mu_{k-1}^0) + \sin 1\right], \quad k = 1, 2, \ldots, N-2, \\ \mu_0^0 &= \sin 1 - \mathrm{Ci}(1), \quad \mu_1^0 = \frac{1}{2}\left[\mathrm{Si}(1) - \frac{\pi}{2} + \sin 1 + \cos 1\right] \end{aligned} \tag{2.71}$$

and the desired moments by (2.70). The variable-precision routine implementing this is (cf. Appendix B.3)

```
function mom=smom_sin0(dig,N)
%SMOM_SIN0 Moments of a special trigonometric weight function.
%   MOM=SMOM_SIN0(DIG,N) generates in DIG-digit arithmetic
%   the 2Nx1 array MOM of the first 2N moments of the weight
%   function W(T)=1+SIN(1/T) on [0,1].

digits(dig); one=vpa(1); pih=vpa(pi/2);
mom=vpa(zeros(2*N,1));
mom(1)=sin(one)-cosint(one);
mom(2)=(sinint(one)-pih+sin(one)+cos(one))/2;
if N==1, return, end
for k=3:2*N
  mom(k)=((cos(one)-mom(k-2))/(k-1)+sin(one))/k;
end
for k=1:2*N
  mom(k)=mom(k)+1/k;
end
```

(b) The routine `dig_sin0.m` can be used to determine the number of digits needed.

```
function [ab,dig]=dig_sin0(N,dig0,dd,nofdig)
%DIG_SIN0 Number of digits needed for orthogonal polynomials
%relative to a special trigonometric weight function.
%   [AB,DIG]=DIG_SIN0(N,DIG0,DD,NOFDIG) determines the number
%   DIG of digits needed to obtain the first N recurrence
%   coefficients for the weight function W(T)=1+SIN(1/T) on
%   [0,1] to an accuracy of NOFDIG decimal digits using the
```

```
%   classical Chebyshev algorithm. It successively increments
%   an initial estimate DIG0 by DD units until the required
%   accuracy is achieved. The recurrence coefficients to
%   NOFDIG digits are placed into the Nx2 array AB.

i=dig0-dd;
maxerr=1;
while maxerr>.5*10^(-nofdig)
  i=i+dd; dig=i
  mom=smom_sin0(dig,N);
  if i==dig0
    ab0=schebyshev(dig,N,mom');
  else
    ab1=schebyshev(dig,N,mom');
    serr=abs((ab1-ab0)./ab1);
    err=double(serr);
    maxerr=max(max(err));
    ab0=ab1;
  end
end
ab=vpa(ab1,nofdig);
```

Run with $N = 80$, `dig0`$=$ 127, `dd`$=$ 1, and `nofdig`$=$ 16, it yields `dig`$=$ 129 and recurrence coefficients, the first and last eight of which are shown below.

```
0.5841029561609566  1.504067061906928
0.4634474607770499  0.07094822535096882
0.4977629714178322  0.07892077774694954
0.535659008862375   0.05547885019105795
0.4669144430825117  0.06259489484316939
0.4951204560106238  0.07262419731845188
0.5406392981446107  0.05561898312698315
0.4856589782121508  0.05738164013582103
        ⋮                    ⋮
0.5042681404311647  0.06102578489812291
0.4923369552841893  0.06291186934809241
0.4955616052055979  0.06478040459915332
0.5079681936575616  0.06179440216769207
0.5051586634352725  0.06041507871232785
0.4999635499313309  0.06214807969960169
0.5013608789849195  0.06230337046586129
0.4943993578328874  0.06271214439653958
```

The α coefficients seem to hover around .5, and the β coefficients around .06, without any indication to ever converge. The complete array `ab` is contained in the file `ab_sin0` and can be uploaded by the routine `loadvpa.m`.

The following routine implements Chebyshev's algorithm (cf. Appendix B.1).

```
function ab=sr_sin0(dig,nofdig,N)
%SR_SIN0 Recurrence coefficients for a special trigonometric
```

```
%weight function.

digits(dig);
mom=smom_sin0(dig,N);
ab=schebyshev(dig,N,mom');
ab=vpa(ab,nofdig);
```

(c) The exact answers can be derived as follows. For I_1, using the change of variables $t \mapsto 1/t$, we get

$$I_1 = \int_0^1 t(1-t)\sin(1/t)\mathrm{d}t = \int_1^\infty \frac{\sin t}{t^3}\,\mathrm{d}t - \int_1^\infty \frac{\sin t}{t^4}\,\mathrm{d}t.$$

Applying integration by parts to both integrals on the right will give, after a short calculation,

$$I_1 = \frac{1}{3}(\sin 1 + \cos 1) - \frac{\pi}{4} + \frac{1}{2}\,\mathrm{Si}(1) - \frac{1}{6}\,\mathrm{Ci}(1).$$

The same change of variables, applied to I_2, gives

$$I_2 = \int_0^1 \frac{\sin(1/t)}{x-t}\,\mathrm{d}t = \int_1^\infty \frac{\sin t}{t(tx-1)}\,\mathrm{d}t = \int_1^\infty \sin t\left[-\frac{1}{t} + \frac{1}{t-1/x}\right]\mathrm{d}t$$
$$= -\int_1^\infty \frac{\sin t}{t}\,\mathrm{d}t + \int_1^\infty \frac{\sin t}{t-1/x}\,\mathrm{d}t.$$

In the last integral we let $t - 1/x = u$ and get

$$\int_1^\infty \frac{\sin t}{t-1/x}\,\mathrm{d}t = \int_{1-1/x}^\infty \frac{\sin(u+1/x)}{u}\,\mathrm{d}u$$
$$= \cos(1/x)\int_{1-1/x}^\infty \frac{\sin u}{u}\,\mathrm{d}u + \sin(1/x)\int_{1-1/x}^\infty \frac{\cos u}{u}\,\mathrm{d}u.$$

Altogether, therefore,

$$I_2 = \cos(1/x)\left[\frac{\pi}{2} - \mathrm{Si}(1-1/x)\right] - \sin(1/x)\mathrm{Ci}(1-1/x) - \left[\frac{\pi}{2} - \mathrm{Si}(1)\right].$$

We now write

$$I_1 = \int_0^1 t(1-t)[1+\sin(1/t)]\mathrm{d}t - \frac{1}{6}, \quad I_2 = \int_0^1 \frac{1+\sin(1/t)}{x-t}\,\mathrm{d}t + \ln(1-1/x)$$

and apply Gaussian quadrature to the integrals in each formula. In the case of I_1, two-point Gauss quadrature should give an exact answer modulo rounding errors. This is confirmed by running the script `I1_sin0.m`. In the case of I_2 the script `I2_sin0.m` determines the number of quadrature nodes required, for selected values of x, to obtain 12 correct decimal digits and prints the relative error achieved. Both scripts are shown below.

```
%I1_SIN0

f0='%24.17f %12.4e\n';
Iexact=(sin(1)+cos(1))/3-pi/4+sinint(1)/2-cosint(1)/6;
```

```
load -ascii ab_sin0;
ab=ab_sin0;
xw=gauss(2,ab); x=xw(:,1); w=xw(:,2);
I1=sum(w.*x.*(1-x))-1/6; err=abs((I1-Iexact)/Iexact);
fprintf(f0,I1,err)

%I2_SIN0

f0='%8.2f %4.0f %16.12f %12.4e\n';
disp('      x      n          I2             err')
load -ascii ab_sin0;
ab=ab_sin0; eps0=.5e-12;
for x=[1.01 1.05 1.1 1.5 2 5 10]
  Iexact=cos(1/x)*(pi/2-sinint(1-1/x))-sin(1/x)*cosint(1-1/x) ...
    -(pi/2-sinint(1));
  n=0; I=1; I0=0;
  while abs((I -I0)/I)>eps0
    n=n+1; I0=I;
    xw=gauss(n,ab);
    I=sum(xw(:,2)./(x-xw(:,1)));
  end
  I2=I+log(1-1/x); err=abs((I2-Iexact)/Iexact);
  fprintf(f0,x,n,I,err)
end
```

The output

```
>> I1_sin0
I1 =
     9.200048152799387e-02
err =
     9.050683807730973e-16

>> I2_sin0
       x     n          I2             err
    1.01    70    3.607638201735    1.7642e-12
    1.05    33    2.269163108094    5.6089e-13
    1.10    24    1.722695077816    3.3074e-13
    1.50    12    0.688918357523    6.9296e-14
    2.00    10    0.406281964773    4.0990e-15
    5.00     6    0.118674569875    3.2743e-15
   10.00     5    0.054502895503    0.0000e+00
>>
```

is as expected.

Exercise 2.31. *Consider the weight function*

$$w(t) = \operatorname{erfc} t \quad \textit{on } [0, \infty],$$

where erfc *is the complementary error function* $\operatorname{erfc} t = (2/\sqrt{\pi})\int_t^\infty e^{-u^2}\mathrm{d}u$.

(a) *Express the moments* $\mu_k = \int_0^\infty t^k w(t)\mathrm{d}t$ *in terms of the gamma function.*

(b) *Use the symbolic MATLAB routine* `schebyshev.m` *to compute the first* $N = 12$ *recurrence coefficients of the orthogonal polynomials* $\pi_k(\cdot\,; w)$ *from the first* $2N$ *moments, accurate to* 32 *decimal digits. What is the precision needed?*

(c) *Use the answer of* (b) *to compute the n-point Gauss formulae for* $n = 2, 3, \ldots, 12$ *to MATLAB double-precision accuracy, and compare with Table* 3 *of* [63].

(d) *Generate the first* 100 *recurrence coefficients to an accuracy of* 32 *decimal digits and determine the precision needed.*

Answer

(a) We have

$$\mu_k = \int_0^\infty t^k \operatorname{erfc} t \, dt = \frac{2}{\sqrt{\pi}} \int_0^\infty t^k dt \int_t^\infty e^{-u^2} du$$
$$= \frac{2}{\sqrt{\pi}} \int_0^\infty e^{-u^2} du \int_0^u t^k dt = \frac{2}{\sqrt{\pi}(k+1)} \int_0^\infty e^{-u^2} u^{k+1} du.$$

Letting $u^2 = x$, $2u\,du = dx$, we get

$$\mu_k = \frac{1}{\sqrt{\pi}(k+1)} \int_0^\infty e^{-x} x^{k/2} dx = \frac{1}{\sqrt{\pi}(k+1)} \Gamma(k/2+1).$$

(b) The number of digits required is determined by the routine `dig_erfc.m`,

```
function [ab,dig]=dig_erfc(N,dig0,dd,nofdig)
%DIG_ERFC Number of digits needed for polynomials orthogonal
%with respect to the coerror function.
%    [AB,DIG]=DIG_ERFC(N,DIGO,DD,NOFDIG) determines the number
%    DIG of digits needed to obtain the first N recurrence
%    coefficients for the weight function W(T)=ERFC(T) on [0,Inf]
%    to an accuracy of NOFDIG decimal digits using the classical
%    Chebyshev algorithm. It successively increments an initial
%    estimate DIGO by DD units until the required accuracy is
%    achieved. The recurrence coefficients to NOFDIG digits
%    are placed into the Nx2 array AB.

i=dig0-dd;
maxerr=1;
while maxerr>.5*10^(-nofdig)
  i=i+dd; dig=i
  mom=smom_erfc(dig,N);
  if i==dig0
    ab0=schebyshev(dig,N,mom');
  else
    ab1=schebyshev(dig,N,mom');
    serr=abs((ab1-ab0)./ab1);
    err=double(serr);
    maxerr=max(max(err));
    ab0=ab1;
  end
end
```

```
ab=vpa(ab1,nofdig);
%ab=double(ab);
%for n=2:12
%  xw=gauss(n,ab)
%end
```

Run with N = 12, dig0 = 35, dd = 1, and nofdig = 32, it produces the output

```
ab =
[ 0.44311346272637900682454187083529, 0.56418958354775627928034964497783]
[ 0.90471467169272446773158918092414, 0.13698379248397125592941812187836]
[  1.2113620086441822956886649944719,  0.2922868336841532736515506172296]
[  1.4569913029741128482893718109762, 0.45354132586269063016076522517975]
[  1.6678006444190714952462701650639, 0.61684149382812875724222712609681]
[  1.8553204773809488972871288841846, 0.78113590190278888030867877650999]
[  2.0258575999911201760375537778177, 0.94600799494371319128954075853875]
[  2.1833071552663532927398988979444,  1.1112520482146228770492716612848]
[  2.3302734878243060393303583532576,  1.2767526793261646501512111240634]
[  2.4685990476556832254568401035065,  1.4424393675912798223674571597561]
[  2.5996437163990289817132244873412,  1.6082662063204072027578015811738]
[  2.7244449239812704960273968854068,  1.7742018274993259779275276127514]

dig =
    37
>>
```

We see that the precision needed is dig=36.

(c) Uncommenting the last four commands in the routine of (b) yields the desired Gauss quadrature rules, where the first column of xw contains the Gaussian nodes, and the second column the Gaussian weights. We display only the rules for $n = 2, 7$, and 12.

```
xw =
        2.377343891939785e-01     4.313627465744712e-01
        1.110093745225125e+00     1.328268369732851e-01

xw =
        5.491230506104055e-02     1.311825703475700e-01
        2.803854593689935e-01     2.117567631417558e-01
        6.556447780228353e-01     1.551840767514100e-01
        1.147813045119681e+00     5.667342430613216e-02
        1.736297739734502e+00     8.934782663909336e-03
        2.424231297939170e+00     4.542108143814861e-04
        3.265875542582317e+00     3.755522597752376e-06

xw =
        2.675963852966877e-02     6.638964910455235e-02
        1.391308241160834e-01     1.309894621129891e-01
        3.342118102564529e-01     1.484378998253909e-01
        6.018439551662339e-01     1.184649327890081e-01
        9.308843830586756e-01     6.698792207842864e-02
        1.311586486634760e+00     2.569882062953836e-02
```

```
1.737140427597650e+00      6.262738409913335e-03
2.204514908177997e+00      8.902612107345368e-04
2.715272299295224e+00      6.575950788467372e-05
3.277540926430984e+00      2.116271860702681e-06
3.912282459957908e+00      2.157537114404687e-08
4.680260379733545e+00      3.208456990236608e-11
```

These agree with those in the cited reference except for the last, or last two, digits, especially when $n=12$.

(d) All we need to do is to rerun the routine in (b) with `N`$=100$, `dig0`$=136$ (found by trial and error), `dd`$=1$, `nofdig`$=32$, to see that the required precision is `dig`$=136$. The 32-digit answers are stored in the file `ab_erfc.txt`.

Exercise 2.32. *Consider the weight function*

$$w_\nu(t)=\frac{2}{\pi}\cos\left(\frac{1}{2}\nu\pi\right)K_\nu(t),\quad 0\le\nu<1,\quad t\in\mathbb{R}_+,$$

where K_ν is the modified Bessel function of the second kind of order ν. Gaussian quadrature relative to this weight function, with $\nu=1/3$ (and a different normalization), is used in [39] to evaluate the "inhomogeneous" Airy function Gi(z).

(a) *Determine the moments $\mu_k=\int_0^\infty t^k w_\nu(t)\mathrm{d}t$ and show that w_ν is normalized, i.e.,* $\mu_0=1$. {Hint: Use formula (6.561.16) in [40].}

(b) *How many digits are needed in the classical Chebyshev algorithm to obtain the first eighty recurrence coefficients for the orthogonal polynomials $\pi_k(\cdot;w_\nu)$, $\nu=1/3$, accurate to* 16 *and* 32 *decimal digits?*

(c) *Write a MATLAB variable-precision routine* `sr_Knu.m` *that uses the classical Chebyshev algorithm in* `dig`*-digit arithmetic to generate the first N recurrence coefficients for w_ν and places them to* `nofdig` *digits into the $N\times 2$ array* `ab`*. Run the routine with* `N`$=80$, `nu`$=1/3$, `nofdig`$=32$, *and* `dig` *as found in* (b), *to produce the array* `ab_modBessel`*. Compare with Table* A.1 *of* [23].

(d) *Use the file* `ab_modBessel` *obtained in* (c) *to check the formula* (cf. [40, Eq. 6.576.1])

$$\int_0^\infty t^{1-\nu}J_0(t)K_\nu(t)\mathrm{d}t=\frac{1}{2}\Gamma(1-\nu),\quad 0\le\nu<1, \tag{2.72}$$

for $\nu=1/3$, using Gaussian quadrature with weight function w_ν. Convergence is expected to be quite slow in this case (why?). How could you speed it up?

Answer

(a) Using the hint, we get

$$\mu_k=\frac{2}{\pi}\cos\left(\frac{1}{2}\nu\pi\right)\cdot 2^{k-1}\Gamma\left(\frac{k+1+\nu}{2}\right)\Gamma\left(\frac{k+1-\nu}{2}\right).$$

This is implemented by the `SOPQ` routine `smom_Knu.m` (cf. Appendix B.3). The reflection formula for the gamma function, $\Gamma(z)\Gamma(1-z)=\pi/\sin(\pi z)$, with $z=(1+\nu)/2$, then yields

$$\Gamma\left(\frac{1+\nu}{2}\right)\Gamma\left(\frac{1-\nu}{2}\right)=\frac{\pi}{\sin(\frac{1}{2}\pi(\nu+1))}=\frac{\pi}{\cos(\frac{1}{2}\pi\nu)},$$

so that

$$\mu_0=\frac{2}{\pi}\cos\left(\frac{1}{2}\nu\pi\right)\cdot\frac{1}{2}\,\frac{\pi}{\cos(\frac{1}{2}\nu\pi)}=1.$$

(b) The routine that answers the question is dig_Knu.m, which is the same as all the other dig_xxx.m routines except for the command generating the moments. Run with $\nu = 1/3$, it yields dig = 86 for 16-digit accuracy, and dig = 100 for 32-digit accuracy.

(c) The routine is (cf. Appendix B.1)

```
function ab=sr_Knu(dig,nofdig,N,nu)
%SR_KNU Recurrence coefficients for the modified Bessel
%weight function.

digits(dig);
mom=smom_Knu(dig,N,nu);
ab=schebyshev(dig,N,mom');
ab=vpa(ab,nofdig);
```

The 80×2 array ab_modBessel produced by this routine, run with dig = 100, nofdig = 32, N = 80, and nu = 1/3, confirms the correctness to all 28 digits of the data given in Table A.1 of [23] for the first 40 recurrence coefficients.

(d) Convergence of Gauss quadrature is expected to be slow because of the fractional power $t^{1-\nu}$ appearing in the integral of (2.72). To restore fast convergence, one has to include this factor as part of the weight function, that is, take $w_\nu^0(t) = t^{1-\nu}K_\nu(t)$. The recurrence coefficients needed can be obtained in the same manner as outlined in (a) and (b). The relevant file, accurate to 16 digits, is ab_modBessel0.txt when $\nu = 1/3$. The script test_modBessel.m,

```
%TEST_MODBESSEL

f0='%8.0f %20.16f %12.4e\n';
ab=loadvpa('ab_modBessel',3,32,80,2);
ab0=double(ab);
disp('        n              I                  err')
exact=sqrt(3)*gamma(2/3)/(2*pi);
for n=10:10:80
  xw=gauss(n,ab0); x=xw(:,1); w=xw(:,2);
  I=sum(w.*x.^(2/3).*besselj(0,x));
  err=abs(I-exact);
  fprintf(f0,n,I,err)
end
fprintf('\n')
load -ascii ab_modBessel0
for n=5:5:25
  xw=gauss(n,ab_modBessel0); x=xw(:,1); w=xw(:,2);
  I=sum(w.*besselj(0,x));
  err=abs(I-exact);
  fprintf(f0,n,I,err)
end
```

compares the performance of the two quadrature procedures, and the output,

```
>> test_modBessel
       n              I                  err
      10    0.3772712581296420    3.9891e-003
      20    0.3748949182003916    1.6127e-003
```

```
    30   0.3742318020740497   9.4963e-004
    40   0.3739342155588093   6.5204e-004
    50   0.3737691923110096   4.8702e-004
    60   0.3736658225583616   3.8365e-004
    70   0.3735957041992177   3.1353e-004
    80   0.3735453879407448   2.6321e-004

     5   0.3733883027683588   1.0613e-004
    10   0.3732822668925997   9.2985e-008
    15   0.3732821738991597   8.2356e-012
    20   0.3732821739073841   1.1158e-014
    25   0.3732821739073953   0.0000e+000
>>
```

demonstrates convincingly the superiority of the second quadrature method.

Exercise 2.33. *Consider the weight function*

$$w(t) = \frac{2^{-1/6}3^{1/6}}{\pi^{1/2}}\, t^{-5/6} e^{-t} \operatorname{Ai}((3t/2)^{2/3}), \quad 0 < t < \infty,$$

where Ai *is the Airy function.* Gauss quadrature relative to this weight function is proposed in [56, §4] to evaluate the complex Airy function. While Eqs. (3.2), (3.3) in this reference are erroneous, the computational work reported in Section 5, in particular Tables II–IV and Table VI, are correct. See also Ex. 5.4.

(a) *Determine the moments* $\mu_k = \int_0^\infty t^k w(t)\mathrm{d}t$ *and show that* w *is normalized, i.e.,* $\mu_0 = 1$. {Hint: Use $\operatorname{Ai}(x) = (1/\pi)\sqrt{x/3}\,K_{1/3}((2/3)x^{3/2})$, cf. [53, Eq. (9.6.1)], in combination with [40, Eq. 6.621.3].}

(b) *How many digits are needed in the classical Chebyshev algorithm to obtain the first eighty, or the first* 200*, recurrence coefficients for the orthogonal polynomials* $\pi_k(\cdot\,; w)$*, accurate to* 16 *and* 32 *digits?*

(c) *Write a MATLAB variable-precision routine* `sr_Airy.m` *that uses the classical Chebyshev algorithm in* `dig`*-digit arithmetic to generate the first* N *recurrence coefficients for* w *and places them to* `nofdig` *digits into the* $N \times 2$ *array* `ab`*. Run the routine with* $N = 200$*,* `nofdig` *=32, and* `dig` *as found in* (b) *to produce the array* `ab_Airy.txt`*.*

Answer

(a) Following the hint, we have

$$w(t) = \frac{2^{-1/6}3^{1/6}}{\pi^{1/2}}\, t^{-5/6} e^{-t} \cdot \frac{t^{1/3}}{2^{1/3}3^{1/6}\pi} K_{1/3}(t) = \frac{2^{-1/2}}{\pi^{3/2}}\, t^{-1/2} e^{-t} K_{1/3}(t), \tag{2.73}$$

so that

$$\mu_k = \frac{2^{-1/2}}{\pi^{3/2}} \int_0^\infty t^{k-1/2} e^{-t} K_{1/3}(t)\mathrm{d}t.$$

Using [40, Eq. 6.621.3] with $\mu = k + 1/2$, $\alpha = \beta = 1$, $\nu = 1/3$, gives

$$\mu_k = \frac{1}{2\pi} 2^{-k} \frac{\Gamma(k+5/6)\Gamma(k+1/6)}{\Gamma(k+1)}, \quad k = 0, 1, 2, \ldots.$$

In particular,

$$\mu_0 = \frac{1}{2\pi}\Gamma(5/6)\Gamma(1/6).$$

By the reflection formula for the gamma function, $\Gamma(z)\Gamma(1-z)=\pi/\sin(\pi z)$, $z=1/6$, one gets

$$\mu_0=\frac{1}{2\pi}\frac{\pi}{\sin(\pi/6)}=\frac{1}{2}\frac{1}{1/2}=1.$$

(b) Similarly to Ex. 2.32(b), the relevant routine is `dig_Airy.m`. Run manually in the MATLAB workspace, it yields

N	nofdig	dig
80	16	84
	32	99
200	16	198
	32	214

where `nofdig` is the number of digits desired, and `dig` the number of digits needed.

(c) The routine is (cf. Appendix B.1)

```
function ab=sr_Airy(dig,nofdig,N)
%SR_AIRY Recurrence coefficients for an Airy weight function.

digits(dig);
mom=smom_Airy(dig,N);
ab=schebyshev(dig,N,mom');
ab=vpa(ab,nofdig);
```

The beginning and end of the array `ab_Airy.txt` produced by this routine with `dig` $=220$, `nofdig` $=32$, `N` $=200$ are shown below.

```
ab =
[ 0.069444444444444444444444444444444,                                1.0]
[   1.0266030956329463792150359314538, 0.032310956790123456790123456790123]
[   2.0202748532320531762368452935313,  0.54020341213402002945166068191076]
[    3.0170748478530614224148283337960,   1.5457945613810474150647315982767]
[   4.0150420680208661497400417733071,   3.0503611575067731728725133181584]
[   5.0136012177952459950203288976878,   5.0543172669999293805537904160359]

                    ⋮                                      ⋮

[   196.00228299607376447709207036289,   9555.2433637961750139091107758986]
[   197.00227721362807008110030023793,   9653.2439347286805700361013679888]
[   198.00227147490453510010857493198,   9751.7445042135857909698983199017]
[   199.00226577935489096755858591548,   9850.7450722618511084136549187284]
```

It will be used in Ex. 5.4 to compute the Airy function Ai in the complex domain.

2.5 ▪ Discretization method

Exercise 2.34. In [62, §2.4] the measure $d\lambda(t)=\exp(-t^2-\sin t)dt$ on $\mathbb{R}$ is considered as an "exotic choice" for illustrating the Riemann–Hilbert approach to Gaussian quadrature. The specific example given is $I=\int_{\mathbb{R}} f(t)d\lambda(t)$ with $f(t)=\sin^2 10t$.

(a) *Use the* SOPQ *routines* `sr_hermite.m` *and* `sgauss.m` *to obtain I accurate to 24 decimal digits.*

(b) *Use the* OPQ *routine* mcdis.m *to generate the first* N *recurrence coefficients* α_k, β_k, $k = 0,1,2,\ldots,N-1$, *for the measure* $\mathrm{d}\lambda$.

(c) *Reproduce* Fig. 5(b) *of* [62] *showing the absolute error of n-point Gaussian quadrature,* $n = 1,2,\ldots,120$, *for the integral* I.

Answer

(a) Write

$$I = \int_{\mathbb{R}} \sin^2(10t)e^{-\sin t} \cdot e^{-t^2}\mathrm{d}t$$

and use Gauss–Hermite quadrature to evaluate I. The script

```
%EXPSIN_EXACT

dig=30; k=0; I=vpa(zeros(9,1),dig);
ab=sr_hermite(dig,237);
for n=[210 220 221:227]
  k=k+1;
  xw=sgauss(dig,n,ab); x=xw(:,1); w=xw(:,2);
  I(k)=sum(w.*(sin(10*x).^2).*exp(-sin(x)));
end
vpa(I,24)
```

and resulting output (slightly edited)

```
 n                  I
210  1.03359634476958320913849
220  1.03359634476958320913882
221  1.03359634476958320913878
222  1.03359634476958320913881
223  1.03359634476958320913879
224  1.03359634476958320913880
225  1.03359634476958320913879
226  1.03359634476958320913880
227  1.03359634476958320913880
```

suggest that the required 24-digit accuracy can be achieved by 224-point Gauss–Hermite quadrature.

(b) We discretize the inner product by Gauss–Hermite quadrature as in (a), and thus employ the quadrature routine quad_expsin.m,

```
function xw=quad_expsin(N,i)
%QUAD_EXPSIN Quadrature routine for discretization used by
%OPexpsin.m in mcdis.m

ab=r_hermite(N);
xw=gauss(N,ab); x=xw(:,1);
xw(:,2)=exp(-sin(x)).*xw(:,2);
```

in mcdis.m. To avoid overflow problems, the recurrence coefficients of the discretized measure are computed using lanczos.m rather than stieltjes.m, thus putting irout = 0 in the routine below.

```
function [ab,Mcap,kount]=OPexpsin(N,eps0)
%OPEXPSIN Recurrence coefficients for an exponential/trigonometric
%weight function.
%   [AB,MCAP,KOUNT]=OPEXPSIN(N,EPS0) generates to an accuracy EPS0
%   the Nx2 array AB of the first N recurrence coefficients for the
%   weight function W(T)=EXP(-T^2-SIN(T)) on [-Inf,Inf]. The output
%   variables MCAP and KOUNT are inherited from the routine mcdis.m.

  global mc mp iq idelta irout DM AB
  mc=1; mp=0; iq=1; idelta=2; irout=0; DM=[0 0];
  AB=[-Inf Inf]; Mmax=600;
  [ab,Mcap,kount]=mcdis(N,eps0,@quad_expsin,Mmax);
```

(c) The script[2]

```
%RUN_OPEXPSIN

f0='%8.0f %20.13f\n';
N=120; eps0=.5e-13;
err=zeros(N,1); exact=1.0335963447695832;
ab=OPexpsin(N,eps0);
for n=1:N
  xw=gauss(n,ab); x=xw(:,1); w=xw(:,2);
  I=sum(w.*(sin(10*x).^2));
  err(n)=abs(I-exact);
end
n=(1:N)';
semilogy(n,err,'LineWidth',1.5);set(gca,'FontSize',14)
axis([0 120 10^(-15) 10])
xlabel('n'), ylabel('error')
```

plots the Gauss quadrature error of $f(t) = \sin^2(10t)$ as a function of the number n of quadrature nodes, by comparing the computed answers with the exact answer of (a) rounded to 16 decimal places after the decimal point. The plot, shown in Fig. 2.18, is practically identical with Fig. 5(b) of [62].

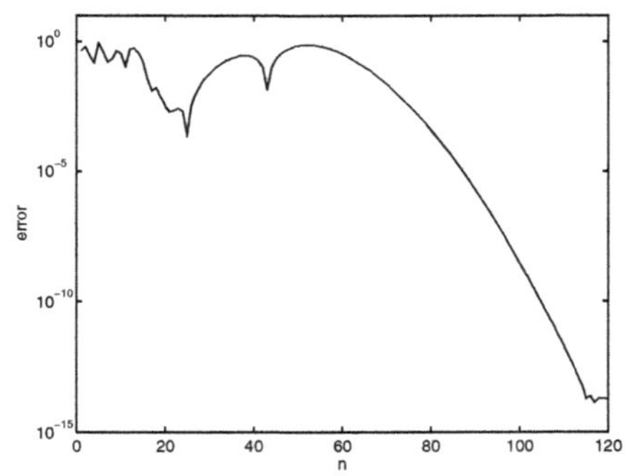

Figure 2.18. *Gauss quadrature error*

[2]This script works well in MATLAB Release R2011a, but does not work at all in MATLAB Release R2014a for reasons indicated in Demo 1.17 of Chapter 1.

Exercise 2.35. The weight function $w(t) = \exp(-e^t)$ on $[0, \infty]$, decreasing "hyperexponentially" at infinity, has been found useful in the evaluation of modified Bessel functions K_ν (also known as Macdonald functions) of complex order ν (see [26]). The three-term recurrence relation for the corresponding orthogonal polynomials has been generated for the first 100 polynomials in Fortran quadruple precision to 30 decimal places, using the subroutine `qmcdis` of the Fortran package `ORTHPOLq`; the results are tabulated in the file `qabmacdonald` of [26, §2], containing the first 100 recurrence coefficients α_k, β_k, $k = 0, 1, 2, \ldots, 99$.

Recompute these coefficients to 32 *decimal places, using the symbolic MATLAB routine* `smcdis.m` *of the package* `SOPQ`. *Specifically, decompose the interval* $[0, \infty]$, *as in* [26, §2], *into four subintervals,* $[0, \infty] = [0, 3/4] \cup [3/4, 3/2] \cup [3/2, 3] \cup [3, \infty]$, *and discretize the inner product* $(u, v) = \int_0^\infty u(t)v(t)\exp(-e^t)\mathrm{d}t$ *by applying* N*-point Fejér quadrature on each of these subintervals, with* N *chosen sufficiently large to achieve 32-digit accuracy* (use 36-digit arithmetic to guard against rounding errors).

Answer. The driver script `sab_hypexp.m` is as follows:

```
%SAB_HYPEXP  Recurrence coefficients for hyperexponential
%    weight function.

global mc mp iq idelta irout DM uv AB dig
dig=36; digits(dig);
N=100; eps0=.5e-32;
mc=4; mp=0; iq=0; idelta=2; irout=1; Mmax=1400;
AB=[[vpa(0) vpa(3/4)];[vpa(3/4) vpa(3/2)];[vpa(3/2) vpa(3)]; ...
    [vpa(3) Inf]];
[ab,Mcap,kount]=smcdis(dig,N,eps0,@squadgp,Mmax);
vpa(ab,32)
```

Be sure to declare `mc uv AB dig` as global variables at the beginning of the general-purpose quadrature routine `squadgp.m` and to append the function statement

```
function y=wf(x,i)
y=exp(-exp(x));
```

at the end thereof. The output `ab` is stored in the file `ab_hypexp.txt` in a form ready to be downloaded by the `loadvpa` command. There is complete agreement to all 30 decimal places with the data stored in the file `qabmacdonald`.

The beginning and end of the file `ab_hypexp.txt` are displayed below.

```
0.44599071252078017094836827453406   0.21938393439552027367716377546012
0.81146076489997903198073055577943   0.12566930523395267920321289748256
0.99871194765081678997834783415999   0.21608925435284678436655910462367
1.1273385336770703576625083922077    0.28926308278963351280666387624614
1.2259121923414294335465722984153    0.35077028482089750945631367962757

 ...        ...          ...        ...          ...           ...

2.5345989882698200285553872839565    1.6030913469339251880863565637734
2.5392853980764249958610507191203    1.6090603090563289598319446900894
2.5439244730029817079941150894754    1.614979606691227044770562753546
2.5485171657539670638677422847815    1.6208501364848626434802184368303
2.5530644004809157454159898611477    1.6266727702110668319464921459056
```

Among the two options in smcdis.m we used Stieltjes's procedure (sstieltjes.m) rather than Lanczos's (slanczos.m), since it is much faster. For $N = 5$, for example, the use of Stieltjes took 30 seconds to run sab_hypexp.m, whereas Lanczos took 13,539 seconds, that is, about 450 times longer!

Exercise 2.36. The weight function $w(t) = \exp(-e^{|t|})$ on $\mathbb{R}$ is a simple example of an *Erdős weight function* [10], meaning that

$$e^{|t|^k} w(t) \to 0 \quad \text{as } |t| \to \infty \tag{2.74}$$

for every $k = 0, 1, 2, \dots$. According to Erdős, "it seems likely that" (2.74) implies that the zeros $-\infty < t_1 < t_2 < \cdots < t_n < \infty$ of $\pi_n(\cdot\,; w)$, asymptotically as $n \to \infty$, are uniformly distributed in the following sense: Transform the interval $[t_1, t_n]$ linearly onto $[-1, 1]$ so that $t_1 < t_2 < \cdots < t_n$ are mapped into $-1 = s_1 < s_2 < \cdots < s_n = 1$. Let $s_k = \cos\theta_k$, $k = 1, 2, \dots, n$. Then for any interval (a, b), $0 < a < b \le \pi$, the number $N_n(a, b)$ of θ_ks on the arc (a, b) satisfies

$$N_n(a,b) = n\,\frac{b-a}{\pi} + o(n), \quad n \to \infty. \tag{2.75}$$

(a) *For $n = 100$ and $n = 200$ generate the orthogonal polynomials π_n and their zeros by a symmetric extension of the procedure in* Ex. 2.35, *and prepare a histogram for the respective θ_k's to illustrate* (2.75).

(b) *For comparison, do the same for the polynomials orthogonal with respect to the weight function $w(t) = \exp(-|t|)$ on $\mathbb{R}$. Use the symbolic version* smcdis.m *of the multicomponent discretization procedure to compute an arbitrary number of the relevant recurrence coefficients to any accuracy. Specifically, prepare a table of the first* 100 *coefficients to an accuracy of* 32 *decimal places.*

Answer

(a) The script sab_hypexpsymm.m,

```
%SAB_HYPEXPSYMM  Recurrence coefficients for symmetric
%   hyperexponential weight function.

global mc mp iq idelta irout DM uv AB dig
tic
dig=48; digits(dig);
N=200; eps0=.5e-32;
mc=8; mp=0; iq=0; idelta=2; irout=1; Mmax=2000;
AB=[[-Inf vpa(-3)];[vpa(-3) vpa(-3/2)];[vpa(-3/2) vpa(-3/4)]; ...
   [vpa(-3/4) vpa(0)];[vpa(0) vpa(3/4)];[vpa(3/4) vpa(3/2)]; ...
   [vpa(3/2) vpa(3)];[vpa(3) Inf]];
[ab,Mcap,kount]=smcdis(dig,N,eps0,@squadgp,Mmax)
vpa(ab,32)
toc
```

generates, and stores in the file sab_hypexpsymm_200.txt, the first 200 recurrence coefficients for the weight function

```
function y=wf(x,i)
y=exp(-exp(abs(x)));
```

which must be appended to the quadrature routine squadgp.m. The histograms in question are produced by the script

```
%ZDIST_HYPEXPSYMM

ab=loadvpa('sab_hypexpsymm_200.txt',3,32,200,2);
n=100;
%n=200;
xw=gauss(n,ab);
x=xw(:,1); xmin=min(x); xmax=max(x);
t=(2*x-(xmin+xmax))./(xmax-xmin);
th=pi-acos(t);
[n,x]=hist(th);
bar(x,n)
```

They are shown in Fig. 2.19. Uniform distribution of the zeros is certainly suggested, although near the endpoints of $[0,\pi]$ uniformity ceases to hold. There may be several reasons for it. For one, n may not be large enough; for another, uniformity may hold only in compact subintervals of $[0,\pi]$.

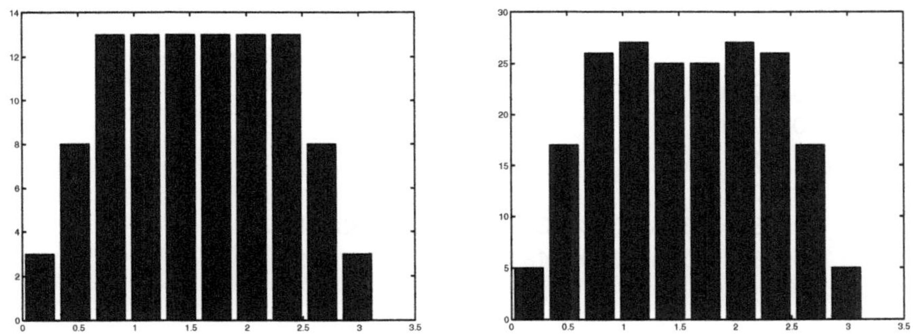

Figure 2.19. *Histograms for the zeros of π_n for $n=100$ (left) and $n=200$ (right)*

(b) For the weight function $w(t)=e^{-|t|}$ on $\mathbb{R}$, we have, given any polynomial p,

$$\int_{\mathbb{R}} p(x)e^{-|x|}\mathrm{d}x = \int_0^\infty [p(-t)+p(t)]e^{-t}\mathrm{d}t. \tag{2.76}$$

Hence, inner products can be easily evaluated by Gauss–Laguerre quadrature, and if p has degree $\le 2n-1$, then n-point Gauss–Laguerre yields exact results. A 2-component discretization procedure, therefore, with the two components corresponding to the two integrals on the right of (2.76), yields exact answers for the first n recurrence coefficients of $\pi_n(\cdot;w)$, if we use n-point Gauss–Laguerre quadrature to do the discretization. This is implemented in variable-precision arithmetic in the script sab_lagsymm.m,

```
%SAB_LAGSYMM

global mc mp iq idelta irout dig
dig=72; digits(dig);
N=100; eps0=.5e-32;
mc=2; mp=0; iq=1; idelta=2; irout=1; Mmax=230;
tic
```

```
[ab,Mcap,kount]=smcdis(dig,N,eps0,@squad_lagsymm,Mmax);
toc
[Mcap kount]
ab=vpa(ab,32)
```

using the quadrature routine squad_lagsymm.m,

```
function xw=squad_lagsymm(N,i)
%SQUAD_LAGSYMM Quadrature routine for discretization used by
%sab_lagsymm.m in smcdis.m.
%
global dig
digits(dig);
ab=sr_laguerre(dig,N);
xw=sgauss(dig,N,ab);
if i==1, xw(:,1)=-xw(:,1); end
```

Apparently, there is a considerable amount of cancellation error, which requires a high working precision to achieve convergence after just one iteration of the discretization process.

The beginning and end of the output file sab_lagsymm.txt are shown below.

```
0  2.0
0  2.0
0  10.0
0  21.6
0  39.73333333333333333333333333333
0  60.96196868008948545861297539149
   ...                  ...

0  22264.62616602098708889776437514 8
0  22742.85219143257298852372979206 4
0  23212.05519328666643718378212489 4
0  23700.25669738873600273861539660 6
0  24179.22366511323443771672582357 1
```

The analogon zdist_lagsymm.m to the script zdist_hypexpsymm.m then produces the histogram shown in Fig. 2.20. It looks anything but uniform!

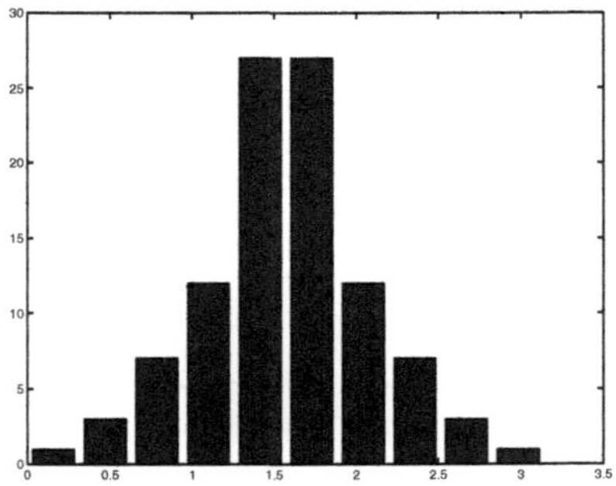

Figure 2.20. *Histogram for the n zeros of* $\pi_n(\cdot; e^{-|t|}dt)$, $n = 100$

Exercise 2.37. The weight function $w(t) = t^t(1-t)^{1-t}$ on $[0,1]$, which is symmetric about the point 1/2, was considered in [24, Example 2.33], where a natural subdivision of the first half of the interval $[0,1]$ was proposed for integrating w over $[0,1/2]$.

(a) *Use this subdivision (extended over the whole interval* $[0,1]$*) in the routine* `mcdis.m`, *in combination with the general-purpose quadrature routine* `quadgp.m`, *to generate the first* N=[5 10 20 40 80] *recurrence coefficients of the orthogonal polynomials* $\pi_k(\cdot;w)$ *to ten decimal digits. Compare the performance with using no subdivision at all.*

(b) *Apply the more efficient of the two routines of* (a) *in symbolic variable-precision arithmetic, using the routines* `smcdis.m` *and* `squadgp.m`, *to generate the first* 100 *recurrence coefficients of* w *to* 16 *decimal places.*

Answer

(a) For the two procedures with and without subdivision the important performance data `Mcap` and `kount` of the routine `mcdis.m` are determined by the script `ab_wfpow.m` below for the requested accuracy of `eps0=.5e-10`, along with the respective run times (in seconds). In the general-purpose quadrature routine `quadgp.m` the variables `mc`, `uv`, and `AB` must be declared as global and the function

```
function y=wf(x,i)
y=exp(x.*log(x)+(1-x).*log(1-x));
```

appended at the end. The routine and results produced are shown below.

```
%AB_WFPOW

global mc mp iq idelta irout DM uv AB
f0='%2.0f %5.0f %5.0f %9.4f %6.0f %6.0f %8.4f %11.4e\n';
disp(' N  Mcap0 kount0     t0      Mcap1  kount1    t1       err')
eps0=.5e-10;
mp=0; iq=0; idelta=2; irout=1; Mmax=700;
for N=[5 10 20 40 80]
  mc=6;
  AB=[[0 0.00115];[0.00115 0.0844];[0.0844 0.5]; ...
     [0.5 0.9156];[0.9156 0.99885];[0.99885 1]];
  tic
  [ab0,Mcap0,kount0]=mcdis(N,eps0,@quadgp,Mmax);
  t0=toc;
  mc=1;
  AB=[0 1];
  tic
  [ab1,Mcap1,kount1]=mcdis(N,eps0,@quadgp,Mmax);
  t1=toc;
  err=max(max(abs((ab1-ab0)./ab0)));
  fprintf(f0,N,Mcap0,kount0,t0,Mcap1,kount1,t1,err)
end
```

The output (edited) is

	with subdivision			without subdivision			
N	Mcap	kount	t	Mcap	kount	t	error
5	41	6	0.0394	371	19	0.0277	6.4315e–11
10	41	4	0.0056	501	16	0.0326	1.8369e–11
20	61	3	0.0058	441	11	0.0255	3.2784e–11
40	121	3	0.0091	481	8	0.0277	2.3242e–11
80	241	3	0.0169	641	6	0.0363	7.7253e–12

Columns 2–4 of the output relate to results obtained with subdivision, columns 5–7 to those without. The last column shows the maximum discrepancy between the two results. It can be seen that subdivision is quite beneficial, not only with regard to the number of iterations, `kount`, but also with regard to run time, `t`.

(b) The driver script is similar to the one in Ex. 2.35,

```
%SAB_WFPOW Recurrence coefficients for the weight function
%    w(t)=t^t*(1-t)^(1-t) on [0,1]

global mc mp iq idelta irout DM uv AB dig
tic
dig=24; digits(dig);
N=100; eps0=.5e-16;
mc=6; mp=0; iq=0; idelta=2; irout=1; Mmax=2000;
AB=[[vpa(0) vpa(115/100000)];[vpa(115/100000) vpa(844/10000)]; ...
    [vpa(844/10000) vpa(1/2)];[vpa(1/2) vpa(9156/10000)]; ...
    [vpa(9156/10000) vpa(99885/100000)];[vpa(99885/100000) 1]];
[ab,Mcap,kount]=smcdis(dig,N,eps0,@squadgp,Mmax)
vpa(ab,16)
toc
```

with the same quadrature routine `squadgp.m` except for the function routine at the end, which is as shown in (a). The output is stored in the file `ab_wfpow.txt`, the beginning and end of which are

```
0.5  0.6178269647290115
0.5  0.09820406964667909
0.5  0.06325486172935019
0.5  0.06381042337979742
0.5  0.06327507962135766
...            ...

0.5  0.06250171326748891
0.5  0.06250167794238825
0.5  0.06250164369866808
0.5  0.06250161049263628
0.5  0.06250157828278533
```

Evidently, all α_k are $1/2$ (by symmetry), and it appears that the β_k tend (slowly!) to $1/16 = 0.0625$ as $k \to \infty$.

The performance data are `Mcap`=1201, `kount`=8, and `t`=1667 sec.

Exercise 2.38. The weight function

$$w(t;\varepsilon)=\frac{\exp[-(t^4/4-t^2/2)/\varepsilon]}{\exp[1/(4\varepsilon)]},\quad \varepsilon>0,\ t\in\mathbb{R},$$

and the associated orthogonal polynomials $\pi_k(\cdot;w)$ are of interest in physics and chemistry in connection with the Fokker–Planck equation for bistable potentials [58, §§2.5.2 and 6.5]. When ε is small, the weight function has sharp peaks at $t=\pm 1$ of magnitude 1 (see Fig. 2.21 for $\varepsilon = .1, .02, .005$, produced by `plot_wbimod.m`), which presents a considerable challenge for generating the respective orthogonal polynomials.

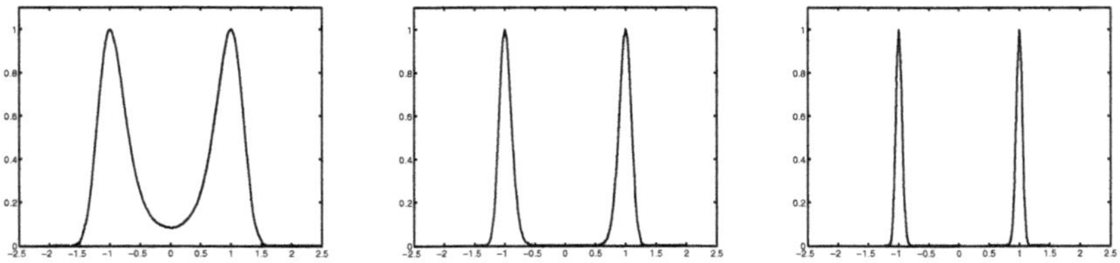

Figure 2.21. *The weight function $w(t)$ for $\varepsilon = .1, .02, .005$ (from left to right)*

(a) *For the values $\varepsilon = 1, .1, .02, .005$ apply a 1-component discretization procedure to obtain the first $N = 12$ recurrence coefficients for $\{\pi_k(\cdot;w)\}$ to an accuracy of* 15 *decimal places, using a suitable quadrature rule to discretize the inner product $\int_{\mathbb{R}} f(t)w(t;\varepsilon)dt$, $f(t)=u(t)v(t)$. Use the results to obtain the respective 12-point Gaussian quadrature rules for the weight function w to* 15 *decimal digits and compare with the* (6–7)*-digit results for the Gaussian nodes given in* [58, Table 2.18]. {Hint: You will have to use the symbolic SOPQ routine `smcdis.m` and manipulate the incrementation of the variable `Mcap` to get precise results about the number of quadrature points needed.}

(b) *Try to get similar results as in* (a) *for larger values of N.*

(c) *Can one do better with c-component discretization, $c>1$? If so, explain how.*

Answer

(a) A natural 1-component discretization of the inner product is in terms of Gauss–Freud quadrature (cf. Ex. 2.23). We write

$$\int_{\mathbb{R}} f(t)w(t;\varepsilon)dt=\frac{1}{e^{1/(4\varepsilon)}}\int_{\mathbb{R}} e^{t^2/(2\varepsilon)}f(t)e^{-t^4/(4\varepsilon)}dt$$

and let

$$t=c\tau,\quad c=(4\varepsilon)^{1/4},$$

to get

$$\int_{\mathbb{R}} f(t)w(t;\varepsilon)dt=ce^{-1/c^4}\int_{\mathbb{R}} e^{2(\tau/c)^2}f(c\tau)e^{-\tau^4}d\tau=c\int_{\mathbb{R}} e^{(2c^2\tau^2-1)/c^4}f(c\tau)e^{-\tau^4}d\tau.$$

Here the integral on the far right can be evaluated by Gauss quadrature relative to the Freud weight function $e^{-\tau^4}$ on $\mathbb{R}$. The respective first 200 recurrence coefficients (in the global array `ab1` of the routine `quad_bimod.m` below) have been generated to an accuracy of 32 decimal places, using the routine `sr_freud.m` of Ex. 2.23 in `dig`$=150$ digits, and are available in the file `ab_freud4.txt`. The quadrature routine to be used in `smcdis.m` thus is

```
function xw=quad_bimod(N,i)
%
global ab1 c dig
xw=sgauss(dig,N,ab1); digits(dig);
xw(:,1)=c*xw(:,1); s1=vpa(1); s4=vpa(4);
xw(:,2)=c*xw(:,2).*exp((2*xw(:,1).^2-s1)/c^s4);
```

Note that symbolic routines are used to execute the quadrature process, since the latter seems to be subject to relatively large rounding errors.

To determine exactly the number M of quadrature points needed for the routine `smcdis.m` to return the first twelve recurrence coefficients accurate to 15 decimal places, we run the routine with `dig`$=32$, `N`$=12$, `eps0`$=.5\times 10^{-15}$, and `Mmax` $=200$, but replace the command `Mcap=floor((2*N-1)/idelta` by `Mcap=Mest`, where `Mest` is a suitable estimate, and increment `Mcap` by 1, replacing the command `if kount>1, iMcap=2^ (floor(kount/5))*N; end` by `if kount>1, iMcap=1; end`. In this way, for $\varepsilon = 1, .1, .02, .005$, we find that M is equal to, resp., 32, 47, 77, and 143. The routines implementing this, and producing the respective Gaussian quadrature points and weights, are

```
%RUN_OPBIMOD

f0='%27.15e %23.15e\n'; f1='%8.0f %23.15e %23.15e\n';
f2='[Mcap kount]= %4.0f %4.0f\n';
dig=32;
for N=12
%for N=48
  fprintf('     N= %3.0f\n',N)
  epsi=1
%  epsi=1/10
%  epsi=2/100
%  epsi=5/1000
  disp('                            ab')
  [ab,Mcap,kount]=OPbimod(N,epsi);
  fprintf(f0,double(ab'));
  fprintf('\n');
  disp('      k             xk                     wk')
  xw=sgauss(dig,N,ab);
  for k=floor(N/2)+1:N
    fprintf(f1,k,double(xw(k,1)),double(xw(k,2)));
  end
  fprintf('\n'); fprintf(f2,Mcap,kount); fprintf('\n')
end
```

and

```
function [ab,Mcap,kount]=OPbimod(N,epsi)
%OPBIMOD Recurrence coefficients for the bimodal weight
%function.
%   [AB,MCAP,KOUNT]=OPBIMOD(N,EPSI) generates the Nx2 array AB of
%   the first N recurrence coefficients for the weight function
%   W(T)=C*EXP(-(T^4/4-T^2/2)/EPSI), C=EXP(-1/(4*EPSI)) on [-Inf,
```

```
%   Inf] using Freud4 quadrature in 32-digit arithmetic. The
%   output variables MCAP and KOUNT are inherited from the routine
%   mcdis.m.

global mc mp iq idelta irout DM AB c ab1 dig
mc=1; mp=0; iq=1; idelta=2; irout=1; DM=[0 0];
AB=[-Inf Inf]; Mmax=200; dig=32; digits(dig);
sepsi=vpa(epsi);
ab1=loadvpa('ab_freud4',3,dig,200,2);
eps0=.5e-15; c=(4*sepsi)^(vpa(1/4));
[ab,Mcap,kount]=smcdis(dig,N,eps0,@quad_bimod,Mmax);
```

The last six Gaussian quadrature nodes xk (in increasing order) and corresponding weights wk (the first six being the same except for the sign of xk) are found to be

```
 k          xk                       wk
 7   3.054319838885385e-01   4.903148891767381e-01   epsilon=1
 8   8.727105106506786e-01   5.191812817617963e-01
 9   1.358944900305454e+00   3.756869810795985e-01
10   1.780640575074145e+00   1.227764823811024e-01
11   2.166333913802086e+00   1.249935930767693e-02
12   2.552701503000944e+00   2.029492361457336e-04

 k          xk                       wk
 7   2.509269622516550e-01   5.299861724224568e-02   epsilon=1/10
 8   6.539266487309450e-01   1.484248509637131e-01
 9   9.463795651942800e-01   2.505116436981626e-01
10   1.181995862875993e+00   1.473581570589986e-01
11   1.391344373622840e+00   2.281564559410119e-02
12   1.598810231565511e+00   5.138514640791632e-04

 k          xk                       wk
 7   5.313424200075479e-01   5.301301224394745e-04   epsilon=2/100
 8   7.651935053423398e-01   2.089293691898918e-02
 9   9.201874815797584e-01   1.024693215018254e-01
10   1.047651293514686e+00   1.064929479684131e-01
11   1.164084164710648e+00   2.374125369461268e-02
12   1.282431167818464e+00   7.008423953364490e-04

 k          xk                       wk
 7   8.130843330110780e-01   3.185701145653734e-04   epsilon=5/1000
 8   8.973338317054135e-01   1.109802068006201e-02
 9   9.658537395073645e-01   5.137162473670748e-02
10   1.027887369306923e+00   5.150940848463445e-02
11   1.088161493587796e+00   1.119103188378906e-02
12   1.152340595880837e+00   3.234839905789916e-04
```

The nodes xk are in agreement with those in [58, Table 2.18] to all (6–7) digits given there, except for two end-figure discrepancies of one unit when $\varepsilon = 1/10$.

(b) For larger values of N we proceed exactly as in (a) as long as the number M of Gauss–Freud quadrature points remains less than, or equal to, Mmax $= 200$. In the most extreme case $\varepsilon = 5/1000$, and for accuracy levels of eps0 $= .5 \times 10^{-16}$, $.5 \times 10^{-12}$, and $.5 \times 10^{-8}$, this was found to be the case for, resp., $N \le 40$, $N \le 48$, and $N \le 58$.

(c) The answer is "yes," there being no longer any restriction on the number of quadrature points required.

We propose two types of multicomponent discretizations, depending on the value of the parameter ε. The first type is used for $\varepsilon \ge 1/10$ and is obtained by drawing the horizontal line $y = h_0 = e^{-1/(4\varepsilon)}$, $x \in \mathbb{R}$, tangent to the minimum of $w(\cdot;\varepsilon)$ at $x = 0$; see Fig. 2.22 (drawn for $\varepsilon = 1$). If we denote by $\mp b$ the abscissae of the first and last points of intersection, this yields the 4-component partition

$$\texttt{AB} = [\,[-\infty - b];[-b\;\;0];[0\;\;b];[b\;\;\infty]\,], \quad b = \sqrt{2}, \tag{2.77}$$

to be used in the OPQ routine mcdis.m. The second type of partition, for $\varepsilon < 1/10$, is obtained similarly by a horizontal line $y = h$, $h = e^{\mu} h_0$, $\mu > 0$, which, if $h < 1$, yields four points of intersection, hence five component intervals,

$$\texttt{AB} = [\,[-\infty - b_1];[-b_1 - b_2];[-b_2\;\;b_2];[b_2\;\;b_1];[b_1\;\;\infty]\,], \tag{2.78}$$

where, for $h > h_0$,

$$\pm b_1 = \pm\sqrt{1+k}, \quad \pm b_2 = \pm\sqrt{1-k},$$
$$k = 2\sqrt{\varepsilon \ln(1/h)}.$$

In the limit $h \downarrow h_0$, we have $k = 1$, that is, $b_2 = 0$, so that the 5-component partition (2.78) shrinks to the 4-component partition (2.77). The choice of $h = e^{\mu}$ in the former is made so as to minimize the number of quadrature points needed for a given accuracy.

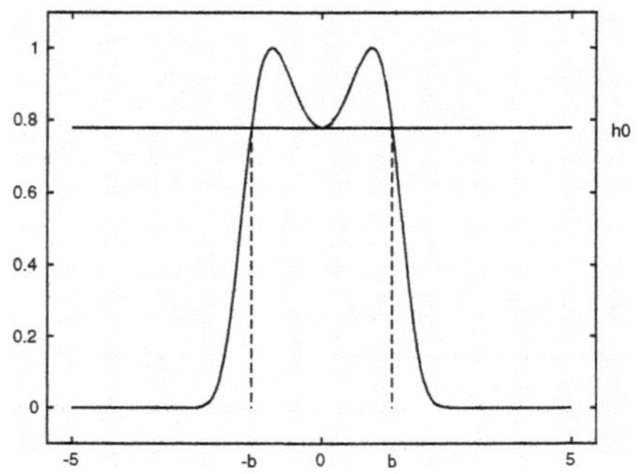

Figure 2.22. *The partition* AB *of* $\mathbb{R}$ *in the case* $\varepsilon \ge 1/10$

Analogous to run_OPbimod.m, OPbimod.m of (a) using Gauss–Freud quadrature on $\mathbb{R}$, we now try the script

```
%RUN_OPBIMOD_GP

global epsi eps0 h
f0='%27.15e %23.15e\n'; f1='%8.0f %23.15e %23.15e\n';
f2='[Mcap kount]= %4.0f %4.0f\n';
eps0=.5e-15;
for N=12
  fprintf('     N= %3.0f\n',N)
  epsi=1
```

```
% epsi=1/10
% epsi=2/100
% epsi=5/1000
  plot_wbimod;
  h0=exp(-1/(4*epsi));
  if epsi<1/10
    mu=input('enter the value of mu')
    h=exp(mu)*h0;
  end
  disp('                          ab')
  [ab,Mcap,kount]=OPbimod_gp(N,epsi);
  fprintf(f0,ab');
  fprintf('\n');
  disp('     k           xk                        wk')
  xw=gauss(N,ab);
  for k=floor(N/2)+1:N
    fprintf(f1,k,xw(k,1),xw(k,2));
  end
  fprintf('\n'); fprintf(f2,Mcap,kount); fprintf('\n')
end
```

and the routine (be sure to include the proper weight function `wf` at the end of `quadgp.m`)

```
function [ab,Mcap,kount]=OPbimod_gp(N,epsi)
%OPBIMOD_GP Recurrence coefficients for the bimodal weight function
%   OPBIMOD_GP(N,EPSI) generates the first N recurrence
%   coefficients for the weight function
%   W(T)=EXP(-(T^4/4-T^2/2)/EPSI) on [-Inf,Inf] using
%   quadgp.m in mcdis.
global mc mp iq idelta irout DM AB eps0 h
mp=0; iq=0; idelta=2; irout=1; DM=[0 0]; Mmax=500;
h0=exp(-1/(4*epsi)); k0=2*sqrt(epsi*log(1/h0));
if epsi<1/10
  k=2*sqrt(epsi*log(1/h));
  a1=-Inf; b1=-sqrt(1+k); a2=b1; b2=-sqrt(1-k);
  a3=b2; b3=sqrt(1-k); a4=b3; b4=sqrt(1+k); a5=b4; b5=Inf;
  mc=5;
  AB=[[a1 b1];[a2 b2];[a3 b3];[a4 b4];[a5 b5]];
else
  a1=-Inf; b1=-sqrt(1+k0); a2=b1; b2=0;
  a3=b2; b3=sqrt(1+k0); a4=b3; b4=Inf;
  mc=4;
  AB=[[a1 b1];[a2 b2];[a3 b3];[a4 b4]];
end
[ab,Mcap,kount]=mcdis(N,eps0,@quadgp,Mmax);
```

using general-purpose Fejér quadrature on appropriate multicomponent partitions of $\mathbb{R}$. The additional routine

```
%MUOPBIMOD_GP

function [Mcapmin,mu]=muOPbimod_gp(N,epsi,mu_low,dmu,mu_high,Mmax)
```

```
global eps0 h
if epsi>=1/10
  disp('epsi too large'); return
end
Mcapmin=Mmax; h0=exp(-1/(4*epsi));
for mu0=mu_low:dmu:mu_high
  h=exp(mu0)*h0;
  [ab,Mcap,kount]=OPbimod_gp(N,epsi);
%  [Mcap mu0]
  if Mcap<Mcapmin
    Mcapmin=Mcap; mu=mu0;
  end
end
```

is an auxiliary routine determining an optimal choice mu of μ to be entered at the input prompt of the script run_OPbimod_gp.m. We used these codes to reproduce the Gaussian quadrature nodes xk and weights wk of (a), for $\varepsilon = 1, .1, .02, .005$, and found almost identical results differing from those in (a) by at most five units in the 15th decimal place after the decimal point. The values of the optimal μ, when $\varepsilon = .02$ and .005, are, resp., mu = 2.5005 and mu = 41.9390 to four decimal places.

Further comparisons between Gauss–Freud and Fejér discretization are made for $N = 48$, eps0 = .5e–12, and implemented in the script compare_OPbimod.m,

```
%COMPARE_OPBIMOD

global epsi eps0 h
N=48; eps0=.5e-12;
epsi=1;
%epsi=1/10;
%epsi=2/100;
%epsi=5/1000;
tic
[ab,Mcap,kount]=OPbimod(N,epsi)
toc
h0=exp(-1/(4*epsi));
if epsi==2/100
  h=h0;
elseif epsi==5/1000
  h=exp(26.8)*h0;
end
tic
[ab_gp,Mcap,kount]=OPbimod_gp(N,epsi)
toc
merr=max(double(abs((ab(:,2)-ab_gp(:,2))./ab(:,2))));
fprintf('merr=%12.4e\n',merr)
```

The number Mcap required in the two cases, along with the maximum relative discrepancies merr and timings, are shown in Table 2.15 for selected values of ε. When $\varepsilon = .02$ and $\varepsilon = .005$, the respective values of μ (determined by mu0Pbimod_gp.m) are $\mu = 0$ and $\mu = 26.8$. It can be seen that, in the case of Gauss–Freud discretization, Mcap increases with decreasing ε, whereas it decreases in the case of Fejér discretization. Rather striking are the differences in timing, Gauss–Freud being as much as eight thousand times slower than Fejér. The reason is the use of multiprecision in the case of Gauss–Freud discretization. The agreement in the results of the two discretizations is well below the error tolerance `eps0`, as shown in the column headed by `merr`.

Table 2.15. *Freud vs Fejér discretization*

	Freud		Fejér		
ε	`Mcap`	Time (in sec)	`Mcap`	Time (in sec)	`merr`
1	69	43.9	318	.0476	3.585(–13)
.1	88	57.1	194	.0495	3.103(–14)
.02	121	94.5	124	.0417	8.787(–14)
.005	197	282	106	.0350	3.206(–15)

In all these computations, the routine `stieltjes.m`, resp., `sstieltjes.m`, was used in `mcdis.m` and `smcdis.m` (`irout = 1`) to obtain the recurrence coefficients from the discrete measures. Using `slanczos.m` instead, in the case of Freud discretization, turns out to be much slower by a factor of as large as 8.1, whereas `lanczos.m` in the case of Fejér discretization yields timings similar to those in Table 2.15. The discrepancies `merr` are always exactly the same.

High-precision results can be obtained by the symbolic analogue `sOPbimod_gp.m` of the routine `OPbimod_gp.m` and the respective driver script `run_sOPbimod_gp.m`. Employing 40-digit arithmetic, for example, we successfully generated the first 100 recurrence coefficients for the respective orthogonal polynomials to an accuracy of 32 decimal places for $\varepsilon = 1, .1, .02$, and $.005$. The values of `Mcap`—the required number of quadrature points in `smcdis.m`—are as high as, resp., 1004, 634, 430, and 354. The values of μ for $\varepsilon = .02$ and $\varepsilon = .005$ are obtained from `muOPbimod_gp.m` to be 0 and 15.3. The results can be found in the files `ab_OPbimod1.txt`, `ab_OPbimod01.txt`, `ab_OPbimod02.txt`, and `ab_OPbimod005.txt`.

The recurrence coefficients β_k (all α_k are zero by symmetry) are depicted in Fig. 2.23 for $\varepsilon = .1, .02, .005$. They start out oscillating, but end up growing monotonically.

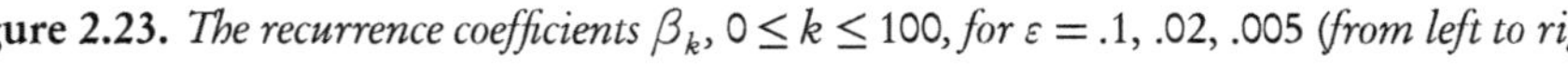

Figure 2.23. *The recurrence coefficients* β_k, $0 \le k \le 100$, *for* $\varepsilon = .1, .02, .005$ *(from left to right)*

2.6 ▪ Modification algorithms

Exercise 2.39. (Ex. 17 of [27, §2].) Let the measure $d\lambda$ be given. Consider modification of the measure by a quadratic factor $(t-x)^2 + y^2 = (t-z)(t-\overline{z})$, where $z = x + iy$, $x \in \mathbb{R}$, $y > 0$. That is, $d\hat{\lambda}(t) = (t-z)(t-\overline{z})d\lambda(t)$. If we denote $\pi_n(t) = \pi_n(t; d\lambda)$,

$\hat{\pi}_n(t)=\pi_n(t;d\hat{\lambda})$, then, by Christoffel's theorem,

$$(t-z)(t-\overline{z})\hat{\pi}_n(t)=\text{const}\begin{vmatrix}\pi_n(t) & \pi_{n+1}(t) & \pi_{n+2}(t)\\ \pi_n(z) & \pi_{n+1}(z) & \pi_{n+2}(z)\\ \pi_n(\overline{z}) & \pi_{n+1}(\overline{z}) & \pi_{n+2}(\overline{z})\end{vmatrix}.$$

To make $\hat{\pi}_n$ monic, the constant must be chosen to be the reciprocal of

$$\begin{vmatrix}\pi_n(z) & \pi_{n+1}(z)\\ \pi_n(\overline{z}) & \pi_{n+1}(\overline{z})\end{vmatrix}.$$

We then have

$$(t-z)(t-\overline{z})\hat{\pi}_n(t)=\pi_{n+2}(t)+s_n\pi_{n+1}(t)+t_n\pi_n(t)$$

for some constants s_n, t_n.

(a) *Determine s_n and t_n in terms of the original orthogonal polynomials.* {Hint: Define $r_n(z)=\pi_{n+1}(z)/\pi_n(z)$ and express the result in terms of $r_n'=\operatorname{Re} r_n(z)$, $r_n''=\operatorname{Im} r_n(z)$, and $|r_n|^2=|r_n(z)|^2$.}

(b) *Supply the details for deriving Algorithm* 4 *of* [27]*:*

Algorithm 4 Modification by a quadratic factor $(t-z)(t-\overline{z})$, $z=x+iy$

initialization:

$$r_0=z-\alpha_0,\ r_1=z-\alpha_1-\beta_1/r_0,\ r_2=z-\alpha_2-\beta_2/r_1,$$
$$\hat{\alpha}_0=\alpha_2+r_2'+\frac{r_2''}{r_1''}r_1'-\left(r_1'+\frac{r_1''}{r_0''}r_0'\right),$$
$$\hat{\beta}_0=\beta_0(\beta_1+|r_0|^2).$$

continuation (if $n>1$):

for $k=1,2,\dots,n-1$ do

$$r_{k+2}=z-\alpha_{k+2}-\beta_{k+2}/r_{k+1},$$
$$\hat{\alpha}_k=\alpha_{k+2}+r_{k+2}'+\frac{r_{k+2}''}{r_{k+1}''}r_{k+1}'-\left(r_{k+1}'+\frac{r_{k+1}''}{r_k''}r_k'\right),$$
$$\hat{\beta}_k=\beta_k\frac{r_{k+1}''r_{k-1}''}{[r_k'']^2}\left|\frac{r_k}{r_{k-1}}\right|^2.$$

Answer

(a) We have

$$-s_n=\frac{\begin{vmatrix}\pi_n(z) & \pi_{n+2}(z)\\ \pi_n(\overline{z}) & \pi_{n+2}(\overline{z})\end{vmatrix}}{\begin{vmatrix}\pi_n(z) & \pi_{n+1}(z)\\ \pi_n(\overline{z}) & \pi_{n+1}(\overline{z})\end{vmatrix}}=\frac{\pi_n(z)\pi_{n+2}(\overline{z})-\pi_n(\overline{z})\pi_{n+2}(z)}{\pi_n(z)\pi_{n+1}(\overline{z})-\pi_n(\overline{z})\pi_{n+1}(z)}.$$

Dividing numerator and denominator by $\pi_n(z)\pi_n(\overline{z})$, one gets

$$
\begin{aligned}
-s_n &= \frac{r_{n+1}(\overline{z})r_n(\overline{z}) - r_{n+1}(z)r_n(z)}{r_n(\overline{z}) - r_n(z)} = \frac{-2\mathrm{i}\,\mathrm{Im}(r_{n+1}(z)r_n(z))}{-2\mathrm{i}\,\mathrm{Im}(r_n(z))} \\
&= \frac{r'_{n+1}r''_n + r''_{n+1}r'_n}{r''_n};
\end{aligned}
$$

thus,

$$
s_n = -\left(r'_{n+1} + \frac{r''_{n+1}}{r''_n} r'_n \right). \tag{2.79}
$$

Similarly,

$$
\begin{aligned}
t_n &= \frac{\pi_{n+1}(z)\pi_{n+2}(\overline{z}) - \pi_{n+1}(\overline{z})\pi_{n+2}(z)}{\pi_n(z)\pi_{n+1}(\overline{z}) - \pi_n(\overline{z})\pi_{n+1}(z)} = \frac{r_n(z)r_{n+1}(\overline{z})r_n(\overline{z}) - r_n(\overline{z})r_{n+1}(z)r_n(z)}{r_n(\overline{z}) - r_n(z)} \\
&= \frac{r_n(z)r_n(\overline{z})[r_{n+1}(\overline{z}) - r_{n+1}(z)]}{r_n(\overline{z}) - r_n(z)} = |r_n|^2 \frac{-2\text{ß}\, r''_{n+1}}{-2\text{ß}\, r''_n};
\end{aligned}
$$

thus,

$$
t_n = \frac{r''_{n+1}}{r''_n} |r_n|^2. \tag{2.80}
$$

(b) Note first that $r_k = \pi_{k+1}(z)/\pi_k(z)$ can be computed recursively by

$$
r_0 = z - \alpha_0, \quad r_k = z - \alpha_k - \beta_k / r_{k-1}, \quad k = 1,2,3,\ldots,
$$

as follows directly from the three-term recurrence relation satisfied by $\{\pi_n(z)\}_{n=0}^{\infty}$ and from $\pi_1(z) = z - \alpha_0$ (cf. Ex. 2.3). This establishes the first lines in the initialization and continuation parts of Algorithm 4, and, by taking real and imaginary parts, yields r'_k, r''_k, $k = 0,1,2,\ldots$. As far as the initialization of $\hat{\beta}_0$ is concerned, we recall that, by definition,

$$
\hat{\beta}_0 = \int_{\mathbb{R}} (t-z)(t-\overline{z})\mathrm{d}\lambda(t) = \int_{\mathbb{R}} [t - \alpha_0 - (z - \alpha_0)][t - \alpha_0 - (\overline{z} - \alpha_0)]\mathrm{d}\lambda(t),
$$

and, since $t - \alpha_0 = \pi_1(t)$ and $\int_{\mathbb{R}} \pi_1^2(t)\mathrm{d}\lambda(t) = \beta_0\beta_1$, we get

$$
\begin{aligned}
\hat{\beta}_0 &= \int_{\mathbb{R}} [\pi_1(t) - (z - \alpha_0)][\pi_1(t) - (\overline{z} - \alpha_0)]\mathrm{d}\lambda(t) \\
&= \int_{\mathbb{R}} \pi_1^2(t)\mathrm{d}\lambda(t) - (\overline{z} + z - 2\alpha_0)\int_{\mathbb{R}} \pi_1(t)\mathrm{d}\lambda(t) + \beta_0|\pi_1(z)|^2 \\
&= \beta_0(\beta_1 + |r_0|^2),
\end{aligned}
$$

as claimed in the initialization part of Algorithm 4.

To complete the verification of Algorithm 4, we note that, on the one hand,

$$
\begin{aligned}
&(t-z)(t-\overline{z})t\hat{\pi}_k(t) = t\pi_{k+2}(t) + s_k t\pi_{k+1}(t) + t_k t\pi_k(t) \\
&\quad = \pi_{k+3}(t) + \alpha_{k+2}\pi_{k+2}(t) + \beta_{k+2}\pi_{k+1}(t) + s_k[\pi_{k+2}(t) + \alpha_{k+1}\pi_{k+1}(t) + \beta_{k+1}\pi_k(t)] \\
&\qquad + t_k[\pi_{k+1}(t) + \alpha_k\pi_k(t) + \beta_k\pi_{k-1}(t)],
\end{aligned}
$$

that is,

$$
\begin{aligned}
&(t-z)(t-\overline{z})t\hat{\pi}_k(t) \\
&\quad = \pi_{k+3}(t)+(\alpha_{k+2}+s_k)\pi_{k+2}(t)+(\beta_{k+2}+s_k\alpha_{k+1}+t_k)\pi_{k+1}(t) \\
&\qquad +(s_k\beta_{k+1}+t_k\alpha_k)\pi_k(t)+t_k\beta_k\pi_{k-1}(t),
\end{aligned}
\tag{2.81}
$$

and on the other hand,

$$
(t-z)(t-\overline{z})t\hat{\pi}_k(t)=(t-z)(t-\overline{z})[\hat{\pi}_{k+1}(t)+\hat{\alpha}_k\hat{\pi}_k(t)+\hat{\beta}_k\hat{\pi}_{k-1}(t)].
$$

Using the identity derived just before (a) at the beginning of the exercise, one gets

$$
\begin{aligned}
&(t-z)(t-\overline{z})t\hat{\pi}_k(t) \\
&\quad = \pi_{k+3}(t)+s_{k+1}\pi_{k+2}(t)+t_{k+1}\pi_{k+1}(t)+\hat{\alpha}_k[\pi_{k+2}(t)+s_k\pi_{k+1}(t)+t_k\pi_k(t)] \\
&\qquad +\hat{\beta}_k[\pi_{k+1}(t)+s_{k-1}\pi_k(t)+t_{k-1}\pi_{k-1}(t)],
\end{aligned}
$$

that is,

$$
\begin{aligned}
&(t-z)(t-\overline{z})t\hat{\pi}_k(t) \\
&\quad = \pi_{k+3}(t)+(s_{k+1}+\hat{\alpha}_k)\pi_{k+2}(t)+(t_{k+1}+s_k\hat{\alpha}_k+\hat{\beta}_k)\pi_{k+1}(t) \\
&\qquad +(t_k\hat{\alpha}_k+s_{k-1}\hat{\beta}_k)\pi_k(t)+t_{k-1}\hat{\beta}_k\pi_{k-1}(t).
\end{aligned}
\tag{2.82}
$$

Comparing the coefficient of $\pi_{k+2}(t)$ in (2.82) with the one in (2.81) yields $s_{k+1}+\hat{\alpha}_k=\alpha_{k+2}+s_k$, that is,

$$
\hat{\alpha}_k=\alpha_{k+2}-s_{k+1}+s_k,
$$

which, using (2.79), for $k=0$ gives the initialization of $\hat{\alpha}_0$ in Algorithm 4, and for $k>0$ the continuation of $\hat{\alpha}_k$. Comparison of the coefficient of $\pi_{k-1}(t)$ in (2.82) with the one in (2.81) yields $t_{k-1}\hat{\beta}_k=t_k\beta_k$, that is,

$$
\hat{\beta}_k=\beta_k\frac{t_k}{t_{k-1}}, \quad k\geq 1,
$$

which, by (2.80), gives

$$
\hat{\beta}_k=\beta_k\frac{r''_{k+1}r''_{k-1}}{(r''_k)^2}\left|\frac{r_k}{r_{k-1}}\right|^2, \quad k=1,2,3,\ldots,
$$

as claimed in the continuation part of Algorithm 4.

Exercise 2.40. (Ex. 18 of [27, §2].) Let the measure $d\lambda$ be given. Consider modification of the measure by a polynomial divisor, $v(t)=\prod_{\mu=1}^{m}(t-v_\mu)$, $m\geq 1$, where the v_μ are distinct numbers. That is, $d\hat{\lambda}(t)=d\lambda(t)/v(t)$. If, as in Ex. 2.39, we denote $\pi_n(t)=\pi_n(t;d\lambda)$, $\hat{\pi}_n(t)=\pi_n(t;d\hat{\lambda})$, there are generalizations of the Christoffel theorem which take different forms depending on whether $m\leq n$ or $m>n$ (cf. [24, §2.4.1]). In the first case,

$$\hat{\pi}_n(t) = \text{const}\begin{vmatrix} \pi_{n-m}(t) & \pi_{n-m+1}(t) & \cdots & \pi_n(t) \\ \rho_{n-m}(v_1) & \rho_{n-m+1}(v_1) & \cdots & \rho_n(v_1) \\ \rho_{n-m}(v_2) & \rho_{n-m+1}(v_2) & \cdots & \rho_n(v_2) \\ \vdots & \vdots & \cdots & \vdots \\ \rho_{n-m}(v_m) & \rho_{n-m+1}(v_m) & \cdots & \rho_n(v_m) \end{vmatrix}, \quad m \le n, \tag{2.83}$$

and in the second case,

$$\hat{\pi}_n(t) = \text{const}\begin{vmatrix} 0 & 0 & 0 & \cdots & 0 & \pi_0(t) & \pi_1(t) & \cdots & \pi_n(t) \\ 1 & v_1 & v_1^2 & \cdots & v_1^{m-n-1} & \rho_0(v_1) & \rho_1(v_1) & \cdots & \rho_n(v_1) \\ 1 & v_2 & v_2^2 & \cdots & v_2^{m-n-1} & \rho_0(v_2) & \rho_1(v_2) & \cdots & \rho_n(v_2) \\ \vdots & \vdots & \vdots & \cdots & \vdots & \vdots & \vdots & \cdots & \vdots \\ 1 & v_m & v_m^2 & \cdots & v_m^{m-n-1} & \rho_0(v_m) & \rho_1(v_m) & \cdots & \rho_n(v_m) \end{vmatrix}, \quad m > n, \tag{2.84}$$

where in both cases $\rho_k(z)$ are the Cauchy transforms of π_k (cf. Ex. 2.9),

$$\rho_k(z) = \rho_k(z; d\lambda) = \int_{\mathbb{R}} \frac{\pi_k(t)}{z-t}\, d\lambda(t), \quad k = 0, 1, 2, \ldots, \quad \rho_{-1}(z) = 1. \tag{2.85}$$

To make the $\hat{\pi}_n$ monic, the constant has to be chosen as the reciprocal of the respective cofactors of $\pi_n(t)$.

The important quantities are now the ratios

$$r_n = \frac{\rho_{n+1}(z)}{\rho_n(z)}, \quad n = -1, 0, 1, 2, \ldots. \tag{2.86}$$

If $m = 1$ and $v_1 = z$, we are in the case (2.83), and thus

$$\hat{\pi}_n(t) = \frac{\begin{vmatrix} \pi_{n-1}(t) & \pi_n(t) \\ \rho_{n-1}(z) & \rho_n(z) \end{vmatrix}}{-\rho_{n-1}(z)} = \frac{\pi_{n-1}(t)\rho_n(z) - \rho_{n-1}(z)\pi_n(t)}{-\rho_{n-1}(z)},$$

that is,

$$\hat{\pi}_n(t) = \pi_n(t) - r_{n-1}\pi_{n-1}(t), \quad n \ge 1. \tag{2.87}$$

Supply the details for deriving Algorithm 5 *of* [27], *which pertains to the linear function* $v(t) = t - z$.

Algorithm 5 Modification by a linear divisor $t - z$

initialization:

$$\hat{\alpha}_0 = \alpha_0 + r_0, \quad \hat{\beta}_0 = -\rho_0(z).$$

continuation (if $n > 1$):

for $k = 1,2,\ldots,n-1$ do
$$\hat{\alpha}_k = \alpha_k + r_k - r_{k-1},$$
$$\hat{\beta}_k = \beta_{k-1} r_{k-1}/r_{k-2}.$$

Answer. Note first that $r_k = \rho_{k+1}(z)/\rho_k(z)$ satisfies, similarly as in Ex. 2.39(b),

$$r_k = z - \alpha_k - \beta_k/r_{k-1}, \quad k = 0,1,2,\ldots, \tag{2.88}$$

where

$$r_{-1} = \rho_0(z)$$

since $\rho_{-1}(z) = 1$.

We first prove the initialization of Algorithm 5. By definition,

$$\hat{\beta}_0 = \int_{\mathbb{R}} \mathrm{d}\hat{\lambda}(t) = \int_{\mathbb{R}} \frac{\mathrm{d}\lambda(t)}{t-z} = -\rho_0(z).$$

Moreover,

$$t\hat{\pi}_0(t) = t\pi_0(t) = \pi_1(t) + \alpha_0\pi_0(t),$$

and also

$$\begin{aligned} t\hat{\pi}_0(t) &= \hat{\pi}_1(t) + \hat{\alpha}_0\hat{\pi}_0(t) \\ &= \pi_1(t) - r_0\pi_0(t) + \hat{\alpha}_0\pi_0(t) \\ &= \pi_1(t) + (\hat{\alpha}_0 - r_0)\pi_0(t), \end{aligned}$$

where, in the second equation, (2.87) for $n = 1$ has been used. Comparison with the first relation for $t\hat{\pi}_0(t)$ gives $\alpha_0 = \hat{\alpha}_0 - r_0$, that is,

$$\hat{\alpha}_0 = \alpha_0 + r_0.$$

To prove the continuation part of Algorithm 5, we note that, on the one hand, by (2.87), for $k \geq 1$,

$$\begin{aligned} t\hat{\pi}_k(t) &= t\pi_k(t) - r_{k-1}t\pi_{k-1}(t) \\ &= \pi_{k+1}(t) + \alpha_k\pi_k(t) + \beta_k\pi_{k-1}(t) - r_{k-1}[\pi_k(t) + \alpha_{k-1}\pi_{k-1}(t) + \beta_{k-1}\pi_{k-2}(t)] \\ &= \pi_{k+1}(t) + (\alpha_k - r_{k-1})\pi_k(t) + (\beta_k - r_{k-1}\alpha_{k-1})\pi_{k-1}(t) - r_{k-1}\beta_{k-1}\pi_{k-2}(t), \end{aligned} \tag{2.89}$$

and on the other hand, again using (2.87),

$$\begin{aligned} t\hat{\pi}_k(t) &= \hat{\pi}_{k+1}(t) + \hat{\alpha}_k\hat{\pi}_k(t) + \hat{\beta}_k\hat{\pi}_{k-1}(t) \\ &= \pi_{k+1}(t) - r_k\pi_k(t) + \hat{\alpha}_k[\pi_k(t) - r_{k-1}\pi_{k-1}(t)] + \hat{\beta}_k[\pi_{k-1}(t) - r_{k-2}\pi_{k-2}(t)] \\ &= \pi_{k+1}(t) + (\hat{\alpha}_k - r_k)\pi_k(t) + (\hat{\beta}_k - \hat{\alpha}_k r_{k-1})\pi_{k-1}(t) - \hat{\beta}_k r_{k-2}\pi_{k-2}(t). \end{aligned} \tag{2.90}$$

Consider first $k = 1$. Comparing the coefficients of $\pi_k(t)$ (when $k = 1$) in (2.89) and (2.90) gives $\alpha_1 - r_0 = \hat{\alpha}_1 - r_1$, that is,

$$\hat{\alpha}_1 = \alpha_1 + r_1 - r_0, \tag{2.91}$$

in agreement with the case $k=1$ of the continuation part. Comparison of the coefficients of $\pi_{k-1}(t)$ (when $k=1$) in (2.89) and (2.90) gives $\beta_1 - r_0\alpha_0 = \hat{\beta}_1 - r_0\hat{\alpha}_1$, that is,

$$\hat{\beta}_1 = \beta_1 + r_0(\hat{\alpha}_1 - \alpha_0). \tag{2.92}$$

Now using (2.91), and (2.88) for $k=1$ and $k=0$, yields

$$\begin{aligned}\hat{\beta}_1 &= \beta_1 + r_0(\alpha_1 + r_1 - r_0 - \alpha_0)\\ &= \beta_1 + r_0\left(\alpha_1 + z - \alpha_1 - \frac{\beta_1}{r_0} - r_0 - \alpha_0\right)\\ &= \beta_1 + r_0(z - r_0 - \alpha_0) - \beta_1\\ &= r_0\left(z - z + \alpha_0 + \frac{\beta_0}{\rho_0(z)} - \alpha_0\right)\\ &= \beta_0 r_0 / r_{-1},\end{aligned}$$

which is the formula for $\hat{\beta}_k$ in the continuation part when $k=1$. For $k \geq 2$, it suffices to compare the second and last terms in (2.89) with those in (2.90) to establish the formulae for $\hat{\alpha}_k$ and $\hat{\beta}_k$ in the continuation part of Algorithm 5.

Exercise 2.41. (Ex. 19 of [27, §2].) Consider the modification of Ex. 2.40, where the divisor is a quadratic polynomial, $v(t) = (t-x)^2 + y^2 = (t-z)(t-\overline{z})$, $z = x + \mathrm{i}y$, $x \in \mathbb{R}$, $y > 0$. The appropriate generalized Christoffel's theorem now yields

$$\hat{\pi}_n(t) = \pi_n(t) + s_n\pi_{n-1}(t) + t_n\pi_{n-2}(t), \quad n \geq 1. \tag{2.93}$$

(a) *Obtain* $\hat{\alpha}_0$ *and* $\hat{\beta}_0$.
(b) *In analogy to* Ex. 2.39, *derive formulae for* s_n *and* t_n *in* (2.93).
(c) *Supply the details for deriving Algorithm* 6 *of* [27]*:*

Algorithm 6 Modification by a quadratic divisor $(t-z)(t-\overline{z})$, $z = x + \mathrm{i}y$

initialization:

$$\begin{aligned}&\hat{\alpha}_0 = x + \rho_0' y/\rho_0'', \quad \hat{\beta}_0 = -\rho_0''/y,\\ &\hat{\alpha}_1 = \alpha_1 - s_2 + s_1, \quad \hat{\beta}_1 = \beta_1 + s_1(\alpha_0 - \hat{\alpha}_1) - t_2,\\ &\hat{\alpha}_2 = \alpha_2 - s_3 + s_2, \quad \hat{\beta}_2 = \beta_2 + s_2(\alpha_1 - \hat{\alpha}_2) - t_3 + t_2.\end{aligned}$$

continuation (if $n > 3$):

for $k = 3, 4, \ldots, n-1$ do

$$\hat{\alpha}_k = \alpha_k - s_{k+1} + s_k, \quad \hat{\beta}_k = \beta_{k-2}\, t_k / t_{k-1}.$$

Answer

(a) We have

$$\frac{1}{(t-z)(t-\overline{z})} = \frac{1}{(z-t)(\overline{z}-t)} = \frac{1}{\overline{z}-z}\left(\frac{1}{z-t} - \frac{1}{\overline{z}-t}\right),$$

that is,

$$\frac{1}{(t-z)(t-\overline{z})}=\frac{\mathrm{i}}{2y}\left(\frac{1}{z-t}-\frac{1}{\overline{z}-t}\right).$$

Multiplying both sides by t gives

$$\frac{t}{(t-z)(t-\overline{z})}=\frac{\mathrm{i}}{2y}\left[-\left(1-\frac{z}{z-t}\right)+\left(1-\frac{\overline{z}}{\overline{z}-t}\right)\right]=\frac{\mathrm{i}}{2y}\left(\frac{z}{z-t}-\frac{\overline{z}}{\overline{z}-t}\right).$$

Integration yields

$$\begin{aligned}\int_{\mathbb{R}}\frac{t\,\mathrm{d}\lambda(t)}{(t-z)(t-\overline{z})}&=\frac{\mathrm{i}}{2y}(z\rho_0(z)-\overline{z}\rho_0(\overline{z}))\\&=\frac{\mathrm{i}}{2y}2\mathrm{i}\,\mathrm{Im}(z\rho_0(z))\\&=-\frac{1}{y}[x\,\mathrm{Im}\,\rho_0(z)+y\,\mathrm{Re}\,\rho_0(z)],\end{aligned}$$

$$\int_{\mathbb{R}}\frac{\mathrm{d}\lambda(t)}{(t-z)(t-\overline{z})}=\frac{\mathrm{i}}{2y}[\rho_0(z)-\rho_0(\overline{z})]=-\frac{1}{y}\mathrm{Im}\,\rho_0(z).$$

Therefore,

$$\hat{\alpha}_0=\frac{\displaystyle\int_{\mathbb{R}}\frac{t\,\mathrm{d}\lambda(t)}{(t-z)(t-\overline{z})}}{\displaystyle\int_{\mathbb{R}}\frac{\mathrm{d}\lambda(t)}{(t-z)(t-\overline{z})}}=x+y\frac{\mathrm{Re}\,\rho_0(z)}{\mathrm{Im}\,\rho_0(z)},\quad \hat{\beta}_0=-\frac{1}{y}\mathrm{Im}\,\rho_0(z).$$

(b) We are now in the case $m=2$ of the generalized Christoffel's theorem (cf. Ex. 2.40, Eqs. (2.83) and Eqs. (2.84)), which gives, with $v_1=z$, $v_2=\overline{z}$,

$$\hat{\pi}_n(t)=\frac{\begin{vmatrix}\pi_{n-2}(t) & \pi_{n-1}(t) & \pi_n(t)\\ \rho_{n-2}(z) & \rho_{n-1}(z) & \rho_n(z)\\ \rho_{n-2}(\overline{z}) & \rho_{n-1}(\overline{z}) & \rho_n(\overline{z})\end{vmatrix}}{\begin{vmatrix}\rho_{n-2}(z) & \rho_{n-1}(z)\\ \rho_{n-2}(\overline{z}) & \rho_{n-1}(\overline{z})\end{vmatrix}},\ n\geq 2;\quad \hat{\pi}_1(t)=\frac{\begin{vmatrix}0 & \pi_0(t) & \pi_1(t)\\ 1 & \rho_0(z) & \rho_1(z)\\ 1 & \rho_0(\overline{z}) & \rho_1(\overline{z})\end{vmatrix}}{\begin{vmatrix}1 & \rho_0(z)\\ 1 & \rho_0(\overline{z})\end{vmatrix}}. \quad (2.94)$$

Assume first $n\geq 2$. From the first relation in (2.94), we have

$$s_n=-\frac{\begin{vmatrix}\rho_{n-2}(z) & \rho_n(z)\\ \rho_{n-2}(\overline{z}) & \rho_n(\overline{z})\end{vmatrix}}{\begin{vmatrix}\rho_{n-2}(z) & \rho_{n-1}(z)\\ \rho_{n-2}(\overline{z}) & \rho_{n-1}(\overline{z})\end{vmatrix}}=\frac{\rho_{n-2}(z)\rho_n(\overline{z})-\rho_{n-2}(\overline{z})\rho_n(z)}{\rho_{n-2}(z)\rho_{n-1}(\overline{z})-\rho_{n-2}(\overline{z})\rho_{n-1}(z)}.$$

Dividing numerator and denominator by $\rho_{n-2}(z)\rho_{n-2}(\overline{z})$ gives

$$\begin{aligned} s_n &= -\frac{r_{n-1}(\overline{z})r_{n-2}(\overline{z})-r_{n-1}(z)r_{n-2}(z)}{r_{n-2}(\overline{z})-r_{n-2}(z)} = \frac{2\mathrm{i}\,\mathrm{Im}(r_{n-1}(z)r_{n-2}(z))}{-2\mathrm{i}\,\mathrm{Im}(r_{n-2}(z)} \\ &= -\frac{r'_{n-1}r''_{n-2}+r''_{n-1}r'_{n-2}}{r''_{n-2}} = -\left(r'_{n-1}+\frac{r''_{n-1}}{r''_{n-2}}r'_{n-2}\right), \quad n\geq 2, \end{aligned}$$

with the r_k defined as in (2.86). Similarly,

$$\begin{aligned} t_n &= \frac{\rho_{n-1}(z)\rho_n(\overline{z})-\rho_{n-1}(\overline{z})\rho_n(z)}{\rho_{n-2}(z)\rho_{n-1}(\overline{z})-\rho_{n-2}(\overline{z})\rho_{n-1}(z)} \\ &= \frac{r_{n-2}(z)r_{n-1}(\overline{z})r_{n-2}(\overline{z})-r_{n-2}(\overline{z})r_{n-1}(z)r_{n-2}(z)}{r_{n-2}(\overline{z})-r_{n-2}(z)} = \frac{|r_{n-2}|^2(r_{n-1}(\overline{z})-r_{n-1}(z))}{r_{n-2}(\overline{z})-r_{n-2}(z)} \\ &= |r_{n-2}|^2\frac{-2\mathrm{i}\,\mathrm{Im}\,r_{n-1}(z)}{-2\mathrm{i}\,\mathrm{Im}\,r_{n-2}(z)} = \frac{r''_{n-1}}{r''_{n-2}}|r_{n-2}|^2, \quad n\geq 2. \end{aligned}$$

Finally, from the second relation in (2.94), when $n=1$,

$$s_1 = -\frac{\begin{vmatrix} 1 & \rho_1(z) \\ 1 & \rho_1(\overline{z}) \end{vmatrix}}{\begin{vmatrix} 1 & \rho_0(z) \\ 1 & \rho_0(\overline{z}) \end{vmatrix}} = -\frac{\rho_1(\overline{z})-\rho_1(z)}{\rho_0(\overline{z})-\rho_0(z)}.$$

Since $\rho_1(z)=r_0(z)\rho_0(z)$ and $r_{-1}(z)=\rho_0(z)$, we can write

$$\begin{aligned} s_1 &= -\frac{r_0(\overline{z})r_{-1}(\overline{z})-r_0(z)r_{-1}(z)}{r_{-1}(\overline{z})-r_{-1}(z)} = \frac{2\mathrm{i}\,\mathrm{Im}(r_0(z)r_{-1}(z))}{-2\mathrm{i}\,\mathrm{Im}\,r_{-1}(z)} = -\frac{r'_0r''_{-1}+r''_0r'_{-1}}{r''_{-1}} \\ &= -\left(r'_0+\frac{r''_0}{r''_{-1}}r'_{-1}\right). \end{aligned}$$

This is the same as the formula above for s_n with $n=1$.

(c) The first line of the initialization of Algorithm 6 has already been established in (a). For the second line, we use, on the one hand, (2.93) with $n=1$,

$$\begin{aligned} t\hat{\pi}_1(t) &= t\pi_1(t)+s_1t\pi_0(t) \\ &= \pi_2(t)+\alpha_1\pi_1(t)+\beta_1\pi_0(t)+s_1(\pi_1(t)+\alpha_0\pi_0(t)) \\ &= \pi_2(t)+(\alpha_1+s_1)\pi_1(t)+(\beta_1+s_1\alpha_0)\pi_0(t), \end{aligned} \tag{2.95}$$

and on the other hand,

$$t\hat{\pi}_1(t)=\hat{\pi}_2(t)+\hat{\alpha}_1\hat{\pi}_1(t)+\hat{\beta}_1\hat{\pi}_0(t).$$

Using here again (2.93), this time with $n=1$ and $n=2$, we get

$$\begin{aligned} t\hat{\pi}_1(t) &= \pi_2(t)+s_2\pi_1(t)+t_2\pi_0(t)+\hat{\alpha}_1[\pi_1(t)+s_1\pi_0(t)]+\hat{\beta}_1\pi_0(t) \\ &= \pi_2(t)+(\hat{\alpha}_1+s_2)\pi_1(t)+(t_2+s_1\hat{\alpha}_1+\hat{\beta}_1)\pi_0(t). \end{aligned} \tag{2.96}$$

Comparison of the coefficients of $\pi_1(t)$ in (2.95) and (2.96) gives $\alpha_1+s_1=\hat{\alpha}_1+s_2$, that is,

$$\hat{\alpha}_1=\alpha_1+s_1-s_2,$$

while comparison of the coefficients of $\pi_0(t)$ gives $\beta_1+s_1\alpha_0=t_2+s_1\hat{\alpha}_1+\hat{\beta}_1$, that is,

$$\hat{\beta}_1=\beta_1+s_1(\alpha_0-\hat{\alpha}_1)-t_2.$$

To prove the third line of the initialization part, as well as the continuation part, we proceed as follows. From (2.93), for $k\geq 2$, we have

$$\begin{aligned}
t\hat{\pi}_k(t)&=t\pi_k(t)+s_kt\pi_{k-1}(t)+t_kt\pi_{k-2}(t)\\
&=\pi_{k+1}(t)+\alpha_k\pi_k(t)+\beta_k\pi_{k-1}(t)+s_k[\pi_k(t)+\alpha_{k-1}\pi_{k-1}(t)+\beta_{k-1}\pi_{k-2}(t)]\\
&\quad+t_k[\pi_{k-1}(t)+\alpha_{k-2}\pi_{k-2}(t)+\beta_{k-2}\pi_{k-3}(t)]\\
&=\pi_{k+1}(t)+(\alpha_k+s_k)\pi_k(t)+(\beta_k+s_k\alpha_{k-1}+t_k)\pi_{k-1}(t)\\
&\quad+(s_k+\beta_{k-1}+t_k\alpha_{k-2})\pi_{k-2}(t)+t_k\beta_{k-2}\pi_{k-3}(t).
\end{aligned}\tag{2.97}$$

On the other hand,

$$t\hat{\pi}_k(t)=\hat{\pi}_{k+1}(t)+\hat{\alpha}_k\hat{\pi}_k(t)+\hat{\beta}_k\hat{\pi}_{k-1}(t),$$

thus, by (2.93),

$$\begin{aligned}
t\hat{\pi}_k(t)&=\pi_{k+1}(t)+s_{k+1}\pi_k(t)+t_{k+1}\pi_{k-1}(t)+\hat{\alpha}_k[\pi_k(t)+s_k\pi_{k-1}(t)+t_k\pi_{k-2}(t)]\\
&\quad+\hat{\beta}_k[\pi_{k-1}(t)+s_{k-1}\pi_{k-2}(t)+t_{k-1}\pi_{k-3}(t)]\\
&=\pi_{k+1}(t)+(s_{k+1}+\hat{\alpha}_k)\pi_k(t)+(t_{k+1}+s_k\hat{\alpha}_k+\hat{\beta}_k)\pi_{k-1}(t)\\
&\quad+(t_k\hat{\alpha}_k+s_{k-1}\hat{\beta}_k)\pi_{k-2}(t)+t_{k-1}\hat{\beta}_k\pi_{k-3}(t).
\end{aligned}\tag{2.98}$$

Comparing the coefficients of $\pi_k(t)$ in (2.97) and (2.98) gives $\alpha_k+s_k=\hat{\alpha}_k+s_{k+1}$, that is,

$$\hat{\alpha}_k=\alpha_k+s_k-s_{k+1},$$

which for $k=2$ establishes the formula for $\hat{\alpha}_2$ in the initialization part of Algorithm 6, and for $k\geq 3$ the first formula in the continuation part. Comparing the coefficients of $\pi_{k-1}(t)$ gives $\beta_k+s_k\alpha_{k-1}+t_k=t_{k+1}+s_k\hat{\alpha}_k+\hat{\beta}_k$, that is,

$$\hat{\beta}_k=\beta_k+s_k(\alpha_{k-1}-\hat{\alpha}_k)+t_k-t_{k+1},$$

which for $k=2$ establishes the formula for $\hat{\beta}_2$ in the initialization part of Algorithm 6. Comparing, finally, the coefficients of $\pi_{k-3}(t)$ for $k\geq 3$ gives $t_k\beta_{k-2}=t_{k-1}\hat{\beta}_k$, that is,

$$\hat{\beta}_k=\beta_{k-2}t_k/t_{k-1},$$

the second formula in the continuation part. For the constants s_k, t_k, see part (b).

Exercise 2.42.* (For this exercise, see also [31].) Modification of a measure $\mathrm{d}\lambda$ by a polynomial divisor,

$$\mathrm{d}\hat{\lambda}(t)=\frac{\mathrm{d}\lambda(t)}{(t-z_1)(t-z_2)\cdots(t-z_m)},\quad m>1,\tag{2.99}$$

where $z_1, z_2, \ldots, z_m$ are distinct real or complex numbers, is dealt with, in theory, by the generalized Christoffel's theorem (2.83), (2.84). (For the case $m = 1$, see Ex. 2.40.) The focus of that theorem, however, is on the orthogonal polynomial itself, not on the respective recurrence coefficients. For the latter, it is more convenient to proceed recursively, that is, to consider a sequence of successive modifications by linear divisors,

$$d\lambda_\mu(t) = \frac{1}{(t - z_\mu)} d\lambda_{\mu-1}(t), \quad \mu = 1, 2, \ldots, m, \quad d\lambda_0(t) = d\lambda(t), \tag{2.100}$$

so that

$$d\hat{\lambda}(t) = d\lambda_m(t).$$

To each modification (2.100) by a linear divisor we can apply Algorithm 5 of Ex. 2.40 to update the recurrence coefficients.

(a) *Denoting by* $\alpha_k^{[\mu]}$, $\beta_k^{[\mu]}$, $k = 0, 1, \ldots, n-1$, *the first* n *recurrence coefficients associated with the measure* $d\lambda_\mu$, *formulate a recursive algorithm, Algorithm* 5bis, *generating in* m *steps the desired recurrence coefficients* $\alpha_k^{[m]}$, $\beta_k^{[m]}$, $k = 0, 1, \ldots, n-1$, *from the given recurrence coefficients* $\alpha_k = \alpha_k^{[0]}$, $\beta_k = \beta_k^{[0]}$. *In particular, explain how* $\rho_0(z)$ *in Algorithm* 5 *must be chosen in each step of Algorithm* 5bis.

(b) *Write a MATLAB routine implementing Algorithm* 5bis.

(c) *Apply Algorithm* 5bis *to compute the recurrence coefficients of the orthogonal polynomials relative to a Szegő–Bernstein measure,*

$$d\lambda(t) = \frac{1}{(c_1^2 - t^2)(c_2^2 - t^2)\cdots(c_{m/2}^2 - t^2)} \frac{dt}{\sqrt{1-t^2}} \quad \text{on } [-1, 1], \tag{2.101}$$

where $m \geq 2$ *is an even integer and* $c_\mu > 1$, $\mu = 1, 2, \ldots, m/2$.

By symmetry, the α-coefficients α_k are all zero, and from [61, Theorem 2.6] it follows that the β_k are all equal to $1/4$ when $k > 1 + m/2$. Thus, the last β-coefficient not equal to $1/4$ is $\beta_{1+m/2}$. *Reproduce Table 6 in* [31] *of these coefficients for* $m = 4:4:48$ *in the case where* $c_\mu = 1 + 1/\mu$, $\mu = 1, 2, \ldots, m/2$.

(d) *Discuss how to deal with the measure* (2.101) *if the minus signs in the factors involving the* c_μ *are replaced by plus signs. In particular, check Table 7 in* [31] *of* $\beta_{1+m/2}$, $m = 4:4:48$, *pertaining to the case* $c_\mu^2 = -1/\mu^2$, $\mu = 1, 2, \ldots, m/2$.

Answer.

(a) Evidently, in place of $\rho_0(z)$ in Algorithm 3, we must take

$$\rho_0^{[\mu]} = \int_{\mathbb{R}} \frac{1}{z_\mu - t} d\lambda_{\mu-1}(t),$$

that is,

$$\rho_0^{[\mu]} = (-1)^{\mu-1} \int_{\mathbb{R}} \frac{d\lambda(t)}{(z_\mu - t)(z_{\mu-1} - t)\cdots(z_1 - t)}, \quad \mu = 1, 2, \ldots, m.$$

Writing

$$\frac{1}{(z_\mu - t)(z_{\mu-1} - t)\cdots(z_1 - t)} = \frac{c_\mu^{[\mu]}}{z_\mu - t} + \frac{c_{\mu-1}^{[\mu]}}{z_{\mu-1} - t} + \cdots + \frac{c_1^{[\mu]}}{z_1 - t},$$

where

$$c_\nu^{[\mu]} = \frac{1}{\prod_{k=1;k\neq\nu}^{\mu}(z_k - z_\nu)}, \quad \nu = 1,2,\ldots,\mu,\ \mu > 1;\ c_1^{[1]} = 1,$$

we get

$$\rho_0^{[\mu]} = (-1)^{\mu-1}\sum_{\nu=1}^{\mu} c_\nu^{[\mu]} \int_{\mathbb{R}} \frac{d\lambda(t)}{z_\nu - t} = (-1)^{\mu-1}\sum_{\nu=1}^{\mu} c_\nu^{[\mu]} \rho_0(z_\nu),$$

where $\rho_0(z)$ is the Cauchy integral (2.85) with $k = 0$. Thus, we have the following:

Algorithm 5bis Modification by m linear divisors $t - z_\mu$, $\mu = 1,2,\ldots,m$

for $\mu = 1,2,\ldots,m$ do
initialization:

$$\alpha_0^{[\mu]} = \alpha_0^{[\mu-1]} + r_0^{[\mu-1]}, \quad \beta_0^{[\mu]} = -\rho_0^{[\mu]}.$$

continuation (if $n > 1$):
for $k = 1,2,\ldots,n-1$ do

$$\alpha_k^{[\mu]} = \alpha_k^{[\mu-1]} + r_k^{[\mu-1]} - r_{k-1}^{[\mu-1]},$$
$$\beta_k^{[\mu]} = \beta_{k-1}^{[\mu-1]} r_{k-1}^{[\mu-1]} / r_{k-2}^{[\mu-1]}.$$

end$k$
endμ

Here, the quantities $r_k^{[\mu-1]} = \rho_{k+1}(z_\mu)/\rho_k(z_\mu)$ are computed by (cf. (2.88))

$$r_k^{[\mu-1]} = z_\mu - \alpha_k^{[\mu-1]} - \beta_k^{[\mu-1]}/r_{k-1}^{[\mu-1]}, \quad k = 0,1,2,\ldots,$$

where

$$r_{-1}^{[\mu-1]} = \rho_0^{[\mu]}.$$

In general, this requires higher-precision arithmetic because of the instability of the recursion involved.

(b) The MATLAB double-precision routine is `mod_multlindiv.m`, shown below. The symbolic variable-precision analogue is `smod_multlindiv.m`.

```
function ab=mod_multlindiv(N,m,z,ab0,rho0)
%MOD_MULTLINDIV  Modification by multiple linear divisors.
%   AB=MOD_MULTLINDIV(N,M,Z,AB0,RHO0) generates the Nx2 array AB
%   of the first N recurrence coefficients for the weight function
%   W(T)=W0(T)/((T-Z_1)*(T-Z_2)*...*(T-Z_M)), where W0 is a given
%   weight function whose Nx2 array of recurrence coefficients is
%   AB0, and  Z=[Z_1;Z_2;...;Z_M]. The routine calls for a MATLAB
%   function r=rho0(z) computing the Cauchy integral int W0(t)dt/
%   (z-t) of W0. The routine rho0.m has to be supplied by the user
%   and its name prefixed by the symbol @ before using it as the
%   last input argument to this routine. Because of inherent
%   instabilities, the use of the symbolic analogue SMOM_MULTLINDIV
%   is preferable.

if m<1, error('m out of range'), end
```

```
r0=zeros(m,1); rmu0=zeros(m,1); r=zeros(N+1,1);
for mu=1:m
  r0(mu)=rho0(z(mu));
end
rmu0(1)=r0(1);
if m>1
  s=1;
  for mu=2:m
    s=-s;
    k=1:mu;
    for nu=1:mu
      c(nu)=1/prod(z(find(k-nu))-z(nu));
    end
    rmu0(mu)=s*sum(c.'.*r0(1:mu));
  end
end
ab1=ab0;
for mu=1:m
  ab=ab1;
  r(1)=rmu0(mu);
  for k=1:N
    r(k+1)=z(mu)-ab(k,1)-ab(k,2)/r(k);
  end
  ab1(1,1)=ab(1,1)+r(2); ab1(1,2)=-rmu0(mu);
  for k=2:N
    ab1(k,1)=ab(k,1)+r(k+1)-r(k);
    ab1(k,2)=ab(k-1,2)*r(k)/r(k-1);
  end
end
ab=ab1;
```

A small technical remark about MATLAB may be in order here: The transposition of the vector c in the assignment statement for rmu0 is effected by the nonconjugate transposition operator .', since when c is complex, as it may well be, we want only transposition, not combined with conjugation.

(c) The measure (2.101) is the same (except for sign, which is unimportant) as (2.99) if we take

$$z_k = c_{(k+1)/2}, \quad z_{k+1} = -z_k, \qquad k = 1,3,\dots,m-1.$$

The Cauchy integral needed here is (cf. [40, Eq. 3.613.1 for $n=0$])

$$\rho_0(z) = \int_{-1}^{1} \frac{1}{z-t} \frac{\mathrm{d}t}{\sqrt{1-t^2}} = \int_0^{\pi} \frac{\mathrm{d}\theta}{z-\cos\theta} = \mathrm{sgn}(z)\frac{\pi}{\sqrt{z^2-1}}, \qquad |z|>1,\ z\in\mathbb{R},$$

and is evaluated by the MATLAB function rho0_SzBe.m for any complex z,

```
function r=rho0_SzBe(z)
%RHO0_SZBE Cauchy integral of the Chebyshev weight function.
%   R=RHO0_SZBE(Z) evaluates R=int_{-1}^1 dt/((Z-t)*SQRT(1-t^2)).

x=real(z); y=imag(z);
```

```
r=pi/sqrt(abs(z^2-1));
if y==0
  if x<0, r=-r; end
else
  if y>0, r=-i*r; else, r=i*r; end
end
```

The symbolic analogue is srho0_SzBe.m.

Reproducing Table 6 in [31] by means of the variable-precision routine smod_multlindiv.m and the driver script sSz_Be.m,

```
%SSZ_BE    Symbolic analogue of Sz_Be.m

f0='%8.0f %21.16f\n';
disp('        m         beta_{1+m/2}')
dig=52; digits(dig);
N=30;
ab0=sr_jacobi(dig,N,-1/2);
for k=1:2:47
  z(k)=vpa(1+2/(k+1)); z(k+1)=-z(k);
%  z(k)=vpa(2*i/(k+1)); z(k+1)=-z(k);
end
for m=4:4:48
  ab=smod_multlindiv(dig,N,m,z,ab0,@srho0_SzBe);
  fprintf(f0,m,double(ab(2+m/2,2)))
end
```

requires 52-digit arithmetic, as was determined by trial and error. The results are shown in Table 2.16. It has been verified that $\beta_k = 1/4$ when $k > 1 + m/2$.

Table 2.16. *The coefficients* $\beta_{1+m/2}$ *for the measure* (2.101)

m	$\beta_{1+m/2}$	m	$\beta_{1+m/2}$
4	0.2526187518831068	28	0.2500000105993838
8	0.2501334100968554	32	0.2500000025371004
12	0.2500122983685047	36	0.2500000006601609
16	0.2500015886586432	40	0.2500000001842993
20	0.2500002570365245	44	0.2500000000546533
24	0.2500000489998413	48	0.2500000000170815

(d) The constants c_μ must be taken in conjugate complex pairs on the imaginary axis, $c_\mu = \pm \mathrm{i}y_\mu, y_\mu > 0$, to obtain $c_\mu^2 - t^2 = -(y_\mu^2 + t^2)$, $\mu = 1, 2, \ldots, m/2$. This can be run by the same routine sSz_Be.m if in the k-loop the second statement (presently commented out) is activated with $z_k = \mathrm{i}y_{(k+1)/2}$, $z_{k+1} = -z_k$, $k = 1, 3, \ldots, m-1$, and the first commented out. Furthermore, the Cauchy integral is now

$$\rho_0(\mathrm{i}y) = \int_{-1}^{1} \frac{1}{\mathrm{i}y - t} \frac{\mathrm{d}t}{\sqrt{1-t^2}},$$

that is,

$$\rho_0(\mathrm{i}y) = \int_0^\pi \frac{\mathrm{d}\theta}{\mathrm{i}y - \cos\theta} = -\int_0^\pi \frac{\mathrm{i}y + \cos\theta}{y^2 + \cos^2\theta}\,\mathrm{d}\theta = -\mathrm{i}y \int_0^\pi \frac{\mathrm{d}\theta}{y^2 + \cos^2\theta}.$$

Since $\cos^2\theta$ is an even function around $\theta=\pi/2$, we have

$$\begin{aligned}\rho_0(\mathrm{i}y) &= -2\mathrm{i}y\int_0^{\pi/2}\frac{d\theta}{y^2+\cos^2\theta} = -2\mathrm{i}y\int_0^{\pi/2}\frac{d\theta}{y^2+1-\sin^2\theta}\\ &= -2\mathrm{i}\frac{y}{y^2+1}\int_0^{\pi/2}\frac{d\theta}{1-a^2\sin^2\theta}, \quad a^2=\frac{1}{y^2+1}.\end{aligned}$$

The last integral is (cf. [40, Eq. 3.615.1 for $n=0$]) $\pi/(2\sqrt{1-a^2})$, so that

$$\rho_0(\mathrm{i}y) = -\mathrm{i}\,\mathrm{sgn}(y)\frac{\pi}{\sqrt{y^2+1}}.$$

The results produced by `smod_multlindiv.m`, when $y_\mu=1/\mu$, are shown in Table 2.17.

Table 2.17. *The coefficients $\beta_{1+m/2}$ for the modified measure* (2.101)

m	$\beta_{1+m/2}$	m	$\beta_{1+m/2}$
4	0.2663837516998524	28	0.2505052273767967
8	0.2551885634664923	32	0.2503902763397229
12	0.2525024432356620	36	0.2503105117367366
16	0.2514669167788029	40	0.2502529137950326
20	0.2509624502661558	44	0.2502099708571666
24	0.2506795558052531	48	0.2501771028004859

We remark that these results do not agree with those in Table 7 of [31]. The reason is that the latter, owing to a programming error, do not correspond to $c_\mu=1/\mu$ as intended, but to $c_\mu=1/(2\mu-1)$, $\mu=1,2,\ldots,m/2$.

Exercise 2.43.* In [24, Example 2.60] the example

$$F(x,y)=\int_{-\infty}^{\infty} e^{-xt}\cos t\,\frac{e^{-t^2}}{t^2+y^2}, \quad y>0, \tag{2.102}$$

was given to illustrate Algorithm 6 of Ex. 2.41 for modifying a measure by a quadratic divisor, t^2+y^2 in this case. *Explore whether incorporating the factor e^{-xt} into the measure and again applying Algorithm 6 with another quadratic divisor* (the integrand then involving only the cosine function) *would result in a more efficient way of evaluating* (2.102).

(a) *Carry out the analysis for the alternative procedure suggested above.*

(b) *Write a MATLAB double-precision routine* `mod_quaddivisor.m` *and a symbolic version thereof,* `smod_quaddivisor.m`, *implementing Algorithm* 6.

(c) *Use the double-precision routine of* (b) *to reproduce Table 2.26 of* [24], *that is, results for $x=[1\ 2\ 5\ 10]$, $y=[.1\ .01\ .001]$, and n (the number of Gauss quadrature nodes) running from* 5 *in steps of* 1 *until fifteen correct decimal places are achieved. Compare with Table 2.26 of* [24].

(d) *Run the symbolic routine of* (b) *in 36-digit arithmetic to determine the smallest value of n that guarantees 32-digit accuracy.*

(e) *Use the symbolic routine* `sEx_Alg6.m` *listed in the answer to part* (d) *to explore larger values of y, say $y=[.5\ 1\ 2\ 3.5\ 5]$. What kind of precision is required to obtain accuracies of at least* 16 *decimal places?*

Answer

(a) Combining the two exponential terms, and completing the square, yields

$$F(x,y)=\int_{-\infty}^{\infty}\cos t\,\frac{e^{-(t^2+xt)}}{t^2+y^2}\,dt=e^{x^2/4}\int_{-\infty}^{\infty}\cos t\,\frac{e^{-(t+x/2)^2}}{t^2+y^2}\,dt.$$

Changing variables, $t+x/2=\tau$, gives

$$F(x,y)=e^{x^2/4}\int_{-\infty}^{\infty}\cos(\tau-x/2)\,\frac{e^{-\tau^2}}{(\tau-x/2)^2+y^2}\,d\tau.$$

Algorithm 6 is again applicable, but now with $z=x/2+\mathrm{i}y$.

The Cauchy integral ρ_0 required is now

$$\rho_0=\int_{-\infty}^{\infty}\frac{e^{-t^2}}{z-t}\,dt=\int_{-\infty}^{\infty}\frac{(\overline{z}-t)e^{-t^2}dt}{|z-t|^2}=\int_{-\infty}^{\infty}\frac{[(x/2-\mathrm{i}y)-t]e^{-t^2}dt}{(t-x/2)^2+y^2}$$

$$=-\mathrm{i}y\int_{-\infty}^{\infty}\frac{e^{-t^2}dt}{(t-x/2)^2+y^2}+\int_{-\infty}^{\infty}\frac{(x/2-t)e^{-t^2}dt}{(t-x/2)^2+y^2},$$

that is, using [1, Eqs. 7.4.13 and 7.4.14],

$$\rho_0=\pi(\mathrm{Im}\,w(z)-\mathrm{i}\,\mathrm{Re}\,w(z)),\quad z=x/2+\mathrm{i}y,$$

where

$$w(z)=e^{-z^2}\mathrm{erfc}(-\mathrm{i}z).$$

(b) The double-precision routine is (see also Section 1.4.4 for an alternative routine)

```
function ab=mod_quaddivisor(N,ab0,x,y,rho)
%MOD_QUADDIVISOR Modification by a quadratic divisor.
%   AB=MOD_QUADDIVISOR(N,AB0,X,Y,RHO) generates the Nx2 array AB
%   of the first N recurrence coefficients for the weight function
%   W(T)=W0(T)/((T-X)^2+Y^2), where X is real, Y>0, and W0 a given
%   weight function whose Nx2 array of recurrence coefficients is
%   AB0. The last input variable RHO is the Cauchy integral
%   int W0(t)dt/(z-t) of W0, with z=X+iY.

if N<2, error('N out of range'), end
N0=size(ab0,1); r=zeros(N+1,1); s=zeros(N,1); t=zeros(N,1);
z=complex(x,y);
if N0<N, error('input array too short'), end
r(1)=rho0;
for k=1:N
  r(k+1)=z-ab0(k,1)-ab0(k,2)/r(k);
end
for k=1:N
  s(k)=-(real(r(k+1))+(imag(r(k+1))/imag(r(k)))*real(r(k)));
  t(k)=(imag(r(k+1))/imag(r(k)))*abs(r(k))^2;
end
ab(1,1)=x+y*real(r(1))/imag(r(1)); ab(1,2)=-imag(r(1))/y;
```

```
ab(2,1)=ab0(2,1)-s(2)+s(1);
ab(2,2)=ab0(2,2)+s(1)*(ab0(1,1)-ab(2,1))-t(2);
if N==2, return, end
ab(3,1)=ab0(3,1)-s(3)+s(2);
ab(3,2)=ab0(3,2)+s(2)*(ab0(2,1)-ab(3,1))-t(3)+t(2);
if N==3, return, end
for k=3:N-1
  ab(k+1,1)=ab0(k+1,1)-s(k+1)+s(k);
  ab(k+1,2)=ab0(k-1,2)*t(k)/t(k-1);
end
```

The symbolic routine `smod_quaddivisor.m` is a straightforward translation from double-precision arithmetic to variable-precision arithmetic.

(c) This is done by the driver script

```
%EX_ALG6

f0='%8.1f %8.3f %4.0f %22.14e\n';
disp('      x        y       n            int')
digits(32);
ab0=r_hermite(80);
for x=[1 2 5 10]
  for y=[1/10 1/100 1/1000]
    z=vpa(x/2+i*y);
    w=exp(-z^2)*(1-erf(-i*z));
    rho0=vpa(pi*(imag(w)-i*real(w)));
    rho0=double(rho0);
    for N=5:15
      ab=mod_quaddivisor(N,ab0,x/2,y,rho0);
      xw=gauss(N,ab); t=xw(:,1); w=xw(:,2);
      F=exp(x^2/4)*sum(w.*cos(t-x/2));
      fprintf(f0,x,y,N,F)
    end
    fprintf('\n')
  end
end
```

It produces the following results:

```
>> Ex_Alg6
     x        y      n             int
    1.0     0.100    5    2.80191141800017e+01
    1.0     0.100    6    2.80191134239162e+01
    1.0     0.100    7    2.80191134381391e+01
    1.0     0.100    8    2.80191134379059e+01
    1.0     0.100    9    2.80191134379093e+01

    1.0     0.010    5    3.10499175911096e+02
    1.0     0.010    6    3.10499175105418e+02
    1.0     0.010    7    3.10499175120478e+02
    1.0     0.010    8    3.10499175120232e+02
    1.0     0.010    9    3.10499175120236e+02
```

```
  1.0    0.010   10    3.10499175120235e+02

  1.0    0.001    5    3.13790449354097e+03
  1.0    0.001    6    3.13790449273014e+03
  1.0    0.001    7    3.13790449274529e+03
  1.0    0.001    8    3.13790449274505e+03

  2.0    0.100    5    2.98880496999992e+01
  2.0    0.100    6    2.98880485214156e+01
  2.0    0.100    7    2.98880485431102e+01
  2.0    0.100    8    2.98880485427605e+01
  2.0    0.100    9    2.98880485427655e+01

  2.0    0.010    5    3.12740541882399e+02
  2.0    0.010    6    3.12740540617102e+02
  2.0    0.010    7    3.12740540640207e+02
  2.0    0.010    8    3.12740540639837e+02
  2.0    0.010    9    3.12740540639842e+02

  2.0    0.001    5    3.14018771170948e+03
  2.0    0.001    6    3.14018771043514e+03
  2.0    0.001    7    3.14018771045839e+03
  2.0    0.001    8    3.14018771045802e+03

  5.0    0.100    5   -2.32414852197157e+01
  5.0    0.100    6   -2.32413378098260e+01
  5.0    0.100    7   -2.32413409820679e+01
  5.0    0.100    8   -2.32413409252200e+01
  5.0    0.100    9   -2.32413409260990e+01
  5.0    0.100   10   -2.32413409260872e+01

  5.0    0.010    5    2.62429728662585e+02
  5.0    0.010    6    2.62429884395975e+02
  5.0    0.010    7    2.62429881047033e+02
  5.0    0.010    8    2.62429881106877e+02
  5.0    0.010    9    2.62429881105954e+02
  5.0    0.010   10    2.62429881105967e+02

  5.0    0.001    5    3.09017251625521e+03
  5.0    0.001    6    3.09017267284074e+03
  5.0    0.001    7    3.09017266947358e+03
  5.0    0.001    8    3.09017266953374e+03
  5.0    0.001    9    3.09017266953280e+03
  5.0    0.001   10    3.09017266953282e+03

 10.0    0.100    5    3.12274885844231e+08
 10.0    0.100    6    3.12274646342174e+08
 10.0    0.100    7    3.12274649101481e+08
 10.0    0.100    8    3.12274649150643e+08
 10.0    0.100    9    3.12274649146792e+08
```

```
        10.0    0.100   10   3.12274649146905e+08

        10.0    0.010    5   3.11965160958971e+08
        10.0    0.010    6   3.11965022406536e+08
        10.0    0.010    7   3.11965014886619e+08
        10.0    0.010    8   3.11965015591293e+08
        10.0    0.010    9   3.11965015565981e+08
        10.0    0.010   10   3.11965015566426e+08
        10.0    0.010   11   3.11965015566419e+08

        10.0    0.001    5   3.11964027864434e+08
        10.0    0.001    6   3.11964826326125e+08
        10.0    0.001    7   3.11964737715587e+08
        10.0    0.001    8   3.11964740969640e+08
        10.0    0.001    9   3.11964740916535e+08
        10.0    0.001   10   3.11964740917149e+08
        10.0    0.001   11   3.11964740917129e+08
        10.0    0.001   12   3.11964740917142e+08
>>
```

The entries in the last line of each group of results are correct to all digits shown, except when $x = 10$ and $y = .1$, in which case the last digits should be 2 instead of 5 (see the answer to part (d) of this exercise). What is remarkable is the good performance of Gauss quadrature; it never requires more than $n = 12$ quadrature nodes to obtain full 15-digit accuracy. This is a lot better than in [24, Table 2.26], where n had to be as large as 80 (for $x = 10$) to get the same accuracy.

(d) The symbolic script for this part of the exercise is as follows:

```
%SEX_ALG6

dig=36; digits(dig);
ab0=sr_hermite(dig,40);
for x=[1 2 5 10]
  for y=[1/10 1/100 1/1000]
    z=vpa(x/2+i*y); x=vpa(x); y=vpa(y);
    w=exp(-z^2)*(1-erf(-i*z));
    rho0=vpa(pi)*(imag(w)-i*real(w));
    for N=12:20
      ab=smod_quaddivisor(dig,N,ab0,x/2,y,rho0);
      xw=sgauss(dig,N,ab); t=xw(:,1); w=xw(:,2);
      F=exp(x^2/4)*sum(w.*cos(t-x/2));
      disp([double(x) double(y) N F])
    end
    fprintf('\n')
  end
end
```

From the results produced in a format similar to those in part (c),

```
>> sEx_Alg6
[ 1, 1/10, 12, 28.0191134379092801243229723153550792]
```

```
[ 1, 1/10, 13, 28.019113437909280124378871865569927 7]
[ 1, 1/10, 14, 28.0191134379092801243783567794577941]
[ 1, 1/10, 15, 28.0191134379092801243783612013880172]
[ 1, 1/10, 16, 28.0191134379092801243783611658560899]
[ 1, 1/10, 17, 28.0191134379092801243783611661244198]
[ 1, 1/10, 18, 28.0191134379092801243783611661225085]
[ 1, 1/10, 19, 28.0191134379092801243783611661225214]

[ 1, 1/100, 12, 310.4991751202353362515730919972870 21]
[ 1, 1/100, 13, 310.4991751202353362516311976374578 66]
[ 1, 1/100, 14, 310.4991751202353362516306631078757 78]
[ 1, 1/100, 15, 310.4991751202353362516306676900279 37]
[ 1, 1/100, 16, 310.4991751202353362516306676532564 75]
[ 1, 1/100, 17, 310.4991751202353362516306676535338 41]
[ 1, 1/100, 18, 310.4991751202353362516306676535318 67]

[ 1, 1/1000, 12, 3137.904492745047864064803570767555 15]
[ 1, 1/1000, 13, 3137.904492745047864064861901948949 16]
[ 1, 1/1000, 14, 3137.904492745047864064861365433535 27]
[ 1, 1/1000, 15, 3137.904492745047864064861370032037 11]
[ 1, 1/1000, 16, 3137.904492745047864064861369995139 26]
[ 1, 1/1000, 17, 3137.904492745047864064861369995417 55]
[ 1, 1/1000, 18, 3137.904492745047864064861369995415 57]

[ 2, 1/10, 12, 29.8880485427654559909856688367596 83]
[ 2, 1/10, 13, 29.8880485427654559910654348402403493]
[ 2, 1/10, 14, 29.8880485427654559910647042341020412]
[ 2, 1/10, 15, 29.8880485427654559910647104733505751]
[ 2, 1/10, 16, 29.8880485427654559910647104234480 26]
[ 2, 1/10, 17, 29.8880485427654559910647104238233265]
[ 2, 1/10, 18, 29.8880485427654559910647104238206632]
[ 2, 1/10, 19, 29.8880485427654559910647104238206 81]

[ 2, 1/100, 12, 312.7405406398422821591944371903224 3]
[ 2, 1/100, 13, 312.7405406398422821592775397034190 84]
[ 2, 1/100, 14, 312.7405406398422821592767799775211 79]
[ 2, 1/100, 15, 312.7405406398422821592767864546764 19]
[ 2, 1/100, 16, 312.7405406398422821592767864029472 51]
[ 2, 1/100, 17, 312.7405406398422821592767864033357 77]
[ 2, 1/100, 18, 312.7405406398422821592767864033330 23]
[ 2, 1/100, 19, 312.7405406398422821592767864033330 42]

[ 2, 1/1000, 12, 3140.187710458022217659638390698316 45]
[ 2, 1/1000, 13, 3140.187710458022217659721834750680 18]
[ 2, 1/1000, 14, 3140.187710458022217659721072047339 24]
[ 2, 1/1000, 15, 3140.187710458022217659721078548795 94]
[ 2, 1/1000, 16, 3140.187710458022217659721078496880 35]
[ 2, 1/1000, 17, 3140.187710458022217659721078497270 23]
[ 2, 1/1000, 18, 3140.187710458022217659721078497267 47]
```

```
[ 5, 1/10, 12, -23.241340926087202438663959760342417 4]
```

```
[ 5, 1/10, 12, -23.2413409260872024386639597603424174]
[ 5, 1/10, 13, -23.2413409260872024552601690650503425]
[ 5, 1/10, 14, -23.2413409260872024551042957973739798]
[ 5, 1/10, 15, -23.2413409260872024551056555756872633]
[ 5, 1/10, 16, -23.2413409260872024551056444991388814]
[ 5, 1/10, 17, -23.2413409260872024551056445837761867]
[ 5, 1/10, 18, -23.2413409260872024551056445831671218]
[ 5, 1/10, 19, -23.2413409260872024551056445831712639]
[ 5, 1/10, 20, -23.2413409260872024551056445831712372]

[ 5, 1/100, 12, 262.4298811059665162758009567630310 65]
[ 5, 1/100, 13, 262.429881105966516258540536633449166]
[ 5, 1/100, 14, 262.429881105966516258702381091535686]
[ 5, 1/100, 15, 262.429881105966516258700971296441041]
[ 5, 1/100, 16, 262.429881105966516258700982765288157]
[ 5, 1/100, 17, 262.429881105966516258700982677757467]
[ 5, 1/100, 18, 262.429881105966516258700982678386673]
[ 5, 1/100, 19, 262.429881105966516258700982678382398]
[ 5, 1/100, 20, 262.429881105966516258700982678382426]

[ 5, 1/1000, 12, 3090.17266953281708924256292934304906]
[ 5, 1/1000, 13, 3090.17266953281708922523457827541272]
[ 5, 1/1000, 14, 3090.17266953281708922539703280737647]
[ 5, 1/1000, 15, 3090.17266953281708922539561790666054]
[ 5, 1/1000, 16, 3090.17266953281708922539562941552091]
[ 5, 1/1000, 17, 3090.17266953281708922539562932769531]
[ 5, 1/1000, 18, 3090.17266953281708922539562932832657]
[ 5, 1/1000, 19, 3090.17266953281708922539562932832228]

[ 10, 1/10, 12, 312274649.146902192474436972077610938]
[ 10, 1/10, 13, 312274649.146902192250625363905197712]
[ 10, 1/10, 14, 312274649.146902192252272352461388462]
[ 10, 1/10, 15, 312274649.146902192252262558270945685]
[ 10, 1/10, 16, 312274649.146902192252262601006627104]
[ 10, 1/10, 17, 312274649.146902192252262600946040209]
[ 10, 1/10, 18, 312274649.146902192252262600944719633]
[ 10, 1/10, 19, 312274649.146902192252262600944739477]
[ 10, 1/10, 20, 312274649.146902192252262600944739286]

[ 10, 1/100, 12, 311965015.566419404944579387368777877]
[ 10, 1/100, 13, 311965015.566419404574692386743991506]
[ 10, 1/100, 14, 311965015.566419404577247387572809163]
[ 10, 1/100, 15, 311965015.566419404577232107539153612]
[ 10, 1/100, 16, 311965015.566419404577232182412941814]
[ 10, 1/100, 17, 311965015.566419404577232182170997462]
[ 10, 1/100, 18, 311965015.566419404577232182170656316]
[ 10, 1/100, 19, 311965015.566419404577232182170671126]
[ 10, 1/100, 20, 311965015.566419404577232182170670959]

[ 10, 1/1000, 12, 311964740.917141596114383123821326818]
```

```
[ 10, 1/1000, 13, 311964740.91714159572446230100469456 2]
[ 10, 1/1000, 14, 311964740.91714159572712860204655247 5]
[ 10, 1/1000, 15, 311964740.91714159572711268849036627 4]
[ 10, 1/1000, 16, 311964740.91714159572711276695631638 4]
[ 10, 1/1000, 17, 311964740.91714159572711276669447373 4]
[ 10, 1/1000, 18, 311964740.91714159572711276669423890 4]
[ 10, 1/1000, 19, 311964740.91714159572711276669425317 2]
[ 10, 1/1000, 20, 311964740.91714159572711276669425300 8]
>>
```

Table 2.18. *16-digit Gaussian quadrature results for the integral in* (2.102)

x	y	n	$F(x,y)$	x	y	n	$F(x,y)$
1.0	0.5	10	3.755544730971707	5.0	0.5	11	-56.63600306221994
	1.0	10	1.268082574826385		1.0	12	-59.90172173201482
	2.0	11	0.3658783418412851		2.0	11	-47.37727163228577
	3.5	11	0.1244908608016984		3.5	11	-28.12528853530108
	5.0	12	0.06163208285296966		5.0	11	-17.25681861160181
2.0	0.5	11	4.649643632129497	10.0	0.5	10	319541804.1135398
	1.0	10	1.680626458283782		1.0	11	339472732.8813645
	2.0	11	0.4954514824494206		2.0	12	388982741.6098690
	3.5	11	0.1665248060445736		3.5	11	414521818.1953617
	5.0	12	0.08164022557372332		5.0	11	376268306.8830812

it is evident that in all cases $n = 19$, and in some cases already $n = 18$ or even $n = 17$, will achieve 32-digit accuracy.

(e) By trial and error one finds that 32-digit arithmetic suffices to obtain results accurate to 16 digits. The number of Gauss quadrature points is between 10 and 12. The results displayed in Table 2.18 were checked by recomputation with 36-digit arithmetic.

Chapter 3

Answers to Exercises on Sobolev Orthogonal Polynomials

In this chapter, polynomials orthogonal with respect to a Sobolev inner product of the type

$$
\begin{aligned}
(u,v)_S = &\int_{\mathbb{R}} u(t)v(t)\mathrm{d}\lambda_0(t) + \int_{\mathbb{R}} u'(t)v'(t)\mathrm{d}\lambda_1(t) + \cdots \\
&+ \int_{\mathbb{R}} u^{(s)}(t)v^{(s)}(t)\mathrm{d}\lambda_s(t), \quad u,v \in \mathbb{P},\ s \geq 1,
\end{aligned}
\tag{3.1}
$$

are considered, where $\mathrm{d}\lambda_0$ is an absolutely continuous "ground measure," and $\mathrm{d}\lambda_\sigma$, $\sigma = 1,2,\ldots,s$, are additional measures, which may be discrete or absolutely continuous. If all measures are positive, the inner product (3.1) is positive definite on $\mathbb{P}$, and therefore defines a unique set of (monic) orthogonal polynomials $\pi_0, \pi_1, \pi_2, \ldots$,

$$
\begin{aligned}
&(\pi_k, \pi_\ell)_S = 0, \quad k \neq \ell,\ k,\ell = 0,1,2,\ldots, \\
&\pi_k(t) = t^k + \text{lower degree terms}, \quad k = 0,1,2,\ldots,
\end{aligned}
$$

the so-called *Sobolev orthogonal polynomials.* Being a sequence of monic polynomials of increasing degrees $0,1,2,\ldots$, they must satisfy a recurrence relation of the form

$$
\begin{aligned}
\pi_{k+1}(t) = t\pi_k(t) - \sum_{j=0}^{k} \beta_j^k \pi_{k-j}(t), \quad k = 0,1,2,\ldots, \\
\pi_0(t) = 1.
\end{aligned}
\tag{3.2}
$$

Placing the coefficients β_j^k into an upper Hessenberg matrix

$$
\boldsymbol{B}_n = \begin{bmatrix}
\beta_0^0 & \beta_1^1 & \beta_2^2 & \cdots & \beta_{n-2}^{n-2} & \beta_{n-1}^{n-1} \\
1 & \beta_0^1 & \beta_1^2 & \cdots & \beta_{n-3}^{n-2} & \beta_{n-2}^{n-1} \\
0 & 1 & \beta_0^2 & \cdots & \beta_{n-4}^{n-2} & \beta_{n-3}^{n-1} \\
\vdots & \vdots & \vdots & \vdots & \vdots & \vdots \\
0 & 0 & 0 & \cdots & \beta_0^{n-2} & \beta_1^{n-1} \\
0 & 0 & 0 & \cdots & 1 & \beta_0^{n-1}
\end{bmatrix},
\tag{3.3}
$$

the zeros $z_1, z_2, \ldots, z_n$ of π_n are the eigenvalues of $\boldsymbol{B}_n$ (cf. [36, §1]).

The majority of cases studied in the literature relate to the case $s = 1$ and frequently involve measures of the special form

$$d\lambda_0(t) = d\lambda(t), \quad d\lambda_1(t) = \gamma\, d\lambda(t), \quad \gamma > 0,$$

where $d\lambda$ is a positive absolutely continuous measure.

The modified Chebyshev algorithm and Stieltjes's algorithm of Chapter 2 have their analogues for Sobolev orthogonal polynomials ([36, cf. §§2 and 4]). The respective MATLAB routines are `chebyshev_sob.m` and `stieltjes_sob.m`. The former is used in Ex. 3.5 and the latter in Exs. 3.6–3.8. The routine for zeros is `sobzeros.m`. There are also symbolic analogues of these routines, which have the same names prefixed by "`s`".

A few elementary formal properties, most notably the "long" recurrence relation (3.2) (Ex. 3.2), are the subject of Section 3.1. Symmetry of the constituent measures leads again to simplifications, which are discussed in Ex. 3.4 of Section 3.2. Zeros of Sobolev orthogonal polynomials, arguably the most frequently studied topic in this area, are dealt with in Section 3.3, which explores reality and nonreality of zeros as well as monotonicity and interlacing properties. There are many conjectures in this area which can be checked computationally.

3.1 ▪ Elementary properties

Exercise 3.1. (Ex. 1 of [27, §3].) *Show that the Sobolev inner product* (3.1) *does not satisfy the shift property* $(tp, q) = (p, tq)$.

Answer. Consider, e.g., $s = 1$. Since

$$(tp, q)_S = \int_{\mathbb{R}} tp(t)q(t)d\lambda_0 + \int_{\mathbb{R}} [p(t) + tp'(t)]q'(t)d\lambda_1(t),$$
$$(p, tq)_S = \int_{\mathbb{R}} tp(t)q(t)d\lambda_0(t) + \int_{\mathbb{R}} p'(t)[q(t) + tq'(t)]d\lambda_1(t),$$

we get

$$(tp, q)_S - (p, tq)_S = \int_{\mathbb{R}} [p(t)q'(t) - p'(t)q(t)]d\lambda_1(t).$$

Now take $p(t) = 1$, $q(t) = t$. Then the right-hand side becomes

$$\int_{\mathbb{R}} d\lambda_1(t) > 0,$$

so $(tp, q)_S \neq (p, tq)_S$ in this case.

Exercise 3.2. (a) *Prove that the Sobolev orthogonal polynomials* $\pi_k(t) = \pi_k(t; S)$ *satisfy the recurrence relation* (3.2).

(b) *Show that the coefficients are given by*

$$\beta_j^k = \frac{(t\pi_k, \pi_{k-j})_S}{(\pi_{k-j}, \pi_{k-j})_S}, \quad j = 0, 1, \ldots, k;\ k = 0, 1, 2, \ldots.$$

Answer

(a) Let $\{p_k\}_{k=0}^{\infty}$ be a sequence of polynomials, where p_k is monic of degree k. Then any subsequence $\{p_k\}_{k=0}^{n}$ is linearly independent since $c_0p_0(t)+c_1p_1(t)+\cdots+c_np_n(t)\equiv 0$, by successively comparing leading coefficients on the left- and right-hand side, implies $c_n=0$, $c_{n-1}=0,\ldots,c_0=0$ in this order.

Now let $q_k(t)=\pi_{k+1}(t)-t\pi_k(t)$. Clearly, q_k is a polynomial of degree k. Therefore, by the linear independence of $\pi_0,\pi_1,\ldots,\pi_k$, it can be expressed as a linear combination of these polynomials, i.e.,

$$\pi_{k+1}(t)-t\pi_k(t)=-\sum_{j=0}^{k}\beta_j^k\pi_{k-j}(t),$$

for some constants β_j^k, which proves the assertion.

(b) Taking the inner product of both sides of (3.2) with π_{k-j} gives, by Sobolev orthogonality and the linearity of the inner product,

$$0=(\pi_{k+1},\pi_{k-j})_S=(t\pi_k,\pi_{k-j})_S-\beta_j^k(\pi_{k-j},\pi_{k-j})_S,\quad j=0,1,\ldots,k,$$

and hence

$$\beta_j^k=\frac{(t\pi_k,\pi_{k-j})_S}{(\pi_{k-j},\pi_{k-j})_S},\quad j=0,1,\ldots,k;\ k=0,1,2,\ldots. \tag{3.4}$$

Exercise 3.3. (a) *For the inner product* (3.1) *with* $s=1$ *and* $d\lambda_0$, $d\lambda_1$ *positive definite, show that* $(t^2p,q)_S=(p,t^2q)_S$ *holds for arbitrary polynomials* p, q *if and only if* $d\lambda_1(t)=\gamma\delta(0)$, *where* $\gamma>0$ *and* δ *is the Dirac delta function.*

(b) *Show that the Sobolev polynomials* π_k *orthogonal with respect to the inner product* $(u,v)_S=\int_{\mathbb{R}}u(t)v(t)d\lambda_0(t)+\gamma u'(0)v'(0)$, $\gamma>0$, *satisfy a five-term recurrence relation of the form*

$$t^2\pi_k(t)=\pi_{k+2}(t)+\sum_{j=k-2}^{k+1}c_j^k\pi_j(t),\quad k\ge 0,$$

where $c_{-2}^0=c_{-1}^0=c_{-1}^1=0$.

(c) *Express the coefficients* c_j^k *in* (b) *in terms of the recurrence coefficients* (3.4) *for the respective Sobolev orthogonal polynomials. Discuss the merits of the five-term recurrence relation compared with the long recurrence relation* (3.2).

Answer

(a) An elementary computation shows that

$$(t^2p,q)_S-(p,t^2q)_S=2\int_{\mathbb{R}}t\,[p(t)q'(t)-p'(t)q(t)]d\lambda_1(t).$$

A sufficient condition for this to be zero for all p, q is $d\lambda_1(t)=\gamma\delta(0)$. This is also necessary since for $p(t)=t^2$, $q(t)=1$, the right-hand side is $-4\int_{\mathbb{R}}t^2d\lambda_1(t)$, which, since $d\lambda_1$ is positive definite, can be zero only if $d\lambda_1(t)=\gamma\delta(0)$, $\gamma>0$.

(b) Since $t^2\pi_k$ is a monic polynomial of degree $k+2$, and thus $t^2\pi_k-\pi_{k+2}\in\mathbb{P}_{k+1}$, it is clear that

$$t^2\pi_k(t)=\pi_{k+2}+\sum_{j=0}^{k+1}c_j^k\pi_j(t),$$

the polynomials $\pi_0, \pi_1, \ldots, \pi_{k+1}$ forming a basis of $\mathbb{P}_{k+1}$ (cf. the answer to Ex. 3.2(a)). By orthogonality,

$$c_j^k = \frac{(t^2\pi_k - \pi_{k+2}, \pi_j)_S}{(\pi_j, \pi_j)_S}, \quad j = 0, 1, 2, \ldots, k+1.$$

But

$$(t^2\pi_k - \pi_{k+2}, \pi_j)_S = (t^2\pi_k, \pi_j)_S,$$

so that

$$c_j^k = \frac{(t^2\pi_k, \pi_j)_S}{(\pi_j, \pi_j)_S}, \quad j = 0, 1, \ldots, k+1.$$

When $k > 2$, then by (a),

$$(t^2\pi_k, \pi_j)_S = (\pi_k, t^2\pi_j)_S = 0 \quad \text{if } j < k-2,$$

proving the assertion.

(c) From the recurrence relation (3.2),

$$t\pi_k(t) = \pi_{k+1}(t) + \sum_{j=0}^{k} \beta_j^k \pi_j(t), \quad k \geq 0,$$

one finds

$$\begin{aligned} t^2\pi_k(t) &= t\pi_{k+1}(t) + \sum_{j=0}^{k} \beta_j^k \cdot t\pi_j(t) \\ &= \pi_{k+2}(t) + \sum_{j=0}^{k+1} \beta_j^{k+1}\pi_j(t) + \sum_{j=0}^{k} \beta_j^k [\pi_{j+1}(t) + \sum_{\ell=0}^{j} \beta_\ell^j \pi_\ell(t)] \\ &= \pi_{k+2}(t) + \sum_{j=0}^{k+1} \beta_j^{k+1}\pi_j(t) + \sum_{j=0}^{k} \beta_j^k \pi_{j+1}(t) + \sum_{j=0}^{k}\sum_{\ell=0}^{j} \beta_j^k \beta_\ell^j \pi_\ell(t). \end{aligned} \tag{3.5}$$

Interchanging the order of summation in the last term, we have

$$\sum_{j=0}^{k}\sum_{\ell=0}^{j} \beta_j^k \beta_\ell^j \pi_\ell(t) = \sum_{\ell=0}^{k} \left(\sum_{j=\ell}^{k} \beta_j^k \beta_\ell^j \right) \pi_\ell(t),$$

so that, if $k > 0$,

$$\begin{aligned} t^2\pi_k(t) = \pi_{k+2}(t) + \beta_0^{k+1}\pi_0(t) &+ \sum_{j=1}^{k+1} (\beta_j^{k+1} + \beta_{j-1}^k)\pi_j(t) \\ &+ \sum_{j=0}^{k} \beta_j^k \beta_0^j \pi_0(t) + \sum_{\ell=1}^{k} \left(\sum_{j=\ell}^{k} \beta_j^k \beta_\ell^j \right) \pi_\ell(t). \end{aligned} \tag{3.6}$$

Thus, when $n \geq 2$,

$$\begin{aligned} t^2\pi_n(t) &= \pi_{n+2}(t) + (\beta_{n+1}^{n+1} + \beta_n^n)\pi_{n+1}(t) \\ &+ \sum_{j=1}^{n} \left(\beta_j^{n+1} + \beta_{j-1}^n + \sum_{i=j}^{n} \beta_i^n \beta_j^i \right) \pi_j(t) + \left(\beta_0^{n+1} + \sum_{j=0}^{n} \beta_j^n \beta_0^j \right) \pi_0(t). \end{aligned}$$

Comparing with the five-term recurrence relation in (b) (where $k=n$) gives

$$\begin{aligned}
c_{n+1}^n &= \beta_{n+1}^{n+1}+\beta_n^n,\\
c_n^n &= \beta_n^{n+1}+\beta_{n-1}^n+(\beta_n^n)^2\\
c_{n-1}^n &= \beta_{n-1}^{n+1}+\beta_{n-2}^n+\beta_{n-1}^n(\beta_{n-1}^{n-1}+\beta_n^n)\\
c_{n-2}^n &= \begin{cases} \beta_0^3+\beta_1^2\beta_0^1+\beta_0^2(\beta_0^0+\beta_2^2) & (n=2)\\ \beta_{n-2}^{n+1}+\beta_{n-3}^n+\beta_{n-1}^n\beta_{n-2}^{n-1}+\beta_{n-2}^n(\beta_{n-2}^{n-2}+\beta_n^n) & (n>2)\end{cases}
\end{aligned}$$

These formulae also hold for $n=0$ and $n=1$ if the β's with negative subscripts are taken to be zero. This follows, resp., from (3.5) and (3.6).

The five-term recurrence relation has the obvious advantage of allowing a more efficient recursive evaluation of the Sobolev polynomials, but has the disadvantage of not providing access to the zeros of the polynomials.

3.2 ▪ Symmetry

Exercise 3.4. (Ex. 3 of [27, §3].) The Sobolev inner product (3.1) is called *symmetric* if each measure $d\lambda_\sigma$ is symmetric in the sense of Ex. 2.12. *For symmetric Sobolev inner products,*

(a) *show that* $\pi_k(-t;S)=(-1)^k\pi_k(t;S)$*;*

(b) *show that* $\beta_{2r}^k=0$ *for* $r=0,1,\dots,\lfloor k/2\rfloor$.

Answer

Let

$$(u,v)_S=\sum_{j=0}^s\int_{-a_j}^{a_j}u^{(j)}(t)v^{(j)}(t)w_j(t)\mathrm{d}t,\quad a_j>0, \tag{3.7}$$

where $w_j(-t)\equiv w_j(t)$ on $[-a_j,a_j]$.

(a) Define $p_k(t):=(-1)^k\pi_k(-t;S)$. Clearly, p_k is monic of degree k. Moreover,

$$\begin{aligned}
(p_k,p_\ell)_S &= \sum_{j=0}^s\int_{-a_j}^{a_j}p_k^{(j)}(t)p_\ell^{(j)}(t)w_j(t)\mathrm{d}t=\sum_{j=0}^s\int_{-a_j}^{a_j}p_k^{(j)}(-t)p_\ell^{(j)}(-t)w_j(-t)\mathrm{d}t\\
&=(-1)^{k+\ell}\sum_{j=0}^s\int_{-a_j}^{a_j}\pi_k^{(j)}(t;S)\pi_\ell^{(j)}(t;S)w_j(t)\mathrm{d}t=0,\quad k\neq\ell.
\end{aligned}$$

By the uniqueness of monic Sobolev orthogonal polynomials, there follows $p_k(t)\equiv\pi_k(t;S)$, that is, $\pi_k(-t;S)=(-1)^k\pi_k(t;S)$, as claimed.

(b) By (3.4),

$$\beta_{2r}^k=\frac{(t\pi_k,\pi_{k-2r})_S}{(\pi_{k-2r},\pi_{k-2r})_S},\quad \pi_m(t)=\pi_m(t;S).$$

Here the numerator is

$$(t\pi_k,\pi_{k-2r})_S=\sum_{j=0}^s\int_{-a_j}^{a_j}(t\pi_k)^{(j)}(t)\pi_{k-2r}^{(j)}(t)w_j(t)\mathrm{d}t. \tag{3.8}$$

By part (a) of this exercise, the polynomials π_m are even or odd functions depending on whether m is even or odd. We now use the simple facts that an even function remains even if differentiated an even number of times, but becomes an odd function if differentiated an odd number of times; and an odd function remains odd if differentiated an even number of times, but becomes an even function if differentiated an odd number of times. Applied to the integrand in (3.8), this implies that, regardless of the parity of the integers k and j, one of the first two factors in the integrands of (3.8) is even and the other odd, so that the integrands are always odd. Therefore, all the integrals in (3.8) vanish, that is, $(t\pi_k, \pi_{k-2r}) = 0$, and thus $\beta_{2r}^k = 0$, as was to be shown.

3.3 ▪ Zeros

Exercise 3.5. Zeros of symmetric Sobolev orthogonal polynomials.

In this exercise, $s = 1$ and

$$d\lambda_0 = d\lambda, \quad d\lambda_1 = \gamma d\lambda, \quad \gamma > 0, \tag{3.9}$$

where $d\lambda$ is a symmetric measure. The zeros, by Ex. 3.4(a), are then located symmetrically with respect to the origin, and it suffices to look at half of them only. We are interested in exploring monotonicity of the zeros as a function of γ.

The matrix $\boldsymbol{B}$ of the recurrence coefficients can be generated by the modified Chebyshev algorithm, using for the polynomials defining the modified moments, $p_k(t) = \pi_k(t; d\lambda)$. The modified moments are then all zero except for the first ones, $m_0 = \mu_0$, resp., $m_0 = \gamma\mu_0$, where $\mu_0 = \int_{\mathbb{R}} d\lambda(t)$. The `OPQ` routine implementing the Chebyshev algorithm is `chebyshev_sob.m`. The zeros of the Sobolev polynomials can then be produced in increasing order (if all are real) by the `OPQ` routine `sobzeros.m`.

(a) *Experiment with* $d\lambda(t) = dt$ *on* $[-1, 1]$: *Althammer polynomials. It is known, in this case, that all zeros are real and contained in* $(-1, 1)$. *Check numerically whether or not the positive zeros of these polynomials are monotonically increasing with* γ. *Identify any computational problems and how to deal with them.*

(b) *Do the same with* $d\lambda(t) = (1-t^2)^\alpha$ *on* $[-1, 1]$. It has been conjectured in [36, §3.2], and proved in [49], that all zeros are real and all are contained in $(-1, 1)$ if $\alpha \geq 0$, and all but two (a symmetric pair) if $-1 < \alpha < 0$.

(c) *Examine the case* $d\lambda(t) = |t|^{2\mu}\exp(-t^2)dt$ *on* $\mathbb{R}$, $\mu > -1/2$.

Answer

(a) To examine the monotonicity of the positive zeros, one needs a routine, `zeros_alth.m`, that calculates the zeros as a function of N, the degree of the polynomial, and γ,

```
function z=zeros_alth(N,g)
%ZEROS_ALTH Zeros of Althammer polynomials.
%   Z=ZEROS_ALTH(N,G) computes the Nx1 array Z of the zeros of
%   a polynomial of degree N orthogonal with respect to the two
%   Sobolev weight functions W_0(T)=1, W_1(T)=G on [-1,1], G>0.

mom=zeros(2,2*N);
mom(1,1)=2; mom(2,1)=2*g;
abm=r_jacobi(2*N-1);
B=chebyshev_sob(N,mom,abm);
z=sobzeros(N,N,B);
```

The monotonicity of the zeros is then checked by the script monzeros_alth.m,

```
%MONZEROS_ALTH

f0='%8.0f %6.2f %12.4e\n';
disp('        N     g        dmin')
for N=5:5:20
  z1=zeros_alth(N,0);
  dmin=100;
  for g=[.2:.2:1 1.5:.5:5 6:10]
    z0=z1;
    z1=zeros_alth(N,g);
    d=min(z1(floor((N+3)/2):N)-z0(floor((N+3)/2):N));
    if d<=0
      disp('monotonicity fails')
      fprintf(f0,N,g,d)
      return
    end
    if d<dmin, dmin=d; end
    if g==1|g==5|g==10
      fprintf(f0,N,g,dmin)
    end
  end
end
```

For $N = 5:5:20$, and for a sequence of increasing values of γ, it calculates the zeros for two successive γ-values and the difference thereof, to check whether it remains positive. The smallest of these differences is printed for three selected values, $\gamma = 1$, $\gamma = 5$, and $\gamma = 10$, to see how close it comes to the machine precision eps. Once it approaches eps, the routine becomes unreliable and could indicate nonmonotonicity by virtue of rounding errors. From the output

```
>> monzeros_alth
       N     g        dmin
       5   1.00   3.2444e-04
           5.00   8.6794e-06
          10.00   2.3172e-06
      10   1.00   1.8197e-09
           5.00   1.2387e-12
          10.00   7.8715e-14
monotonicity fails
      15   0.60  -4.9072e-14
>>
```

it can be seen that this happens when $N = 15$ and $\gamma = 0.6$.

To avoid uncertainties caused by rounding, we prepared symbolic versions, szeros_alth.m and smonzeros_alth.m, of the two preceding programs,

```
function z=szeros_alth(dig,N,g)
%SZEROS_ALTH  Symbolic analogue of zeros_alth.m.

digits(dig);
```

```
sg=vpa(g,dig);
mom=vpa(zeros(2,2*N));
mom(1,1)=vpa(2); mom(2,1)=2*sg;
abm=sr_jacobi(dig,2*N-1);
B=schebyshev_sob(dig,N,mom,abm);
z=ssobzeros(dig,N,N,B);

%SMONZEROS_ALTH  Symbolic analogue of monzeros_alth.m.

f0='%8.0f %6.2f %12.4e\n';
dig=48;
disp('       N     g        dmin')
for N=5:5:20
  g=0;
  z1=szeros_alth(dig,N,g);
  dmin=100;
  for g=[.2:.2:1 1.5:.5:5 6:10]
    z0=z1;
    z1=szeros_alth(dig,N,g);
    d=min(double(z1(floor((N+3)/2):N)-z0(floor((N+3)/2):N)));
    if d<0
      disp('monotonicity fails')
      fprintf(f0,N,g,d)
      return
    end
    if d<dmin, dmin=d; end
    if g==1|g==5|g==10
      fprintf(f0,N,g,dmin)
    end
  end
end
```

and ran them with the same input data in 48-digit arithmetic. The results,

```
>> smonzeros_alth
       N     g        dmin
       5   1.00  3.2444e-004
           5.00  8.6794e-006
          10.00  2.3172e-006
      10   1.00  1.8197e-009
           5.00  1.2346e-012
          10.00  7.8591e-014
      15   1.00  3.5842e-016
           5.00  2.0151e-021
          10.00  1.6637e-023
      20   1.00  1.6680e-023
           5.00  2.4416e-030
          10.00  4.8551e-033
>>
```

suggest that the zeros are indeed growing monotonically with increasing γ.

(b) The relevant routines are now `zeros_jac_sob.m`, evaluating the zeros of Gegenbauer–Sobolev polynomials, and `monzeros_jac_sob.m` checking the monotone growth of their positive zeros as functions of γ. The symbolic analogues are `szeros_jac_sob.m` and `smonzeros_jac_sob.m`.

Running `monzeros_jac_sob.m` for $\alpha = 3/4$ and $N = 5:5:40$, and the same values of g as before, yields

```
>> monzeros_jac_sob
         N      g         dmin
         5    1.00   3.6604e-004
              5.00   2.7759e-005
             10.00   1.3451e-005
        10    1.00   6.6835e-006
              5.00   5.9167e-007
             10.00   2.9570e-007
        15    1.00   7.9633e-007
              5.00   7.0802e-008
             10.00   3.5402e-008
        20    1.00   1.8034e-007
              5.00   1.6038e-008
             10.00   8.0191e-009
        25    1.00   5.7507e-008
              5.00   5.1081e-009
             10.00   2.5562e-009
        30    1.00   2.2664e-008
              5.00   1.8883e-009
             10.00   9.7446e-010
        35    1.00   1.1540e-008
monotonicity fails
        35    3.00  -4.2891E-009
>>
```

showing a probably spurious failure when $N = 35$ and $g = 3$. Indeed, running the same routine in 48-digit arithmetic, yields (after more than ten hours run time!)

```
>> smonzeros_jac_sob
         N      g         dmin
         5    1.00   3.6604e-004
              5.00   2.7759e-005
             10.00   1.3451e-005
        10    1.00   6.6835e-006
              5.00   5.9167e-007
             10.00   2.9570e-007
        15    1.00   7.9633e-007
              5.00   7.0802e-008
             10.00   3.5402e-008
        20    1.00   1.8034e-007
              5.00   1.6038e-008
             10.00   8.0196e-009
        25    1.00   5.7486e-008
```

```
          5.00   5.1121e-009
         10.00   2.5562e-009
   30     1.00   2.2684e-008
          5.00   2.0170e-009
         10.00   1.0086e-009
   35     1.00   1.0359e-008
          5.00   9.2107e-010
         10.00   4.6055e-010
   40     1.00   5.2617e-009
          5.00   4.6781e-010
         10.00   2.3391e-010

>>
```

thus confirming monotone growth. The same supporting result was observed for other positive values of α, specifically for $\alpha = 0.25 : 0.25 : 1$, $1.5 : 0.5 : 3, 4, 5$, providing strong evidence for the conjecture that all positive zeros increase monotonically when $\alpha > 0$. When $-1 < \alpha < 0$, similar testing reveals that monotone growth takes place only for the positive zeros < 1.

(c) If $\mu = 0$, that is, $\mathrm{d}\lambda(t) = e^{-x^2}\mathrm{d}t$, we note that the property $H_k'(t) = 2kH_{k-1}(t)$ of Hermite polynomials implies that these polynomials are also Hermite–Sobolev polynomials with respect to the special Sobolev inner product (3.9). So the zeros of Hermite–Sobolev polynomials are identical with the zeros of Hermite polynomials and thus do not depend on γ. Monotonicity, therefore, is not an issue in this case.

With regard to the generalized Hermite measure $\mathrm{d}\lambda(t) = |t|^{2\mu}e^{-t^2}\mathrm{d}t$, we found that already for $N = 5$, $\mu = -1/4$, the first positive zero of the generalized Hermite–Sobolev polynomial of degree N decreases, while the next zero increases as we go from $g = .2$ to $g = .4$.

Exercise 3.6. Atomic Sobolev orthogonal polynomials.

These are orthogonal with respect to an inner product (3.1) in which all measures $\mathrm{d}\lambda_\sigma$, $1 \le \sigma \le s$, are 1-point atomic measures, $\mathrm{d}\lambda_\sigma = m_\sigma \delta(t - c_\sigma)$, the masses m_σ being nonnegative and not all equal to zero. The ground measure $\mathrm{d}\lambda_0$ is usually absolute continuous but may also have a 1-point atomic component.

(a) *Assuming that the absolute continuous part of the ground measure is specified by the array* `ab` *of recurrence coefficients, write a MATLAB routine* `sob_atom.m` *that uses the* `OPQ` *routine* `stieltjes_sob.m` *to compute the recurrence matrix* $\boldsymbol{B}_n$ *of atomic Sobolev orthogonal polynomials.*

(b) *Use your routine of* (a) *together with the* `OPQ` *routine* `sobzeros.m` *to check Tables* 2–4 *in* [20], *relating to the Hermite measure* $\mathrm{d}\lambda_0(t) = \exp(-t^2)\mathrm{d}t$ *and a single atomic measure involving the* r*th derivative at the point* $c = 1$ *or* $c = 5$ *and mass* $m = 1$. In the cited reference, the results were obtained by a different ad hoc method.

(c) *In the case of the Laguerre measure* $\mathrm{d}\lambda_0(t) = \exp(-t)\mathrm{d}t$ *on* $\mathbb{R}_+$ *and* $c_\sigma = 0$, $m_\sigma = 1$, $\sigma = 1, 2 \dots, s$, *it may be conjectured that any complex zero that occurs has negative real part. Use your routine and* `sobzeros.m` *to check out this conjecture.*

(d) *For* $\mathrm{d}\lambda_0$, c_σ, m_σ, *as in* (c), *determine the pattern of occurrence of complex zeros. Cover the range* $1 \le s \le 10$, $1 \le n \le 40$.

(e) *Repeat* (c) *and* (d) *with* $\mathrm{d}\lambda_0$ *the Laguerre measure plus an atomic measure with mass* 1 *at the origin.*

Answer

(a) The core of this part of the exercise is to set up the matrix xw in the routine stieltjes_sob.m for computing the matrix $\boldsymbol{B}_n$ of recurrence coefficients. It must contain in the first half the abscissae of the discretized measures $d\lambda_\sigma$, $\sigma = 0, 1, \dots, s$, and in the second half the corresponding weights, and is therefore of size $N_d \times (2s+2)$, where N_d is the maximum number of points in the individual discretizations of the measures. In the present application, the matrix xw is quite sparse, having in column 1 the n Gaussian nodes of the ground measure $d\lambda_0$ and in column $s+2$ the corresponding weights. The elements in the first row and columns 2 to $s+1$ are occupied by the abscissae c_σ, and those in the first row and columns $s+3$ to $2s+2$ by the corresponding masses m_σ. All other elements are zero.

The routine sob_atom.m implementing this thus looks as follows:

```
function B=sob_atom(n,s,c,m,ab0)
%SOB_ATOM Atomic Sobolev polynomials.
%   B=SOB_ATOM(N,S,C,M,AB0) uses Stieltjes's procedure to
%   compute the NxN recurrence matrix B for the Sobolev
%   polynomials orthogonal with respect to an inner product
%   having 1-point atomic measures associated with the S
%   derivatives, the locations of which are given by the
%   elements of the 1xS vector C, and the respective masses
%   by the elements of the 1xS vector M. The ground measure
%   is specified by its Nx2 recurrence array AB0.

xw0=gauss(n,ab0);
xw=zeros(n,2*s+2);
xw(:,1)=xw0(:,1); xw(:,s+2)=xw0(:,2);
xw(1,2:s+1)=c; xw(1,s+3:2*s+2)=m;
nd=[n ones(1,s)];
a0=ab0(1,1); same=0;
B=stieltjes_sob(n,s,nd,xw,a0,same);
```

(b) This is the case where $s = r$, $c_\sigma = \delta_{\sigma s}$ or $5\delta_{\sigma s}$, and $m_\sigma = \delta_{\sigma s}$, $\sigma = 1, 2, \dots, s$. The scripts applying sob_atom.m to generate the special atomic Sobolev orthogonal polynomials of [20] and their zeros, and to check Tables 2–4 of that reference, are run_sob_atom_herm.m, run_sob_atom_lag.m, and run_sob_atom_leg.m, each of them calling on auxiliary print routines zB_herm.m, zB_lag.m, and zB_leg.m, resp. We display only the first of these, the others being similar.

```
function z=zB_herm(N,B,c)
%ZB_HERM  Auxiliary print function for run_sob_atom_herm.m

f0='%10.0f %5.0f %12.8f %+10.8f*i\n';
z=sobzeros(N,N,B);
for n=1:N
  if imag(z(n))~=0
    fprintf(f0,c,N,real(z(n)),imag(z(n)))
  end
end
fprintf('\n')
```

The 2-digit results in Tables 3–4 are confirmed by these routines, except for a printing error in Table 4, where the imaginary part in the entry for $k = 8$, $r = 3$ should be .04 instead of .42. The results for Table 2, however, are less satisfactory. There are discrepancies of one or two units in the terminal digit when $r = 3$ and $c = 5$, and completely erroneous results when $r = 4$ and $c = 5$ that even suggest the presence of many more than one pair of conjugate complex zeros. The problem here, evidently, is instability of Stieltjes's procedure, wiping out all but two digits in the answers when $r = 3$ and $c = 5$, and all of them when $r = 4$ and $c = 5$.

In order to verify this, we prepared symbolic routines `ssob_atom.m`, `szB_herm.m`, and a symbolic script `run_ssob_atom_herm.m`, and used them in 48-digit arithmetic to reproduce the result for $r = 4$, $c = 5$, and $N = 40$ (run time = 11.4 hours!). Comparison with the result obtained in 40-digit arithmetic suggests an accuracy of 39 digits after the decimal point for the 48-digit results.

(c),(d) Experimentation with `run_sob_atom_lag0.m` , `zB_lag0.m` led to the following conjectures: (1) If $s = 1$ and $N > 2$, there are N distinct real zeros, exactly one being negative and tending to zero as $N \to \infty$. This was checked for all $N \leq 40$. (2) If $s > 1$ and $N > s$, there is either at least one negative zero, or at least one pair of conjugate complex zeros having a negative real part, or both. This was checked for $2 \leq s \leq 10$ and all $N \leq 40$.

(e) The same conjectures as in (c, d) can be made by running `run_specsob_lag00.m` and `zB_lag0.m` with the same values of s and N, in the case where $d\lambda_0(t)$ includes a 1-point measure supported at $t = 0$ with mass $m = 1$. This requires the use of two additional routines, `r_lagplusdis.m` and `quadlp.m`. The support point of the extra measure in $d\lambda_0$ could in fact be $t = t_0$, with any $t_0 > 0$, but not $t_0 < 0$. Very likely, though this has not been checked carefully, the masses in all the discrete measures could be arbitrary positive and the conjectures still holding.

Exercise 3.7* Sobolev orthogonal polynomials relative to the Gegenbauer-like inner product

$$(u,v)_S = \int_{-1}^{1} u(t)v(t)(1-t^2)^{\alpha-1}dt + \gamma \int_{-1}^{1} u'(t)v'(t)\frac{(1-t^2)^{\alpha}}{t^2+y^2}\,dt, \quad \alpha > 0,\ y > 0 \quad (3.10)$$

and their zeros.

The zeros have been studied by Groenevelt [42, §5]. Among the results proved are the following. (i) All zeros of the Sobolev orthogonal polynomial $\pi_n(\cdot\,; S)$ are real except, possibly, for one pair of conjugate complex zeros [42, Theorem 5.7]. (ii) If n is odd, all zeros of π_n are real and distinct [42, Theorem 5.8]. (iii) If $n \geq 4$ is even, and $\gamma \to \infty$, then π_n has a pair of conjugate complex zeros if $y \leq y_0$ and y_0 is "sufficiently small" [42, Theorem 5.9].

(a) *If $n = 2$, show that $\pi_n(\cdot\,; S)$ has the zeros $\pm 1/\sqrt{2\alpha+1}$, regardless of the values of y and γ.*

(b) Generating the recurrence coefficients of the modified measure in the second integral of (3.10) requires the use of Algorithm 6 of Ex. 2.41, which in turn calls for the Cauchy integral

$$\rho_0(z) = \int_{-1}^{1} \frac{(1-t^2)^{\alpha}}{z-t}\,dt, \quad z = \mathrm{i}y,\ y > 0.$$

Devise a scheme of calculating $\rho_0(\mathrm{i}y)$ by Gaussian quadrature.

(c) *Write a double-precision (for small y) and a symbolic variable-precision (for large y) MATLAB routine for calculating the zeros of $\pi_n(\cdot\,; S)$ in increasing order if they are real, and* *in increasing absolute value otherwise.* (By symmetry, the zeros are symmetric with respect to the origin; cf. Ex. 3.4(a).)

For the remaining part of this exercise, assume, for simplicity, that $\alpha = 1/2$.

(d) *Try to quantify property* (iii) *when* γ *is finite, say for* $\gamma = [.25, .5, 1, 2, 4, 8]$. *Specifically, for given* $\gamma > 0$, *find* $y_0 = y_0(n)$, $4 \le n(\text{even}) \le 40$ *to an accuracy of five decimal digits such that for* $y < y_0(n)$ *the polynomial* $\pi_n(\cdot\,; S)$ *has a pair of complex zeros, and for* $y > y_0(n)$ *near* $y_0(n)$ *all zeros of* $\pi_n(\cdot\,; S)$ *are real. For each* γ, *produce a curve* $\mathscr{C}_\gamma$ *showing the dependence of* $y_0(n)$ *on* n. {Hint: Let y increase from $y = .01$ in steps of $.01$ until for the first time all zeros of $\pi_n(\cdot\,; S)$ are real. Then use a bisection-type procedure to determine y_0 to five significant decimal digits.}

(e) *Produce evidence to suggest that above the curves* $\mathscr{C}_\gamma$ *of* (d) *all zeros of* $\pi_n(\cdot\,; S)$ *are indeed real.*

(f) *Show that any complex zero has to be purely imaginary.*

Answer

(a) Using (3.4) of Ex. 3.2(b), with $k = 0$, gives

$$\beta_0^0 = \frac{(t\pi_0, \pi_0)_S}{\|\pi_0\|_S^2}.$$

Since

$$(t\pi_0, \pi_0)_S = \int_{-1}^{1} t(1-t^2)^{\alpha-1}\,\mathrm{d}t = 0,$$

we have $\beta_0^0 = 0$. Also $\beta_0^1 = 0$ by symmetry (cf. Ex. 3.4(b) for $r = 0$, $k = 1$). To compute β_1^1, by (3.4) we need

$$\begin{aligned}(t\pi_1, \pi_0)_S &= (t^2, 1)_S = 2\int_0^1 t^2(1-t^2)^{\alpha-1}\,\mathrm{d}t \\ &= 2\int_0^{\pi/2} \cos^2\theta \sin^{2\alpha-1}\theta\,\mathrm{d}\theta = 2\int_0^{\pi/2} (1-\sin^2\theta)\sin^{2\alpha-1}\theta\,\mathrm{d}\theta.\end{aligned}$$

Using (cf. [40, Eq. 3.621.1])

$$\int_0^{\pi/2} \sin^{\mu-1}\theta\,\mathrm{d}\theta = 2^{\mu-2}B(\mu/2, \mu/2) = 2^{\mu-2}\frac{\Gamma^2(\mu/2)}{\Gamma(\mu)},$$

we find by a short computation

$$(t\pi_1, \pi_0)_S = \frac{2^{2\alpha-1}}{2\alpha+1}\frac{\Gamma^2(\alpha)}{\Gamma(2\alpha)}.$$

Since, furthermore,

$$(\pi_0, \pi_0)_S = 2\int_0^{\pi/2} \sin^{2\alpha-1}\theta\,\mathrm{d}\theta = 2\cdot 2^{2\alpha-2}B(\alpha, \alpha) = 2^{2\alpha-1}\frac{\Gamma^2(\alpha)}{\Gamma(2\alpha)},$$

we get, again by (3.4),

$$\beta_1^1 = \frac{(t\pi_1, \pi_0)_S}{(\pi_0, \pi_0)_S} = \frac{1}{2\alpha+1}.$$

Thus (cf. (3.3)),

$$B_2 = \begin{bmatrix} 0 & 1/(2\alpha+1) \\ 1 & 0 \end{bmatrix},$$

and since the zeros z_1, z_2 of π_2 are the eigenvalues of B_2,

$$z_{1,2} = \pm\frac{1}{\sqrt{2\alpha+1}}, \qquad \alpha > 0.$$

(b) We have

$$\begin{aligned}\rho_0(\mathrm{i}y) &= \int_{-1}^{1} \frac{(1-t^2)^\alpha}{\mathrm{i}y - t}\,dt = -\int_{-1}^{1} \frac{\mathrm{i}y+t}{y^2+t^2}(1-t^2)^\alpha dt \\ &= -\int_{-1}^{1} \frac{t(1-t^2)^\alpha}{y^2+t^2}\,dt - \mathrm{i}y\int_{-1}^{1} \frac{(1-t^2)^\alpha}{y^2+t^2}\,dt,\end{aligned}$$

and since the first integral on the far right vanishes,

$$\rho_0(\mathrm{i}y) = -2\mathrm{i}y\int_0^1 \frac{(1+t)^\alpha}{y^2+t^2}(1-t)^\alpha dt. \tag{3.11}$$

The integral in (3.11) can be computed accurately by Gauss–Jacobi quadrature (on $[0,1]$) with Jacobi parameters α and 0, even for y fairly small. The respective MATLAB routine is `rho0_groen.m`.

```
function [r,n]=rho0_groen(y,a,eps0)
%RHO0_GROEN Gauss quadrature evaluation of a Gegenbauer type
%integral.
%   [R,N]=RHO0_GROEN(Y,A,EPS0) evaluates R=-2iY*int_0^1 (1-t^2)^A
%   dt/(Y^2+t^2) by Gauss-Jacobi quadrature to a relative accuracy
%   EPS0. The output variable N is the number of Gauss quadrature
%   points needed.

N=400;
ab=r_jacobi01(N,a,0);
n=0;
r0=0; r1=1;
while abs((r1-r0)/r1)>eps0
  n=n+1; r0=r1;
  xw=gauss(n,ab); x=xw(:,1); w=xw(:,2);
  r1=sum(w.*(1+x).^a./(y^2+x.^2));
end
r=-2*i*y*r1;
```

The performance of the routine, for $\alpha = 1/2$ and `eps0=.5e-12`, is shown in Table 3.1.

For later use (in (e)) we also prepared a symbolic version, `srho0_groen.m`, of the routine, but in both routines we let n increase in steps of 10 to reduce running time.

(c) The routines are `groen.m` and `sgroen.m`. We show only the latter.

Table 3.1. *12-digit Gauss quadrature approximations to the integral in* (3.11) *when* $\alpha = 1/2$, *and the number* n *of Gauss quadrature points needed*

y	n	Integral
1.000	10	6.50645142284e–001
0.100	33	1.42155113836e+001
0.010	101	1.55516690138e+002
0.001	288	1.56922631584e+003

```
function z=sgroen(dig,n,a,y,g)
%SGROEN Zeros of a Gegenbauer-like Sobolev orthogonal polynomial.
%   Z=SGROEN(DIG,N,A,Y,G) computes in DIG-digit arithmetic
%   the Nx1 array Z of the zeros of a polynomial of degree
%   N orthogonal with respect to the two Sobolev weight
%   functions W_0(T)=(1-T^2)^(A-1), W_1(T)=G*(1-T^2)^A/(T^2+Y^2)
%   on [-1,1], A>0, G>0.

digits(dig);
s=1; same=0; eps0=.5e-12; g=vpa(g);
nd=[n n];
ab=sr_jacobi(dig,n,a-1); zw=sgauss(dig,n,ab);
a0=ab(1,1);
ab0=sr_jacobi(dig,n,a); x=vpa(0);
rho0=srho0_groen(dig,y,a,eps0);
ab1=smod_quaddivisor(dig,n,ab0,x,y,rho0); zw1=sgauss(dig,n,ab1);
xw=[zw(:,1) zw1(:,1) zw(:,2) g*zw1(:,2)];
B=sstieltjes_sob(dig,n,s,nd,xw,a0,same);
z=ssobzeros(dig,n,n,B); z=sort(z);
```

(d) The procedure outlined in the hint is implemented in the following MATLAB script:

```
%RUN_GROEN

f0='%8.0f %12.4e\n';
a=1/2; eps1=.5e-5; ymax=1;
for g=[.25 .5 1 2 4 8]
  yp=zeros(19,1);
  for k=1:19
    n=2*(k+1);
    z=i*ones(n,1); y=0;
    while any(abs(imag(z)))>0
      y=y+.01; if y>ymax, error('always complex zeros'); end
      z=groen(n,a,y,g);
    end
    y0=y-.01; y1=y;
    while abs((y1-y0)/y1)>eps1
      y=(y0+y1)/2;
      z=groen(n,a,y,g);
      if any(abs(imag(z)))>0
```

```
          y0=y;
        else
          y1=y;
        end
      end
      yp(k)=y;
      fprintf(f0,n,y)
    end
    n=(4:2:40)';
    plot(n,yp,'LineWidth',1.5);set(gca,'FontSize',14)
    xlabel('n'), ylabel('y0')
    hold on
  end
```

For each n, the first `while` statement checks whether there are any complex zeros (when y is small, say $y = .01$). If this turns out to be true for some $y > 1$, the routine terminates with an error message indicating that for all y tested in the range $.01 \leq y \leq 1$ there is always a pair of conjugate complex zeros (something we do not expect).

The second `while` statement implements the bisection-type procedure mentioned in the hint.

When run, the script does not produce any error message as long as $n \leq 40$, meaning that our expectation is confirmed at least for the y-values tested in the range $0 < y < 1$. The curves $\mathscr{C}_\gamma$ it produced are shown in Fig. 3.1. As can be seen, they are practically independent of γ except for the first few values of n. The values of $y_0(n)$ (to five decimal digits) for $n = 4:2:40$ are shown in Table 3.2 for the case $\gamma = 1$. It appears that at $y = y_0$ the polynomial $\pi_n(\cdot;S)$, n(even)≥ 4, has a zero at the origin of multiplicity 2 which, when y passes downward through y_0 splits into a pair of conjugate complex zeros on the imaginary axis (cf. part (f) of this exercise).

(e) Verifying that all zeros are real if $y > y_0$, for y up to, say, $y = 5$, requires high-precision calculation, more precision the larger y and/or n. An idea of what precision is required can be obtained by running the script `test_groen.m`, say for $n = 40$, $y = 5$, $\gamma = 1$, and $\alpha = 1/2$. (This may take awhile.)

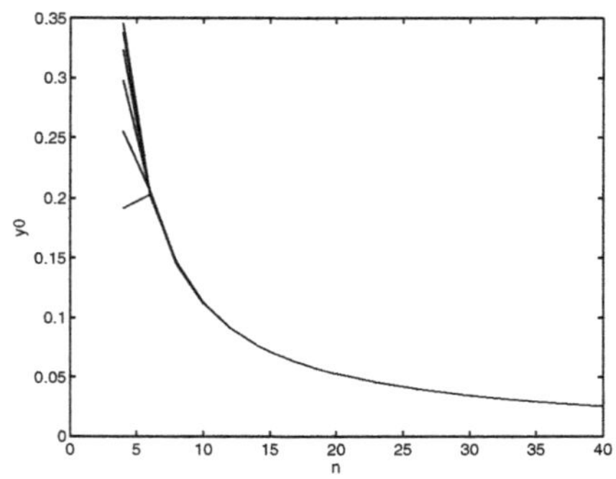

Figure 3.1. *The curves $\mathscr{C}_\gamma$ for selected values of γ*

```
%TEST_GROEN

f0='%20.4e %12.4e\n';
a=1/2; g=1; y=5; n=40;
```

```
z1=zeros(n,1);
for dig=[32 48 64]
  z0=z1;
  realz0=sort(real(z0));
  tic
  z1=sgroen(dig,n,a,y,g);
  toc
  realz1=sort(real(z1));
  err=max(abs(double((realz1-realz0)./realz1)));
  imagmax=max(abs(double(imag(z1))));
  fprintf(f0,err,imagmax)
end
```

Table 3.2. *5-digit values of* $y_0(n)$ *when* $\gamma = 1$

n	y_0	n	y_0	n	y_0	n	y_0
4	.29755	14	.077180	24	.043522	34	.030317
6	.20653	16	.066831	26	.040034	36	.028583
8	.14494	18	.058934	28	.037064	38	.027033
10	.11197	20	.052710	30	.034504	40	.025630
12	.091350	22	.047677	32	.032276		

It is found that `dig`$=40$ is a reasonable choice. The script

```
%CHECK_GROEN

a=1/2; g=1; dig=40;
for n=5:5:20
  for y=[.5 1 3 5]
    z=sgroen(dig,n,a,y,g);
    z0=double(z)
  end
end
```

successfully spot-checks the reality of the zeros above the curve $\mathscr{C}_\gamma$ for $\gamma = 1$. No complex zeros are found.

(f) Since the inner product (3.10) is symmetric, the Sobolev orthogonal polynomials associated with it, according to Ex. 3.4(a), are either even or odd. Moreover, their coefficients are all real. Therefore, if $z = x + iy$ is a zero, then so are $\overline{z}$, $-z$, and $-\overline{z}$. If x were nonzero, we would have four distinct zeros associated with z instead of two. Therefore, $x = 0$ and $z = iy$.

Exercise 3.8. Multiatomic Sobolev orthogonal polynomials.

These are orthogonal with respect to an inner product (3.1) in which all measures $d\lambda_\sigma$, $1 \le \sigma \le s$, are K_σ-point atomic measures $d\lambda_\sigma(t) = \sum_{k=1}^{K_\sigma} m_k^\sigma \delta(t - c_k^\sigma)$, the masses m_k^σ, $k = 1, 2, \ldots, K_\sigma$, being nonnegative and not all equal to zero. The ground measure $d\lambda_0$ is usually absolutely continuous, but may also have a K_0-point atomic component. *Consider the special multiatomic Sobolev polynomial* π_n *of degree* n *with* $s = 1$, $K_0 = K$, $K_1 = L$, *that is,*

$$\mathrm{d}\lambda_0(t)=\mathrm{d}\lambda(t)+\sum_{k=1}^{K} m_k^0\,\delta(t-c_k^0),\quad \mathrm{d}\lambda_1(t)=\sum_{\ell=1}^{L} m_\ell^1\,\delta(t-c_\ell^1),$$

where δ *is the Dirac delta function. Assume, for simplicity, that* $L\le n+K$. (For the special case $K=L=1$, $c_1^0=c_1^1$, see [2].) *To approximate* $(u,v)_S$ *by a discrete measure, it suffices to discretize the measure* $\mathrm{d}\lambda_0$. *This can be done in two ways, both yielding* π_n *exactly* (cf. [24, p. 143, (1) and (2)]). *Use Stieltjes's procedure to obtain the matrix* $\boldsymbol{B}_n$ *of recurrence coefficients.*

(a) *Approximate* $\mathrm{d}\lambda_0$ *by an* $(n+K)$*-point discrete measure obtained by discretizing* $\mathrm{d}\lambda$ *by an* n*-point Gaussian quadrature formula and adding on the given* K*-point multiatomic measure.*

(b) *Discretize the entire measure* $\mathrm{d}\lambda_0$ *by an* n*-point Gaussian quadrature formula.* *Use the routine* `stieltjes_sob.m` *to implement both approaches.*

(c) *Verify that both approaches produce the same matrix* $\boldsymbol{B}_n\in\mathbb{R}^{n\times n}$ *of recurrence coefficients by taking, for example,* $K=2$, $L=3$,

$$\boldsymbol{c}^0=\begin{bmatrix}-1\\1\end{bmatrix},\ \boldsymbol{m}^0=\begin{bmatrix}1\\1\end{bmatrix};\quad \boldsymbol{c}^1=\begin{bmatrix}-1\\0\\1\end{bmatrix},\ \boldsymbol{m}^1=\begin{bmatrix}1\\2\\1\end{bmatrix},$$

and $\mathrm{d}\lambda(t)=\mathrm{d}t$ *on* $[-1,1]$. *Write a script that for this particular example, and for* $n=5:5:40$, *prints the maximum absolute discrepancy between the two* $\boldsymbol{B}_n$*-matrices and between the two vectors of zeros. At the same time determine whether or not all zeros are real.*

Answer

In both (a) and (b), one needs the input variables $\mathtt{same}=0$ and $\mathtt{a0}=\alpha_0(\mathrm{d}\lambda_0)$, that is,

$$\mathtt{a0}=\frac{(t,1)_{\mathrm{d}\lambda_0}}{(1,1)_{\mathrm{d}\lambda_0}}=\frac{\int_{\mathbb{R}}t\,\mathrm{d}\lambda(t)+\sum_{k=1}^{K}m_k^0c_k^0}{\int_{\mathbb{R}}\mathrm{d}\lambda(t)+\sum_{k=1}^{K}m_k^0}=\frac{\mu_0\alpha_0(\mathrm{d}\lambda)+\sum_{k=1}^{K}m_k^0c_k^0}{\mu_0+\sum_{k=1}^{K}m_k^0},$$

where $\mu_0=\int_{\mathbb{R}}\mathrm{d}\lambda(t)$. What differs in (a) and (b) are the input parameters `nd` and `xw`.

(a) Here, $\mathtt{nd}=[n+K,L]$ and

$$\mathtt{xw}=\begin{bmatrix}\boldsymbol{\tau} & \boldsymbol{c}^1 & \boldsymbol{\omega} & \boldsymbol{m}^1\\ \boldsymbol{c}^0 & 0 & \boldsymbol{m}^0 & 0\end{bmatrix},$$

where $\boldsymbol{\tau}^{\mathrm{T}}=[\tau_1,\tau_2,\dots,\tau_n]$, $\boldsymbol{\omega}^{\mathrm{T}}=[\omega_1,\omega_2,\dots,\omega_n]$ are the n-vectors of nodes and weights of the Gaussian quadrature rule for $\mathrm{d}\lambda$, and $\boldsymbol{c}^0$, $\boldsymbol{m}^0$ the K-vectors of the abscissae and weights of the multiatomic measure in $\mathrm{d}\lambda_0$, while $\boldsymbol{c}^1$, $\boldsymbol{m}^1$ are the L-vectors of the abscissae and weights of the multiatomic measure $\mathrm{d}\lambda_1$. The relevant routine is

```
function [B,z]=sob_matom_d1(n,K,L,c0,m0,c1,m1,ab)
%SOB_MATOM_D1 Multiatomic Sobolev orthogonal polynomials.
%   [B,Z]=SOB_MATOM_D1(N,K,L,C0,M0,C1,M1,AB) uses Stieltjes's
%   procedure to compute the NxN recurrence matrix B and the
%   Nx1 array Z of the zeros of Sobolev polynomials orthogonal
%   with respect to an inner product having an L-point atomic
%   measure, with nodes C1 and masses M1, associated with the
%   first derivative. The ground measure DLAMBDA_0 is the sum
```

```
%    of an absolutely continuous measure DLAMBDA and a K-point
%    atomic measure, the former being specified by its Nx2
%    recurrence matrix AB, and the latter by its nodes C0 and
%    masses M0. The discretization is effected by discretizing
%    the measure DLAMBDA using N-point Gaussian quadrature.

s=1; nd=[n+K L]; same=0; mu0=ab(1,2);
a0=(mu0*ab(1,1)+sum(m0.*c0))/(mu0+sum(m0));
tm=gauss(n,ab); t=tm(:,1); w=tm(:,2);
xw=zeros(n+K,4);
xw(1:n,1)=t; xw(n+1:n+K,1)=c0;
xw(1:L,2)=c1;
xw(1:n,3)=w; xw(n+1:n+K,3)=m0;
xw(1:L,4)=m1;
B=stieltjes_sob(n,s,nd,xw,a0,same);
z=sobzeros(n,n,B);
```

(b) Here we first need to apply the routine `mcdis.m` to obtain the $n\times 2$ array `ab0` of recurrence coefficients for the orthogonal polynomials $\pi_k(\cdot;d\lambda_0)$ in order to obtain the discretization of $d\lambda_0$ by the n-point Gauss quadrature rule `tm=gauss(n,ab0)` for $d\lambda_0$. Then `nd=[n,L]` and

$$\mathtt{xw}=[\mathtt{tm}(1:\mathtt{n},1)\;\; c^1\;\; \mathtt{tm}(1:\mathtt{n},2)\;\; \boldsymbol{m}^1].$$

The relevant routine now is

```
function [B,z]=sob_matom_d2(n,K,L,c0,m0,c1,m1,ab)
%SOB_MATOM_D2 Multiatomic Sobolev orthogonal polynomials.
%    [B,Z]=SOB_MATOM_D2(N,K,L,C0,M0,C1,M1,AB) has the same
%    objective as SOB_MATOM_D1, but attains it by discretizing
%    the whole measure DLAMBDA_0 by an N-point atomic measure.

global mc mp iq irout idelta AB DM
s=1; nd=[n L]; same=0; mu0=ab(1,2);
a0=(mu0*ab(1,1)+sum(m0.*c0))/(mu0+sum(m0));
mc=1; mp=K; iq=1; irout=1; idelta=2;
eps0=.5e-12; Mmax=100;
AB=[-1 1];
DM=[c0(1) m0(1)];
for k=2:K
  DM=[DM;[c0(k) m0(k)]];
end
[ab0,Mcap,kount]=mcdis(n,eps0,@quad_sob_matom_d2,Mmax);
tm=gauss(n,ab0);
t=tm(:,1); w=tm(:,2);
xw=zeros(n,4);
xw(1:n,1)=t; xw(1:L,2)=c1;
xw(1:n,3)=w; xw(1:L,4)=m1;
B=stieltjes_sob(n,s,nd,xw,a0,same);
z=sobzeros(n,n,B);
```

(c) The script asked for is as follows.

```
%RUN_SOB_MATOM_D12

f0='%8.0f %12.4e %12.4e\n';
K=2; L=3;
c0=[-1;1]; m0=[1;1];
c1=[-1;0;1]; m1=[1;2;1];
disp('       n      errB          errz')
for n=5:5:40
  ab=r_jacobi(n);
  [B1,z1]=sob_matom_d1(n,K,L,c0,m0,c1,m1,ab);
  [B2,z2]=sob_matom_d2(n,K,L,c0,m0,c1,m1,ab);
  im1=max(abs(imag(z1))); im2=max(abs(imag(z2)));
  c=0;
  if im1~=0 | im2~=0
    c=1;
  end
  errB=max(max(abs(B1-B2)));
  errz=max(z1-z2);
  fprintf(f0,n,errB,errz)
end
fprintf('\n')
if c==0
  disp('      all zeros are real')
else
  disp('      there are complex zeros')
end
```

The output

```
>> run_sob_matom_d12
       n      errB          errz
       5  7.0158e-016   1.9984e-015
      10  4.6684e-015   1.1324e-014
      15  1.0405e-014   2.5400e-012
      20  1.1803e-014   2.0932e-012
      25  2.7007e-014   2.1163e-008
      30  4.1146e-014   3.4240e-009
      35  6.1753e-014   1.1444e-005
      40  1.3331e-013   1.6871e-006

      all zeros are real
>>
```

confirms the agreement (modulo rounding errors) between the two B-matrices and similarly, though somewhat distorted by increasingly larger rounding errors, the agreement between the two sets of zeros. Also confirmed (but not shown) is the symmetry of the zeros about the origin and the fact that all zeros are contained in the interior of the interval $[-1,1]$ except for one pair of zeros outside thereof. The latter seem to approach the endpoints of the interval monotonically as $n \to \infty$.

Exercise 3.9. *Consider the special multiatomic Sobolev orthogonal polynomials of* Ex. 3.8 *with*

$$K = L = c_1^0 = c_1^1 = 1 \quad \textit{and} \quad \mathrm{d}\lambda(t) = \mathrm{d}t \textit{ on } [-1,1]$$

(cf. [2, Proposition 3.2]). *Test the following conjectures.*

(a) Conjecture 1: *for* $m0 = m_1^0 \geq 0$, $m1 = m_1^1 > 0$ *and for all* $n \geq 3$, *the zeros of* π_n *are all in the interior of* $[-1,1]$ *except for one, which is larger than* 1.

(b) Conjecture 2: *The exceptional zeros* $z_n > 1$ *of* π_n *tend monotonically to* 1 *for* $n \geq 4$,

$$z_4 > z_5 > \cdots > 1, \quad \lim_{n\to\infty} z_n = 1.$$

(c) Conjecture 3: *The zeros of* π_n *in the interior of* $[-1,1]$ *are all simple zeros.*

(d) *Investigate the interlacing of the zeros of the Sobolev polynomials* $\pi_k(\cdot\,;S)$ *among themselves and with the zeros of the orthogonal polynomials* $\pi_k(\cdot\,;\mathrm{d}\lambda)$. *Specifically, test the following conjectures.*

Conjecture 4: *If* $n = 2$, *the zero of* $\pi_1(\cdot\,;S)$ *interlaces with the two zeros of* $\pi_2(\cdot\,;S)$; *if* $n > 2$, *only the first* $n-2$ *zeros of* $\pi_{n-1}(\cdot\,;S)$ *interlace with those of* $\pi_n(\cdot\,;S)$.

Conjecture 5: *If* $n = 2$, *the zero of* $\pi_1(\cdot\,;\mathrm{d}\lambda)$ *interlaces with the two zeros of* $\pi_2(\cdot\,;S)$; *if* $n > 2$, *only the first* $n-2$ *zeros of* $\pi_{n-1}(\cdot\,;\mathrm{d}\lambda)$ *interlace with those of* $\pi_n(\cdot\,;S)$.

The statements for $n > 2$ in Conjectures 4 and 5 is all that can be expected, given the monotonicity property in Conjecture 2 for the exceptional zeros.

Answer. All conjectures will be tested for all values of $m0$ from 0 in steps of .2 to 10, for $m1 = .2:.2:10$, and for n as large as $n = 40$. To make sure the inequalities expressing the conjectures are not the result of spurious phenomena caused by rounding errors, the minimum "strength" of the inequalities is measured by a quantity "`dmin`," which in the case of spurious effects of rounding would be near machine precision, and larger by several orders of magnitude otherwise.

(a)–(c) The scripts for testing Conjectures 1–3 are `conj1_sob_matom.m`, `conj2_sob_matom.m`, `conj3_sob_matom.m`, and have the following form:

```
%CONJ1-3_SOB_MATOM

K=1; L=1; c0=1; c1=1;
ab=r_jacobi(40);
for m0=0:.2:10
  dmin=1e5;
  for m1=.2:.2:10
    Conjecture
  end
  [m0 dmin]
end
```

where "`Conjecture`" is to be replaced, resp., by `conj1`, `conj2`, and `conj3`, where

```
%CONJ1

for n=1:40
  nc=0; nlo=0; nhi=0;
  [B,z]=sob_matom_d1(n,K,L,c0,m0,c1,m1,ab);
```

```
    for k=1:n
      if imag(z(k))~=0, nc=nc+1; end
      d=1+z(k); if abs(d)<dmin, dmin=abs(d); end
      if d<=0, nlo=nlo+1; end
      d=z(k)-1; if abs(d)<dmin, dmin=abs(d); end
      if d>=0, nhi=nhi+1; end
    end
    if (nc~=0|nlo~=0|nhi~=1) & n>=3
      m0, m1, n, z
      error('unexpected')
    end
  end


  %CONJ2

  zprec=1e5;
  for n=4:40
    [B,z]=sob_matom_d1(n,K,L,c0,m0,c1,m1,ab);
    d=z(n)-zprec; if abs(d)<dmin, dmin=abs(d); end
    if ~(d<0)
      m0, m1, n, zprec, z(n)
      error('unexpected')
    end
    zprec=z(n);
  end


  %CONJ3

  for n=2:40
    [B,z]=sob_matom_d1(n,K,L,c0,m0,c1,m1,ab);
    for k=2:n
      d=z(k)-z(k-1); if abs(d)<dmin, dmin=abs(d); end
      if dmin<1e-12
        m0, m1, z(k), z(k-1)
        error('unexpected')
      end
    end
  end
```

When run, no error message ever occurred and `dmin` was always larger than 10^{-6}, lending strong support for the validity of the conjectures. The same positive result was obtained for $d\lambda(t)=(1-t^2)^{\pm 1/2}$ and $d\lambda(t)=(1-t)^{-1/2}(1+t)^{1/2}$ on $[-1,1]$, although we had to lower the maximum value of n to 35 in order to avoid deceptive negative results due to an accumulation of rounding errors (in the zeros).

(d) It is found that Conjecture 4 tests positively, but Conjecture 5 only when $n>2$. The latter is false for $n=2$, first when $m0=1$ and $m1=5.4$, in which case the first zero of $\pi_2(\cdot;S)$ is positive, while $\pi_1(\cdot;d\lambda)$ vanishes at the origin. The scripts for testing Conjectures 4–5 therefore are as follows.

```
%CONJ4-5_SOB_MATOM

K=1; L=1; c0=1; m1=1;
ab=r_jacobi(40);
for m0=0:.2:10
  dmin=1e5;
  for m1=.2:.2:10
    Conjecture
  end
  [m0 dmin]
end
```

where "Conjecture" is to be replaced, resp., by `conj4` and `conj5`, where

```
%CONJ4

for n=1:40
  if n>1, zprec(1:n-1)=z; end
  [B,z]=sob_matom_d1(n,K,L,c0,m0,c1,m1,ab);
  if n==2
    d=zprec(1)-z(1); if abs(d)<dmin, dmin=abs(d); end
    d=z(2)-zprec(1); if abs(d)<dmin, dmin=abs(d); end
    if ~(z(1)<zprec(1) & zprec(1)<z(2))
      n, z(1), zprec(1), z(2)
      error('not expected')
    end
  elseif n>2
    for k=1:n-2
      d=zprec(k)-z(k); if abs(d)<dmin, dmin=abs(d); end
      d=z(k+1)-zprec(k); if abs(d)<dmin, dmin=abs(d); end
      if ~(z(k)<zprec(k) & zprec(k)<z(k+1))
        n, k, z(k), zprec(k), z(k+1)
        error('not expected')
      end
    end
  end
end


%CONJ5

for n=1:40
  if n>1, zprec(1:n-1)=xw(:,1); end
  [B,z]=sob_matom_d1(n,K,L,c0,m0,c1,m1,ab);
  if n>2
    for k=1:n-2
      d=zprec(k)-z(k); if abs(d)<dmin, dmin=abs(d); end
      d=z(k+1)-zprec(k); if abs(d)<dmin, dmin=abs(d); end
      if ~(z(k)<zprec(k) & zprec(k)<z(k+1))
        m0, m1, n, k, z(k), zprec(k), z(k+1)
        error('not expected')
```

```
            end
          end
        end
        xw=gauss(n,ab);
      end
```

For the values of $m0$ and $m1$ indicated at the beginning of the answer, and for n as large as $n = 40$, no error message was ever triggered and `dmin` was always returned larger than 10^{-6}.

Chapter 4

Answers to Exercises on Quadrature

The approximation of a weighted integral with positive measure $\mathrm{d}\lambda$ by a linear combination of n values of the integrand,

$$\int_{\mathbb{R}} f(t)\mathrm{d}\lambda(t) = \sum_{\nu=1}^{n} \lambda_\nu f(\tau_\nu) + R_n(f), \tag{4.1}$$

is called a *Gaussian quadrature formula* if $R_n(f) = 0$ whenever $f \in \mathbb{P}_{2n-1}$, i.e., whenever (4.1) has (polynomial) degree of exactness $d = 2n-1$, which is the largest possible. It is achieved if and only if (4.1) is interpolatory, i.e., has degree of exactness $n-1$, and $\tau_\nu = \tau_\nu^{(n)}$ are the zeros of the nth-degree orthogonal polynomial $\pi_n(\cdot;\mathrm{d}\lambda)$ (Ex. 4.1). Moreover, all Gaussian nodes $\tau_\nu = \tau_\nu^G$ are simple and the Gaussian weights $\lambda_\nu = \lambda_\nu^G$ are all positive (Ex. 4.2). These, and other elementary properties, notably a pretty "circle theorem" (Ex. 4.5) for the Gaussian weights, are the subjects of Section 4.1.

There are interesting connections between Gaussian quadrature and *Cauchy integrals* and *Cauchy principal value integrals*, which are explored in Section 4.2. *Gauss–Radau* and *Gauss–Lobatto quadrature rules* are slight modifications of the Gauss quadrature rule, which like the latter are also optimal with regard to degree of exactness, but have prescribed nodes at one or both endpoints of the interval of integration. These, and some generalizations of them, are dealt with in Section 4.3. A more substantial extension is the *Gauss–Kronrod quadrature rule*,

$$\int_{\mathbb{R}} f(t)\mathrm{d}\lambda(t) = \sum_{\nu=1}^{n} \lambda_\nu^K f(\tau_\nu^G) + \sum_{\mu=1}^{n+1} \lambda_\mu^{*K} f(\tau_\mu^K) + R_n^K(f), \tag{4.2}$$

in which τ_ν^G are the n Gauss nodes of $\mathrm{d}\lambda$ and the additional $n+1$ "Kronrod nodes" τ_μ^K and all $2n+1$ weights are chosen so as to optimize the degree of exactness, i.e., to make it as large as, or larger than, $3n+1$. The node polynomial of the Kronrod nodes,

$$\omega_{n+1}^K(t) = \prod_{\mu=1}^{n+1}(t-\tau_\mu^K), \tag{4.3}$$

of degree $n+1$, also known as the *Stieltjes polynomial*, must satisfy (cf. [24, §3.1.2])

$$\int_{\mathbb{R}} \omega_{n+1}^K(t)p(t)\pi_n(t;\mathrm{d}\lambda)\mathrm{d}\lambda(t) = 0 \quad \text{for all } p \in \mathbb{P}_n. \tag{4.4}$$

Thus, it is orthogonal to all lower-degree polynomials, but relative to the oscillatory measure $d\lambda^K = \pi_n(\cdot;d\lambda)d\lambda$. Although ω_{n+1}^K exists uniquely (Example 4.20), some of its zeros may well be complex. The study of the zeros of ω_{n+1}^K indeed is a currently active area of research.

Still another type of orthogonality—"power orthogonality"—

$$\int_{\mathbb{R}} [\omega_{n,s}^T(t)]^{2s+1} p(t) d\lambda(t) = 0 \quad \text{for all } p \in \mathbb{P}_{n-1}, \tag{4.5}$$

where $s > 0$ is a natural number and $\omega_{n,s}^T$ a monic polynomial of degree n, arises when one considers Gauss-type quadrature rules involving the first $2s$ derivatives,

$$\int_{\mathbb{R}} f(t) d\lambda(t) = \sum_{\nu=1}^{n} \sum_{\sigma=0}^{2s} \lambda_\nu^{(\sigma)} f^{(\sigma)}(\tau_\nu) + R_{n,s}(f). \tag{4.6}$$

If $R_{n,s}(f) = 0$ for all $f \in \mathbb{P}_{(2s+2)n-1}$, which is the maximum possible degree of exactness (Ex. 4.21), they are called *Gauss–Turán quadrature rules.* Indeed, the respective nodes $\tau_\nu = \tau_\nu^T$ are the zeros of $\omega_{n,s}^T$, the so-called s-orthogonal polynomial $\omega_{n,s}^T = \pi_{n,s}$.

In cases of integrands that have poles outside the interval of integration, special quadrature rules

$$\int_{\mathbb{R}} g(t) d\lambda(t) = \sum_{\nu=1}^{n} \lambda_\nu g(\tau_\nu) + R_n(g) \tag{4.7}$$

may be constructed in which R_n vanishes not only for polynomials, but for a combination of polynomials and rational functions, that is, for

$$g \in \mathbb{Q}_m \oplus \mathbb{P}_{2n-m-1}, \quad 0 \le m \le 2n,$$

where

$$\mathbb{Q}_m = \text{span}\left\{ r(t) = \frac{1}{1+\zeta_\mu t}, \ \mu = 1,2,\ldots,m \right\}$$

and $\zeta_\mu \in \mathbb{C}$, $\zeta_\mu \neq 0$, $1 + \zeta_\mu t \neq 0$ on $\operatorname{supp} d\lambda$. The poles $-1/\zeta_\mu$ are chosen to match the pole(s) of g closest to the support interval of $d\lambda$. The construction of such special quadrature rules (4.7) is relatively simple (cf. [24, §3.1.4.1]): With

$$\omega_m(t) = \prod_{\mu=1}^{m} (1 + \zeta_\mu t),$$

construct the n-point (polynomial) Gaussian formula

$$\int_{\mathbb{R}} f(t) \frac{d\lambda(t)}{\omega_m(t)} = \sum_{\nu=1}^{n} \lambda_\nu^G f(\tau_\nu^G), \quad f \in \mathbb{P}_{2n-1}, \tag{4.8}$$

for the modified measure $d\lambda(t)/\omega_m(t)$. Then

$$\tau_\nu = \tau_\nu^G, \quad \lambda_\nu = \omega_m(\tau_\nu^G)\lambda_\nu^G, \quad \nu = 1,2,\ldots,n. \tag{4.9}$$

For an example, see Ex. 4.23.

In Section 4.7 we discuss how to estimate, from above and below, matrix functionals $\boldsymbol{u}^{\mathrm{T}} f(A)\boldsymbol{u}$, $\|\boldsymbol{u}\| = 1$, where A is a symmetric positive definite matrix, by means of Lanczos's algorithm and Gauss-type quadratures applied to a discrete measure.

4.1 ▪ Elementary properties

Exercise 4.1. *Show that the n-point quadrature formula* (4.1) *is Gaussian if and only if it is interpolatory and* $\tau_\nu = \tau_\nu^{(n)}$ *are the zeros of* $\pi_n(\cdot;\mathrm{d}\lambda)$.

Answer. Necessity: The fact that (4.1) has to be interpolatory is trivial. Furthermore, if $\omega_n(t) = \prod_{\nu=1}^{n}(t-\tau_\nu)$ and $p \in \mathbb{P}_{n-1}$ is an arbitrary polynomial of degree $\le n-1$, then, since $p\omega_n \in \mathbb{P}_{2n-1}$ and (4.1) is assumed Gaussian, hence exact for all $f \in \mathbb{P}_{2n-1}$, we have

$$\int_{\mathbb{R}} p(t)\omega_n(t)\mathrm{d}\lambda(t) = \sum_{\nu=1}^{n} \lambda_\nu p(\tau_\nu)\omega_n(\tau_\nu) = 0,$$

since $\omega_n(\tau_\nu) = 0$. Thus ω_n is orthogonal to $\mathbb{P}_{n-1}$, that is, $\omega_n(t) = \pi_n(t;\mathrm{d}\lambda)$, as was to be shown.

Sufficiency: Let $f \in \mathbb{P}_{2n-1}$ be an arbitrary polynomial of degree $\le 2n-1$. Divide f by $\pi_n = \pi_n(\cdot;\mathrm{d}\lambda)$ to get

$$f(t) = q(t)\pi_n(t) + r(t),$$

where $q \in \mathbb{P}_{n-1}$ is the quotient polynomial and $r \in \mathbb{P}_{n-1}$ the remainder. Integrating, and using the fact that (4.1) is assumed interpolatory and that π_n is orthogonal to $\mathbb{P}_{n-1}$, one obtains

$$\begin{aligned}\int_{\mathbb{R}} f(t)\mathrm{d}\lambda(t) &= \int_{\mathbb{R}} q(t)\pi_n(t)\mathrm{d}\lambda(t) + \int_{\mathbb{R}} r(t)\mathrm{d}\lambda(t)\\ &= \int_{\mathbb{R}} r(t)\mathrm{d}\lambda(t) = \sum_{\nu=1}^{n} \lambda_\nu r(\tau_\nu)\\ &= \sum_{\nu=1}^{n} \lambda_\nu [f(\tau_\nu) - q(\tau_\nu)\pi_n(\tau_\nu)] = \sum_{\nu=1}^{n} \lambda_\nu f(\tau_\nu),\end{aligned}$$

that is, (4.1) is Gaussian.

Exercise 4.2. Let (4.1) be Gaussian, i.e.,

$$\int_{\mathbb{R}} f(t)\mathrm{d}\lambda(t) = \sum_{\nu=1}^{n} \lambda_\nu^G f(\tau_\nu^G), \quad f \in \mathbb{P}_{2n-1}.$$

(a) *Prove that the nodes* τ_ν^G *are mutually distinct and contained in the interior of the support interval* $[a,b]$ *of* $\mathrm{d}\lambda$.

(b) *Show that* $\lambda_\nu^G > 0$, $\nu = 1,2,\ldots,n$.

Answer.

(a) Since $\pi_n = \pi_n(\cdot;\mathrm{d}\lambda)$ by orthogonality satisfies $\int_{\mathbb{R}} \pi_n(t)\mathrm{d}\lambda(t) = 0$, $n \ge 1$, there must be at least one point in the interior of $[a,b]$ at which π_n changes sign. Let $t_1, t_2, \ldots, t_k$, $1 \le k \le n$, be all such points. If there were $k < n$, orthogonality would imply

$$\int_{\mathbb{R}} \pi_n(t)\prod_{\kappa=1}^{k}(t - t_\kappa)\mathrm{d}\lambda(t) = 0.$$

But this is impossible since the integrand has constant sign. Therefore, $k = n$.

(b) Let $\ell_\mu \in \mathbb{P}_{n-1}$ be the elementary Lagrange interpolation polynomials relative to the nodes $\tau_1^G, \tau_2^G, \dots, \tau_n^G$, that is, $\ell_\mu(\tau_\mu^G) = 1$, $\ell_\mu(\tau_\nu^G) = 0$ for $\nu \neq \mu$. Since $\ell_\mu^2 \in \mathbb{P}_{2n-2}$ and (4.1) is exact for $f \in \mathbb{P}_{2n-1}$, we have that

$$0 < \int_{\mathbb{R}} \ell_\mu^2(t)\mathrm{d}\lambda(t) = \sum_{\nu=1}^{n} \lambda_\nu^G \ell_\mu^2(\tau_\nu^G) = \lambda_\mu^G, \quad \mu = 1,2,\dots,n,$$

as claimed.

Exercise 4.3. Let $J_n = J_n(\mathrm{d}\lambda)$ be the nth-order Jacobi matrix of the measure $\mathrm{d}\lambda$ (cf. Eq. (1.4)) and $p^{\mathrm{T}}(t) = [\tilde{\pi}_0(t), \tilde{\pi}_1(t), \dots, \tilde{\pi}_{n-1}(t)]$, where $\tilde{\pi}_k = \tilde{\pi}_k(\,\cdot\,;\mathrm{d}\lambda)$ are the orthonormal polynomials for the measure $\mathrm{d}\lambda$.

(a) *Show that the Gaussian nodes* τ_ν^G, $\nu = 1,2,\dots,n$, *are the eigenvalues of* J_n *and* $p(\tau_\nu^G)$ *corresponding eigenvectors.*

(b) *Show that the Gaussian weights* λ_ν^G *are*

$$\lambda_\nu^G = \beta_0 v_{\nu,1}^2, \quad \nu = 1,2,\dots,n,$$

where $v_{\nu,1}$ *are the first components of the normalized eigenvectors* v_ν *of* J_n, *and* $\beta_0 = \int_{\mathbb{R}} \mathrm{d}\lambda(t)$.

Answer

(a) Since (cf. Ex. 2.5(c))

$$t p(t) = J_n p(t) + \sqrt{\beta_n}\,\tilde{\pi}_n(t) e_n, \quad e_n^{\mathrm{T}} = [0,0,\dots,1] \in \mathbb{R}^n,$$

letting $t = \tau_\nu^G$ yields

$$J_n p(\tau_\nu^G) = \tau_\nu^G p(\tau_\nu^G),$$

which proves the assertion since $p \neq 0$.

(b) By (a), we have

$$v_\nu = \frac{1}{\|p(\tau_\nu^G)\|}\, p(\tau_\nu^G), \tag{4.10}$$

where

$$\|p(\tau_\nu^G)\|^2 = p(\tau_\nu^G)^T p(\tau_\nu^G) = \sum_{k=0}^{n-1} [\tilde{\pi}_k(\tau_\nu^G)]^2.$$

Taking the square of the first component on either side of (4.10) and multiplying by β_0 gives

$$\beta_0 v_{\nu,1}^2 = \frac{1}{p(\tau_\nu^G)^T p(\tau_\nu^G)} = \frac{1}{\sum_{k=0}^{n-1}[\tilde{\pi}_k(\tau_\nu^G)]^2}$$

since $\tilde{\pi}_0 = \beta_0^{-1/2}$. We claim that the right-hand side is λ_ν^G. Indeed, we have, by orthonormality,

$$\beta_0^{1/2}\delta_{k,0} = \int_{\mathbb{R}} \tilde{\pi}_k(t)\mathrm{d}\lambda(t) = \sum_{\mu=1}^{n} \lambda_\mu^G \tilde{\pi}_k(\tau_\mu^G), \quad k = 0,1,\dots,n-1,$$

that is, in vector form,

$$\beta_0^{1/2} e_1 = \sum_{\mu=1}^{n} \lambda_\mu^G p(\tau_\mu^G).$$

Since eigenvectors of a symmetric real matrix are mutually orthogonal, and $\tilde{\pi}_0 = \beta_0^{-1/2}$, we get, left-multiplying by $p(\tau_\nu^G)^T$,

$$1 = \lambda_\nu^G \|p(\tau_\nu^G)\|^2, \qquad \lambda_\nu^G = \frac{1}{\|p(\tau_\nu^G)\|^2} = \beta_0 v_{\nu,1}^2,$$

as claimed.

Exercise 4.4. (Ex. 1 of [27, §4].) *Prove that the n-point Gaussian quadrature sum for a measure* $d\lambda$ *can be written as*

$$\sum_{\nu=1}^{n} \lambda_\nu^G f(\tau_\nu^G) = \beta_0 e_1^{\mathrm{T}} f(J_n(d\lambda)) e_1, \quad e_1^{\mathrm{T}} = [1, 0, \ldots, 0], \tag{4.11}$$

where J_n *is the nth-order Jacobi matrix of* $d\lambda$ (cf. Eq. (1.4)).

Answer. Since V, the matrix of normalized eigenvectors $v_1, v_2, \ldots, v_n$ of the Jacobi matrix $J_n = J_n(d\lambda)$, is an orthogonal matrix, we have, by the spectral resolution of J_n,

$$J_n(d\lambda) = V D_\tau V^{\mathrm{T}}, \quad D_\tau = \mathrm{diag}(\tau_1^G, \tau_2^G, \ldots, \tau_n^G).$$

Therefore,

$$e_1^{\mathrm{T}} f(J_n) e_1 = e_1^{\mathrm{T}} f(V D_\tau V^{\mathrm{T}}) e_1 = e_1^{\mathrm{T}} V f(D_\tau) V^{\mathrm{T}} e_1.$$

Since $V = [v_1, v_2, \ldots, v_n]$, we can write

$$e_1^{\mathrm{T}} V f(D_\tau) V^{\mathrm{T}} e_1 = [v_{1,1}, v_{2,1}, \ldots, v_{n,1}] \begin{bmatrix} f(\tau_1^G) & & & 0 \\ & f(\tau_2^G) & & \\ & & \ddots & \\ 0 & & & f(\tau_n^G) \end{bmatrix} \begin{bmatrix} v_{1,1} \\ v_{2,1} \\ \vdots \\ v_{n,1} \end{bmatrix}$$

$$= [v_{1,1}, v_{2,1}, \ldots, v_{n,1}] \begin{bmatrix} f(\tau_1^G) v_{1,1} \\ f(\tau_2^G) v_{2,1} \\ \vdots \\ f(\tau_n^G) v_{n,1} \end{bmatrix} = \sum_{\nu=1}^{n} f(\tau_\nu^G) v_{\nu,1}^2,$$

that is,

$$e_1^{\mathrm{T}} f(J_n) e_1 = \sum_{\nu=1}^{n} f(\tau_\nu^G) v_{\nu,1}^2.$$

Since (cf. Ex. 4.3(b)) $\lambda_\nu^G = \beta_0 v_{\nu,1}^2$, that is, $v_{\nu,1}^2 = \lambda_\nu^G / \beta_0$, this yields

$$\frac{1}{\beta_0} \sum_{\nu=1}^{n} \lambda_\nu^G f(\tau_\nu^G) = e_1^{\mathrm{T}} f(J_n) e_1,$$

which, after multiplying both sides by β_0, gives (4.11).

Exercise 4.5*. (Ex. 3* of [27, §4].) Circle theorems for Gaussian quadrature weights (cf. [8]). For this exercise, consult also [28].

(a) Gauss–Jacobi quadrature.

Let $w(t)=(1-t)^\alpha(1+t)^\beta$ be the Jacobi weight function. It is known [61, §15.3.10] that the nodes $\tau_\nu = \tau_\nu^{(n)}$ and weights $\lambda_\nu = \lambda_\nu^{(n)}$ of the n-point Gauss–Jacobi quadrature formula satisfy

$$\lambda_\nu \sim \frac{\pi}{n} w(\tau_\nu)\sqrt{1-\tau_\nu^2}, \quad n \to \infty,$$

for τ_ν on any compact interval contained in $(-1,1)$. Thus, suitably normalized weights, plotted against the nodes, lie asymptotically on the unit circle. *Use MATLAB to demonstrate this graphically.*

(b) Gauss quadrature for the algebraic/logarithmic weight function $w(t)=t^\alpha \ln(1/t)$ on $[0,1]$ (cf. [24, Example 2.27]).

Try, numerically, to find a circle theorem in this case also, and experiment with different values of the parameter $\alpha > -1$. (Use the `OPQ` routine `r_jaclog.m` to generate the recurrence coefficients of the orthogonal polynomials for the weight function w.)

(c) Gauss–Kronrod quadrature.

With w as in (a), the analogous result for the $2n+1$ nodes τ_ν and weights λ_ν of the $(2n+1)$-point Gauss–Kronrod formula is expected to be

$$\lambda_\nu \sim \frac{\pi}{2n} w(\tau_\nu)\sqrt{1-\tau_\nu^2}, \quad n \to \infty.$$

That this indeed is the case, when $\alpha,\beta \in [0,\frac{5}{2})$, follows from [54, Theorem 2]. *Use MATLAB to illustrate this graphically.*

(d) *Experiment with the Gauss–Kronrod formula for the algebraic/logarithmic weight function of* (b), *when* $\alpha = 0$.

Answer

(a) The script `circle_jac.m` plots the points $[\tau_\nu^{(N)}, N\lambda_\nu^{(N)}/(\pi w(\tau_\nu^{(N)}))]$, $\nu = 1,2,\ldots,N$, for selected values of α, β with $\beta \geq \alpha$.

```
%CIRCLE_JAC  Circle theorem for Gauss-Jacobi quadrature

N=20
%N=200;
c=zeros(N,1);
for a=[-.9:.1:1 1.5:.5:5 6:10]
  for b=[-.9:.1:1 1.5:.5:5 6:10]
    if b>=a
      ab=r_jacobi(N,a,b);
      xw=gauss(N,ab); x=xw(:,1); w=xw(:,2);
      c=N*w./(pi*(1-x).^a.*(1+x).^b);
      plot(x,c,'.');set(gca,'FontSize',14)
      plot([-1.2 1.2],[0 0])
      plot([0 0],[0 1.2])
      axis('equal')
      axis([-1.3 1.3 -.2 1.3])
      if a==-.9 & b==-.9, hold on; end
    end
  end
end
```

The plot on the left of Fig. 4.1 is for $N=20$, the one on the right for $N=200$.

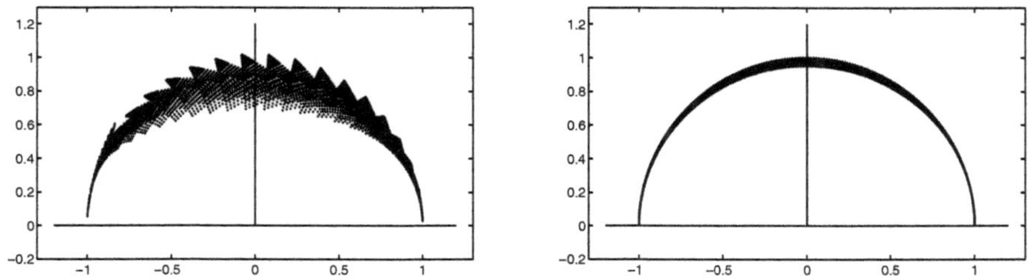

Figure 4.1. *Circle theorem for Gauss–Jacobi quadrature*

(b) For the weight function $w(t) = t^\alpha \log(1/t)$ on $[0,1]$, the routine circle_jac log.m attempts to do the same as the routine circle_jac.m does in part (a) for the Jacobi weight function. It turns out, however, that for large values of N, say $N = 40$, instabilities emerge, producing negative values for Gaussian weights, if they are small, or inaccurate values otherwise. We therefore wrote a symbolic routine scircle_jaclog.m that uses the symbolic routine sr_jaclog.m of Ex. 2.27*(c).

```
%SCIRCLE_JACLOG Circle theorem for Gauss quadrature with
%algebraic/logarithmic weight function.

dig=48; digits(dig);
N=20;
%N=80;
c=vpa(zeros(N,1));
for a=[-.9:.1:1 1.5:.5:5 6:10]
  ab=sr_jaclog(dig,N,a);
  xw=sgauss(dig,N,ab); x=xw(:,1); w=xw(:,2);
  c=vpa(N*w./(pi*x.^a.*log(1./x)));
  plot(double(x),double(c),'.');set(gca,'FontSize',14)
  plot([-0.2 1.2],[0 0])
  plot([0 0],[0 0.6])
  axis('equal')
  axis([-.3 1.3 -.2 .7])
  if a==-.9, hold on; end
end
```

Run in 48-digit arithmetic with $N = 20$ and $N = 80$, it produces the results shown in Fig. 4.2 on the left and right, resp.

(c) The routine circle_kron.m tries to verify the circle theorem for Gauss–Kronrod quadrature with Jacobi weight function $w^{(\alpha,\beta)}(x) = (1-x)^\alpha(1+x)^\beta$, $\alpha, \beta \in [0, 5/2)$. Theoretically, the circle theorem should hold; numerically, the Gauss–Kronrod formula is found to exist only when α and/or β are somewhat smaller than $5/2$. Varying α, β from 0 in steps of .1 to 2.5 produces a nonexistence message when α and/or β are equal to 2.4 in the case $N = 20$, or 2.3 in the case $N = 80$. For values of α, $\beta \leq 2$, Fig. 4.3 shows the plots obtained by circle_kron.m for $N = 20$ (on the left) and for $N = 80$ (on the right). The jaggedness of the "circles" is probably due to numerical inaccuracies near the upper boundary 2.5 of the existence interval. It disappears when the limit is lowered to 1.5.

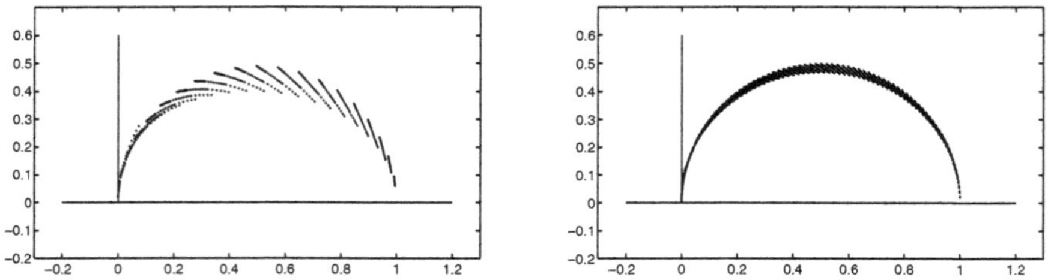

Figure 4.2. *Circle theorem for algebraic/logarithmic weight functions*

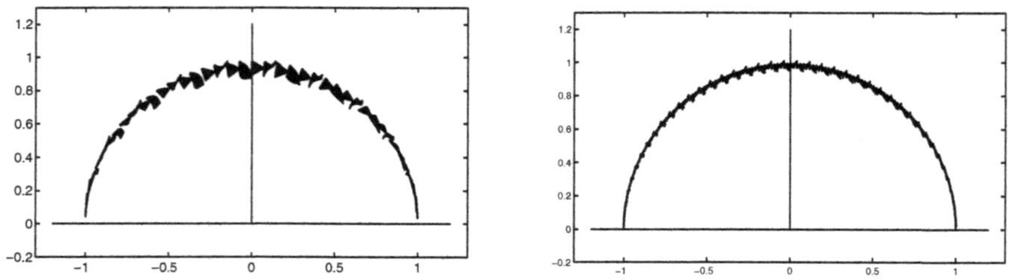

Figure 4.3. *Circle theorem for Gauss-Kronrod quadrature and Jacobi weight functions*

(d) Experimenting with the algebraic/logarithmic weight function $w(x) = x^\alpha \log(1/x)$ on $[0,1]$ shows that the circle theorem is still valid, not only for $\alpha = 0$, but also for $0 \le \alpha \le 1.5$, at least. Indeed, the results produced by `circle_kronlog.m` are shown in Fig. 4.4 for $N = 20$ (on the left) and for $N = 80$ (on the right).

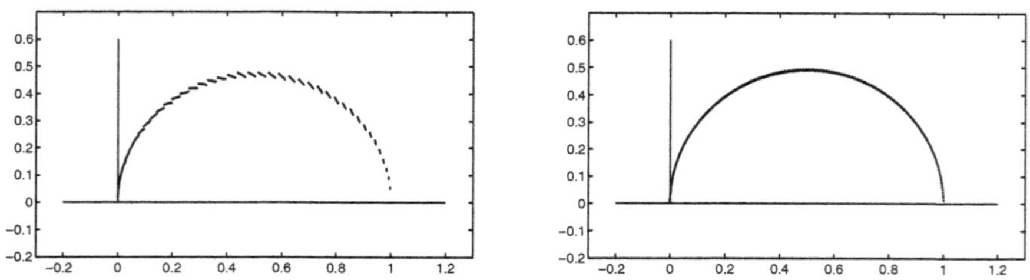

Figure 4.4. *Circle theorem for Gauss-Kronrod quadrature and algebraic/logarithmic weight functions*

4.2 ▪ Cauchy and Cauchy principal value integrals

Exercise 4.6. (Ex. 5 of [27, §4].) Cauchy integral.

Let

$$F(z) = F(z; d\lambda) = \int_{\mathbb{R}} \frac{d\lambda(t)}{z - t} \tag{4.12}$$

be the Cauchy transform of the measure $d\lambda$ and, more generally,

$$\rho_n(z) = \rho_n(z; d\lambda) = \int_{\mathbb{R}} \frac{\pi_n(t)}{z - t}\, d\lambda(t) \tag{4.13}$$

the Cauchy transform (or integral) of the orthogonal polynomial. Then

$$\sigma_n(z) = \int_{\mathbb{R}} \frac{\pi_n(z) - \pi_n(t)}{z - t} \, \mathrm{d}\lambda(t)$$

(cf. Ex. 2.8) can be given the form

$$\sigma_n(z) = \pi_n(z) F(z) - \rho_n(z),$$

so that

$$\frac{\sigma_n(z)}{\pi_n(z)} = F(z) - \frac{\rho_n(z)}{\pi_n(z)}. \tag{4.14}$$

(a) *Consider the Cauchy integral*

$$\rho_n(z) = \rho_n(z; \mathrm{d}\lambda) = \int_{\mathbb{R}} \frac{\pi_n(t; \mathrm{d}\lambda)}{z - t} \, \mathrm{d}\lambda(t).$$

Show that

$$\rho_n(z) = O(z^{-n-1}) \quad \textit{as } z \to \infty.$$

{Hint: Expand the integral defining $\rho_n(z)$ in descending powers of z.}

(b) *Show that*

$$\int_{\mathbb{R}} \frac{\mathrm{d}\lambda(t)}{z - t} - \frac{\sigma_n(z)}{\pi_n(z)} = \frac{\rho_n(z)}{\pi_n(z)} = O(z^{-2n-1}) \quad \textit{as } z \to \infty.$$

{Hint: Use (4.14).}

(c) *Consider the partial fraction decomposition*

$$\frac{\sigma_n(z)}{\pi_n(z)} = \sum_{\nu=1}^{n} \frac{\lambda_\nu}{z - \tau_\nu^G}$$

of $\sigma_n(z)/\pi_n(z)$ in (4.14). *Use* (b) *to show that $\lambda_\nu = \lambda_\nu^G$ are the weights of the n-point Gaussian quadrature formula for* $\mathrm{d}\lambda$. *In particular, show that*

$$\lambda_\nu^G = \frac{\sigma_n(\tau_\nu^G)}{\pi_n'(\tau_\nu^G)}.$$

(d) *Discuss what happens if $z \to x$, $x \in$* $\mathrm{supp}(\mathrm{d}\lambda)$.

Answer

(a) Since

$$\frac{1}{z - t} = \frac{1}{z\left(1 - \frac{t}{z}\right)} = \frac{1}{z}\left(1 + \frac{t}{z} + \frac{t^2}{z^2} + \cdots\right),$$

we have by orthogonality

$$\begin{aligned}
\rho_n(z) &= \int_{\mathbb{R}} \frac{\pi_n(t; \mathrm{d}\lambda)}{z - t} \, \mathrm{d}\lambda(t) = \int_{\mathbb{R}} \pi_n(t; \mathrm{d}\lambda) \left(\frac{1}{z} + \frac{t}{z^2} + \frac{t^2}{z^3} + \cdots\right) \mathrm{d}\lambda(t) \\
&= \int_{\mathbb{R}} \pi_n(t; \mathrm{d}\lambda) \left(\frac{t^n}{z^{n+1}} + \cdots\right) \mathrm{d}\lambda(t) = \frac{1}{z^{n+1}} \int_{\mathbb{R}} \pi_n^2(t; \mathrm{d}\lambda) \mathrm{d}\lambda(t) + \cdots \\
&= \frac{\|\pi_n\|_{\mathrm{d}\lambda}^2}{z^{n+1}} + \cdots,
\end{aligned}$$

which is $O(z^{-n-1})$ since $\|\pi_n\|_{\mathrm{d}\lambda}^2 > 0$.

(b) The first equality is a direct consequence of (4.14) and the definition (4.12) of $F(z)$. The second equality follows from $\rho_n(z) = O(z^{-n-1})$, proved in part (a), and $1/\pi_n(z) = O(z^{-n})$.

(c) From part (b) we have

$$\frac{\sigma_n(z)}{\pi_n(z)} = \int_{\mathbb{R}} \frac{\mathrm{d}\lambda(t)}{z-t} - \frac{\rho_n(z)}{\pi_n(z)} = \frac{1}{z}\int_{\mathbb{R}} \frac{\mathrm{d}\lambda(t)}{1-t/z} + O(z^{-2n-1})$$
$$= \frac{\mu_0}{z} + \frac{\mu_1}{z^2} + \cdots + \frac{\mu_{2n-1}}{z^{2n}} + O(z^{-2n-1}),$$

where

$$\mu_k = \int_{\mathbb{R}} t^k \mathrm{d}\lambda(t), \quad k = 0, 1, 2, \ldots$$

are the moments of $\mathrm{d}\lambda$.

Comparing the above expansion with

$$\sum_{\nu=1}^{n} \frac{\lambda_\nu}{z-\tau_\nu^G} = \frac{\sum_{\nu=1}^n \lambda_\nu}{z} + \frac{\sum_{\nu=1}^n \lambda_\nu \tau_\nu^G}{z^2} + \cdots + \frac{\sum_{\nu=1}^n \lambda_\nu (\tau_\nu^G)^{2n-1}}{z^{2n}} + \cdots,$$

we get

$$\sum_{\nu=1}^{n} \lambda_\nu (\tau_\nu^G)^k = \mu_k, \quad k = 0, 1, \ldots, 2n-1,$$

which characterizes the λ_ν as the weights of the n-point Gaussian quadrature formula for $\mathrm{d}\lambda$.

Multiplying the identity

$$\frac{\sigma_n(z)}{\pi_n(z)} = \sum_{\nu=1}^{n} \frac{\lambda_\nu^G}{z-\tau_\nu^G}$$

by $z-\tau_\mu^G$ and then letting $z \to \tau_\mu^G$ gives

$$\lambda_\mu^G = \frac{\sigma_n(\tau_\mu^G)}{\pi_n'(\tau_\mu^G)}, \quad \mu = 1, 2, \ldots, n.$$

(d) It is known [24, §2.3.1] that if the integral defining $\rho_n(z)$ is interpreted as a Cauchy principal value integral when $z = x \in \operatorname{supp}(\mathrm{d}\lambda)$, the sequence $\{\rho_n(x)\}$ is the solution of the basic three-term recurrence relation for the orthogonal polynomials $\{\pi_k(\cdot;\mathrm{d}\lambda)\}$ with initial values

$$\rho_{-1}(x) = 1, \quad \rho_0(x) = ⨍_{\mathbb{R}} \frac{\mathrm{d}\lambda(t)}{x-t}.$$

Alternatively (cf. [24, Eq. (2.3.9)]),

$$\rho_n(x) = \lim_{z\downarrow x} \operatorname{Re} \rho_n(z).$$

Exercise 4.7. (Ex. 18 of [27, §4].) Cauchy principal value integral.

The Cauchy integral

$$(Cf)(z;\mathrm{d}\lambda) = \int_{\mathbb{R}} \frac{f(t)}{z-t} \mathrm{d}\lambda(t),$$

when $z=x$ is a point in the interior of the support interval of $\mathrm{d}\lambda$, must be taken as a Cauchy principal value integral (cf. Ex. 4.6(d))

$$(Cf)(x;\mathrm{d}\lambda) := \fint_{\mathbb{R}} \frac{f(t)}{x-t}\,\mathrm{d}\lambda(t) = \lim_{\varepsilon\downarrow 0}\left(\int_{-\infty}^{x-\varepsilon} + \int_{x+\varepsilon}^{\infty}\right)\frac{f(t)}{x-t}\,\mathrm{d}\lambda(t). \tag{4.15}$$

Show that

$$(Cf)(x;\mathrm{d}\lambda) = \frac{\rho_n(x)}{\pi_n(x)}f(x) + \sum_{\nu=1}^{n}\lambda_\nu \frac{f(\tau_\nu)}{x-\tau_\nu} + R_n(f;x), \quad n\ge 1, \tag{4.16}$$

where (cf. Ex. 4.6(d))

$$\rho_n(x) = \fint_{\mathbb{R}} \frac{\pi_n(t)}{x-t}\,\mathrm{d}\lambda(t), \quad \pi_n(t) = \pi_n(t;\mathrm{d}\lambda), \tag{4.17}$$

and $\tau_\nu = \tau_\nu^G$, $\lambda_\nu = \lambda_\nu^G$ *are the nodes and weights of the n-point Gaussian quadrature rule for* $\mathrm{d}\lambda$. *What can be said about the remainder term* R_n*?* {Hint: Use Ex. 4.6(c) and (4.14).}

Answer. Let

$$F(x) = \fint_{\mathbb{R}} \frac{\mathrm{d}\lambda(t)}{x-t}.$$

Then (4.15) can be written in the form

$$(Cf)(x;\mathrm{d}\lambda) = f(x)F(x) - \int_{\mathbb{R}} \frac{f(x)-f(t)}{x-t}\,\mathrm{d}\lambda(t).$$

Applying the n-point Gaussian quadrature rule for $\mathrm{d}\lambda$ to the integral on the right,

$$\int_{\mathbb{R}} \frac{f(x)-f(t)}{x-t}\,\mathrm{d}\lambda(t) = \sum_{\nu=1}^{n}\lambda_\nu \frac{f(x)-f(\tau_\nu)}{x-\tau_\nu} + R_n(f;x), \tag{4.18}$$

we get

$$\begin{aligned}(Cf)(x;\mathrm{d}\lambda) &= f(x)F(x) - \sum_{\nu=1}^{n}\frac{\lambda_\nu}{x-\tau_\nu}f(x) + \sum_{\nu=1}^{n}\lambda_\nu\frac{f(\tau_\nu)}{x-\tau_\nu} + R_n(f;x)\\ &= \left(F(x) - \sum_{\nu=1}^{n}\frac{\lambda_\nu}{x-\tau_\nu}\right)f(x) + \sum_{\nu=1}^{n}\lambda_\nu\frac{f(\tau_\nu)}{x-\tau_\nu} + R_n(f;x).\end{aligned}$$

The coefficient of $f(x)$, by (4.14) and Ex. 4.6(c), both for $z=x$, equals

$$F(x) - \frac{\sigma_n(x)}{\pi_n(x)} = \frac{\rho_n(x)}{\pi_n(x)}$$

with $\rho_n(x)$ as in (4.17). Therefore,

$$(Cf)(x;\mathrm{d}\lambda) = \frac{\rho_n(x)}{\pi_n(x)}f(x) + \sum_{\nu=1}^{n}\lambda_\nu\frac{f(\tau_\nu)}{x-\tau_\nu} + R_n(f;x),$$

as claimed in (4.16). Here, $R_n(f;x)$ is the remainder term in (4.18), which vanishes whenever $f\in\mathbb{P}_{2n}$.

Exercise 4.8. (Ex. 19 of [27, §4].) *Show that the approximation on the right-hand side of* (4.16) *is equal to*

$$\fint_{\mathbb{R}} \frac{p_n(f;t)}{x-t}\,\mathrm{d}\lambda(t), \tag{4.19}$$

where $p_n(f;\cdot)$ is the polynomial of degree n interpolating f at the n Gaussian nodes τ_ν and at x.

Answer. By the Lagrange interpolation formula,

$$p_n(f;t)=\sum_{\nu=1}^{n}\ell_\nu(t)f(\tau_\nu)+\ell_0(t)f(x), \quad \tau_\nu=\tau_\nu^G,$$

where

$$\ell_\nu(t)=\frac{t-x}{\tau_\nu-x}\prod_{\substack{\mu=1\\ \mu\neq\nu}}^{n}\frac{t-\tau_\mu}{\tau_\nu-\tau_\mu}=\frac{t-x}{\tau_\nu-x}\,\frac{\pi_n(t)}{(t-\tau_\nu)\pi_n'(\tau_\nu)}, \quad \nu>0,$$

$$\ell_0(t)=\prod_{\nu=1}^{n}\frac{t-\tau_\nu}{x-\tau_\nu}=\frac{\pi_n(t)}{\pi_n(x)}.$$

Thus,

$$p_n(f;t)=\sum_{\nu=1}^{n}\frac{(t-x)\pi_n(t)}{(\tau_\nu-x)(t-\tau_\nu)\pi_n'(\tau_\nu)}f(\tau_\nu)+\frac{\pi_n(t)}{\pi_n(x)}f(x). \tag{4.20}$$

By virtue of the well-known formula for λ_ν,

$$\lambda_\nu=\int_{\mathbb{R}}\ell_\nu(t)\mathrm{d}\lambda(t)=\int_{\mathbb{R}}\frac{\pi_n(t)\mathrm{d}\lambda(t)}{(t-\tau_\nu)\pi_n'(\tau_\nu)},$$

expressing the interpolatory nature of Gaussian quadrature, the Cauchy integral of p_n then becomes, using (4.17),

$$\begin{aligned}\fint_{\mathbb{R}}\frac{p_n(f;t)}{x-t}\mathrm{d}\lambda(t)&=\sum_{\nu=1}^{n}f(\tau_\nu)\int_{\mathbb{R}}\frac{\pi_n(t)\mathrm{d}\lambda(t)}{(x-\tau_\nu)(t-\tau_\nu)\pi_n'(\tau_\nu)}\\ &\quad+\frac{f(x)}{\pi_n(x)}\fint_{\mathbb{R}}\frac{\pi_n(t)}{x-t}\mathrm{d}\lambda(t)=\frac{\rho_n(x)}{\pi_n(x)}f(x)+\sum_{\nu=1}^{n}\frac{\lambda_\nu}{x-\tau_\nu}f(\tau_\nu),\end{aligned}$$

which is precisely the approximation on the right of (4.16).

Exercise 4.9. (Ex. 20 of [27, §4].) *Derive an alternative and more stable form of* (4.16),

$$(Cf)(x;\mathrm{d}\lambda)=\sum_{k=0}^{n}a_k\rho_k(x)+R_n(f;x), \quad n\geq 1, \tag{4.21}$$

where the ρ_k are defined in (4.17) *and, with τ_ν^G, λ_ν^G denoting the nodes and weights of the n-point Gaussian quadrature rule for $\mathrm{d}\lambda$ and $\pi_k(\cdot)=\pi_k(\cdot;\mathrm{d}\lambda)$,*

$$a_k=\frac{1}{\|\pi_k\|^2}\sum_{\nu=1}^{n}\lambda_\nu^G f(\tau_\nu^G)\pi_k(\tau_\nu^G),\ k<n; \quad a_n=\sum_{\nu=1}^{n}\frac{f(x)-f(\tau_\nu^G)}{(x-\tau_\nu^G)\pi_n'(\tau_\nu^G)}. \tag{4.22}$$

{Hint: Expand p_n in (4.19) in the orthogonal polynomials π_k. Also use the fact that the elementary Lagrange interpolation polynomials sum up to 1.}

Answer. One has, using the expansion suggested in the hint and Ex. 4.8,

$$(Cf)(x;\mathrm{d}\lambda) = \fint_{\mathbb{R}} \frac{p_n(f;t)}{x-t}\mathrm{d}\lambda(t) + R_n(f;x) = \fint_{\mathbb{R}} \sum_{k=0}^{n} a_k \frac{\pi_k(t)}{x-t}\mathrm{d}\lambda(t) + R_n(f;x)$$
$$= \sum_{k=0}^{n} a_k \rho_k(x) + R_n(f;x),$$

where

$$a_k = \frac{1}{\|\pi_k\|^2}\int_{\mathbb{R}} p_n(f;t)\pi_k(t)\mathrm{d}\lambda(t). \tag{4.23}$$

If $k < n$, applying the n-point Gauss quadrature rule to the integral on the right, we obtain

$$a_k = \frac{1}{\|\pi_k\|^2}\sum_{\nu=1}^{n} \lambda_\nu^G p_n(f;\tau_\nu^G)\pi_k(\tau_\nu^G)$$
$$= \frac{1}{\|\pi_k\|^2}\sum_{\nu=1}^{n} \lambda_\nu^G f(\tau_\nu^G)\pi_k(\tau_\nu^G), \quad k < n,$$

which is the first relation in (4.22).

If $k = n$, we use the formula (4.20) for p_n in the right-hand side of (4.23) to get

$$a_n = \frac{1}{\|\pi_n\|^2}\int_{\mathbb{R}} \left(\frac{\pi_n(t)}{\pi_n(x)} f(x) + \sum_{\nu=1}^{n} \frac{(t-x)\pi_n(t)}{(t-\tau_\nu^G)(\tau_\nu^G - x)\pi_n'(\tau_\nu^G)} f(\tau_\nu^G)\right)\pi_n(t)\mathrm{d}\lambda(t). \tag{4.24}$$

Since the sum of the coefficients of $f(x)$ and $f(\tau_\nu^G)$, i.e., the sum of the elementary Lagrange interpolation polynomials, is identically equal to 1, we have

$$\frac{\pi_n(t)}{\pi_n(x)} + \sum_{\nu=1}^{n} \frac{(t-x)\pi_n(t)}{(t-\tau_\nu^G)(\tau_\nu^G - x)\pi_n'(\tau_\nu^G)} \equiv 1.$$

Multiplying this by $\pi_n(t)$ and integrating (with measure $\mathrm{d}\lambda$) gives

$$\frac{\|\pi_n\|^2}{\pi_n(x)} + \sum_{\nu=1}^{n} \frac{1}{(\tau_\nu^G - x)\pi_n'(\tau_\nu^G)} \int_{\mathbb{R}} \frac{(t-x)\pi_n^2(t)}{t-\tau_\nu^G}\mathrm{d}\lambda(t) = 0,$$

since the integral over $\pi_n(t)$ is zero by orthogonality. To evaluate the integral on the right-hand side, we note that

$$\frac{(t-x)\pi_n(t)}{t-\tau_\nu^G} = (t-x)(t^{n-1}+\cdots) = t^n + \cdots = \pi_n(t) + \cdots,$$

so that, again by orthogonality,

$$\int_{\mathbb{R}} \frac{(t-x)\pi_n^2(t)}{t-\tau_\nu^G}\mathrm{d}\lambda(t) = \|\pi_n\|^2. \tag{4.25}$$

There follows

$$\frac{1}{\pi_n(x)}=\sum_{\nu=1}^{n}\frac{1}{(x-\tau_\nu^G)\pi_n'(\tau_\nu^G)}.$$

Inserting this in (4.24) and using again (4.25) gives

$$a_n=\frac{1}{\|\pi_n\|^2}\int_{\mathbb{R}}\pi_n^2(t)\sum_{\nu=1}^{n}\frac{1}{(x-\tau_\nu^G)\pi_n'(\tau_\nu^G)}[f(x)-f(\tau_\nu^G)]\mathrm{d}\lambda(t),$$

that is,

$$a_n=\sum_{\nu=1}^{n}\frac{f(x)-f(\tau_\nu^G)}{(x-\tau_\nu^G)\pi_n'(\tau_\nu^G)},$$

which is the second relation in (4.22).

Exercise 4.10. (Ex. 21 of [27, §4].) *Obtain the polynomial p_{n-1} of degree $n-1$ interpolating f at the n Gaussian nodes $\tau_\nu=\tau_\nu^G$ and integrate it in the sense of* (4.15) *to get a quadrature rule "in the strict sense"* (not containing x as a node),

$$(Cf)(x;\mathrm{d}\lambda)=\sum_{\nu=1}^{n}\frac{\rho_n(x)-\rho_n(\tau_\nu)}{(x-\tau_\nu)\pi_n'(\tau_\nu)}f(\tau_\nu)+R_n^*(f;x), \tag{4.26}$$

for the Cauchy integral (4.15) *of Ex.* 4.7. *What can be said about the remainder term R_n^*?*

Answer. The interpolation formula in question is

$$f(t)=p_{n-1}(f;t)+E_{n-1}(f;t),\quad p_{n-1}(f;t)=\sum_{\nu=1}^{n}\frac{\pi_n(t)}{(t-\tau_\nu)\pi_n'(\tau_\nu)}f(\tau_\nu), \tag{4.27}$$

where $E_{n-1}(f;\cdot)$ is the interpolation error. Integrating the formula, as suggested in the exercise, gives

$$\begin{aligned}(Cf)(x;\mathrm{d}\lambda)&=\fint_{\mathbb{R}}\frac{p_{n-1}(t)}{x-t}\mathrm{d}\lambda(t)+\fint_{\mathbb{R}}\frac{E_{n-1}(f;t)}{x-t}\mathrm{d}\lambda(t)\\&=\sum_{\nu=1}^{n}\frac{f(\tau_\nu)}{\pi_n'(\tau_\nu)}\fint_{\mathbb{R}}\frac{\pi_n(t)}{(x-t)(t-\tau_\nu)}\mathrm{d}\lambda(t)+R_n^*(f;x),\end{aligned}$$

where

$$R_n^*(f;x)=\fint_{\mathbb{R}}\frac{E_{n-1}(f;t)}{x-t}\mathrm{d}\lambda(t). \tag{4.28}$$

Using

$$\frac{1}{(x-t)(t-\tau_\nu)}=\frac{1}{x-\tau_\nu}\left[\frac{1}{x-t}-\frac{1}{\tau_\nu-t}\right],$$

we get

$$\begin{aligned}(Cf)(x;\mathrm{d}\lambda)&=\sum_{\nu=1}^{n}\frac{f(\tau_\nu)}{\pi_n'(\tau_\nu)}\frac{1}{x-\tau_\nu}\left[\fint_{\mathbb{R}}\frac{\pi_n(t)}{x-t}\mathrm{d}\lambda(t)-\fint_{\mathbb{R}}\frac{\pi_n(t)}{\tau_\nu-t}\mathrm{d}\lambda(t)\right]+R_n^*(f;x)\\&=\sum_{\nu=1}^{n}\frac{\rho_n(x)-\rho_n(\tau_\nu)}{(x-\tau_\nu)\pi_n'(\tau_\nu)}f(\tau_\nu)+R_n^*(f;x),\end{aligned}$$

as claimed. Since the interpolation error E_{n-1} vanishes for polynomials of degree $n-1$, so does the remainder R_n^* by (4.28).

Exercise 4.11. (Ex. 22 of [27, §4].) *In analogy to* Ex. 4.9, *develop an alternative and more stable form of* (4.26).

Answer. Using the relation (4.27) and expanding p_{n-1} similarly as in Ex. 4.9 yields

$$\begin{aligned}(Cf)(x;\mathrm{d}\lambda) &= \sum_{k=0}^{n-1} a_k' \rho_k(x) + R_n^*(f;x),\\ a_k' &= \frac{1}{\|\pi_k\|^2}\int_{\mathbb{R}} p_{n-1}(f;t)\pi_k(t)\mathrm{d}\lambda(t).\end{aligned} \tag{4.29}$$

Here,

$$\begin{aligned}a_k' &= \frac{1}{\|\pi_k\|^2}\int_{\mathbb{R}} \sum_{\nu=1}^{n} \frac{\pi_n(t)}{(t-\tau_\nu)\pi_n'(\tau_\nu)} f(\tau_\nu)\pi_k(t)\mathrm{d}\lambda(t)\\ &= \frac{1}{\|\pi_k\|^2}\sum_{\nu=1}^{n} \frac{f(\tau_\nu)}{\pi_n'(\tau_\nu)}\int_{\mathbb{R}} \frac{\pi_n(t)}{t-\tau_\nu}\pi_k(t)\mathrm{d}\lambda(t), \quad k<n.\end{aligned}$$

The integrand on the far right is a polynomial of degree $\le 2n-2$; hence the integral can be evaluated by n-point Gaussian quadrature. The resulting quadrature sum consists of only one term corresponding to the Gauss node τ_ν, so that

$$\int_{\mathbb{R}} \frac{\pi_n(t)}{t-\tau_\nu}\pi_k(t)\mathrm{d}\lambda(t) = \lambda_\nu \pi_n'(\tau_\nu)\pi_k(\tau_\nu).$$

Therefore,

$$a_k' = \frac{1}{\|\pi_k\|^2}\sum_{\nu=1}^{n} \lambda_\nu f(\tau_\nu)\pi_k(\tau_\nu), \quad k<n,$$

and hence $a_k' = a_k$, with a_k as in the first equation of (4.22).

4.3 ▪ Gauss–Radau and Gauss–Lobatto quadrature

Exercise 4.12. (Ex. 6 of [27, §4].) The $(n+1)$-point left- and right-hand *Gauss-Radau rule* is given by (cf. [24, §3.1.1.2])

$$\int_a^\infty f(t)\mathrm{d}\lambda(t) = \lambda_0^a f(a) + \sum_{\nu=1}^{n} \lambda_\nu^a f(\tau_\nu^a) + R_n^a(f), \quad n\ge 1, \tag{4.30}$$

resp.,

$$\int_{-\infty}^b f(t)\mathrm{d}\lambda(t) = \sum_{\nu=1}^{n} \lambda_\nu^b f(\tau_\nu^b) + \lambda_{n+1}^b f(b) + R_n^b(f), \quad n\ge 1, \tag{4.31}$$

where in both cases the remainder term vanishes whenever f is a polynomial of degree $d=2n$. For the left-hand formula, the nodes τ_ν^a can be characterized as zeros of the orthogonal polynomial $\pi_n(\cdot;(t-a)d\lambda)$ and the Jacobi–Radau matrix as

$$J_{n+1}^{R,a}(d\lambda)=\begin{bmatrix} J_n(d\lambda) & \sqrt{\beta_n}e_n \\ \sqrt{\beta_n}e_n^T & \alpha_n^R \end{bmatrix}, \tag{4.32}$$

where

$$e_n^T=[0,0,\dots,1]\in\mathbb{R}^n,\quad \alpha_n^R=a-\beta_n\frac{\pi_{n-1}(a)}{\pi_n(a)}.$$

Obtain the analogous characterizations for the nodes τ_ν^b *in* (4.31) *and the corresponding Jacobi–Radau matrix.*

Answer. For any quadrature rule of the type

$$\int_{\mathbb{R}} f(t)d\lambda(t)=\sum_{\nu=1}^{n}\lambda_\nu f(\tau_\nu)+R_n(f), \tag{4.33}$$

the following theorem is well known (cf. [24, Theorem 1.45]): *Given an integer* k *with* $0\le k\le n$, *the quadrature rule* (4.33) *has degree of exactness* $d=n-1+k$ *if and only if both of the following conditions are satisfied*:

(i) *Formula* (4.33) *is interpolatory.*

(ii) *The node polynomial* $\omega_n(t)=\prod_{\nu=1}^n(t-\tau_\nu)$ *satisfies*

$$\int_{\mathbb{R}}\omega_n(t)p(t)d\lambda(t)=0 \text{ for all } p\in\mathbb{P}_{k-1}.$$

Applying this to (4.31) (replace n by $n+1$ in the theorem above and take $k=n$), it follows that (4.31) must be interpolatory and the node polynomial

$$\omega_{n+1}(t)=(t-b)\prod_{\nu=1}^{n}(t-\tau_\nu^b)$$

must be orthogonal to all polynomials of degree $\le n-1$. If we let $\pi_n(t)=\prod_{\nu=1}^n(t-\tau_\nu^b)$, this implies

$$\int_{-\infty}^{b}\pi_n(t)p(t)(b-t)d\lambda(t)=0\quad\text{for all } p\in\mathbb{P}_{n-1},$$

and characterizes the τ_ν^b as the zeros of $\pi_n(\cdot;(b-t)d\lambda(t))$.

The relevant Jacobi–Radau matrix is obtained from (4.32) by replacing a in each occurrence by b (cf. [24, Remark (b) to Theorem 3.2]), that is,

$$J_{n+1}^{R,b}(d\lambda)=\begin{bmatrix} J_n(d\lambda) & \sqrt{\beta_n}e_n \\ \sqrt{\beta_n}e_n^T & \alpha_n^R \end{bmatrix},$$

where

$$e_n^T=[0,0,\dots,1]\in\mathbb{R}^n,\quad \alpha_n^R=b-\beta_n\frac{\pi_{n-1}(b)}{\pi_n(b)}.$$

Exercise 4.13. (Ex. 7 of [27, §4].) *Show that the remainders $R_n^a(f)$, $R_n^b(f)$ in* (4.30), *resp.,* (4.31), *if $f \in C^{2n+1}(I)$, $I = \mathrm{supp}(\mathrm{d}\lambda)$, have the property*

$$R_n^a(f) > 0, \quad R_n^b(f) < 0 \quad \text{if } \operatorname{sgn} f^{(2n+1)} = 1 \text{ on } I, \tag{4.34}$$

with the inequalities reversed if $\operatorname{sgn} f^{(2n+1)} = -1$.

Answer. The formula (4.30) is interpolatory in the sense of being the result of integrating the Hermite-type interpolation formula in which a is a point of multiplicity 1 and all the τ_ν^a points of multiplicity 2. The remainder term $R_n^a(f)$ therefore is the integral of the error term in the interpolation formula, that is,

$$R_n^a(f) = \frac{1}{(2n+1)!} \int_a^\infty (t-a)\pi_n^2(t) f^{(2n+1)}(t)\mathrm{d}\lambda(t),$$

where $\pi_n(t) = \prod_{\nu=1}^n (t - \tau_\nu^a)$. From this, it follows immediately that $R_n^a(f) \gtrless 0$ if $f^{(2n+1)}(t) \gtrless 0$ on $[a, \infty]$. Similarly,

$$R_n^b(f) = \frac{1}{(2n+1)!} \int_{-\infty}^b (t-b)\pi_n^2(t) f^{(2n+1)}(t)\mathrm{d}\lambda(t),$$

implying $R_n^b(f) \lessgtr 0$ if $f^{(2n+1)}(t) \gtrless 0$ on $[-\infty, b]$.

Exercise 4.14. (Ex. 8 of [27, §4].) The quantity α_n^R in the Jacobi–(left-hand)Radau matrix (4.32) is explicitly known for the Jacobi measure $\mathrm{d}\lambda^{(\alpha,\beta)}(t) = (1-t)^\alpha(1+t)^\beta \mathrm{d}t$ on $[-1, 1]$, $\alpha > -1$, $\beta > -1$, and for the generalized Laguerre measure $\mathrm{d}\lambda^{(\alpha)}(t) = t^\alpha e^{-t}\mathrm{d}t$ on $[0, \infty]$, $\alpha > -1$.

(a) *For the Jacobi measure, prove that*

$$\alpha_n^R = -1 + \frac{2n(n+\alpha)}{(2n+\alpha+\beta)(2n+\alpha+\beta+1)}.$$

{Hint: Use the relation between the Jacobi polynomials $P_k = P_k^{(\alpha,\beta)}$ customarily defined and the monic Jacobi polynomials $\pi_k = \pi_k^{(\alpha,\beta)}$, expressed by $P_k(t) = 2^{-k}\binom{2k+\alpha+\beta}{k}\pi_k(t)$. You also need $P_k(-1) = (-1)^k\binom{k+\beta}{k}$ and the β-coefficient for monic Jacobi polynomials, $\beta_n^J = 4n(n+\alpha)(n+\beta)(n+\alpha+\beta)/((2n+\alpha+\beta)^2(2n+\alpha+\beta+1)(2n+\alpha+\beta-1))$, $n \geq 1$.}

(b) *For the generalized Laguerre measure, prove that*

$$\alpha_n^R = n.$$

{Hint: With $\pi_k^{(\alpha)}$ and $L_k^{(\alpha)}$ denoting the monic, resp., conventional, generalized Laguerre polynomials, use $L_k^{(\alpha)}(t) = ((-1)^k/k!)\pi_k^{(\alpha)}(t)$. You also need $L_k^{(\alpha)}(0) = \binom{k+\alpha}{k}$, and $\beta_n^{\mathrm{L}} = n(n+\alpha)$ for monic generalized Laguerre polynomials.}

Answer

(a) Since, from the information given in the exercise,

$$P_k(t) = 2^{-k}\binom{2k+\alpha+\beta}{k}\pi_k(t) = 2^{-k}\frac{\Gamma(2k+\alpha+\beta+1)}{\Gamma(k+\alpha+\beta+1)k!}\pi_k(t),$$

and using $\Gamma(x+1)=x\Gamma(x)$, we have

$$\begin{aligned}\frac{P_{n-1}(t)}{P_n(t)}&=\frac{2\Gamma(2n+\alpha+\beta-1)\Gamma(n+\alpha+\beta+1)n!}{\Gamma(n+\alpha+\beta)(n-1)!\Gamma(2n+\alpha+\beta+1)}\,\frac{\pi_{n-1}(t)}{\pi_n(t)}\\&=\frac{2n(n+\alpha+\beta)}{(2n+\alpha+\beta)(2n+\alpha+\beta-1)}\,\frac{\pi_{n-1}(t)}{\pi_n(t)},\end{aligned}$$

that is,

$$\frac{\pi_{n-1}(t)}{\pi_n(t)}=\frac{(2n+\alpha+\beta)(2n+\alpha+\beta-1)}{2n(n+\alpha+\beta)}\,\frac{P_{n-1}(t)}{P_n(t)}.$$

For $t=-1$, since by the information given,

$$P_k(-1)=(-1)^k\binom{k+\beta}{k}=(-1)^k\frac{\Gamma(k+\beta+1)}{\Gamma(\beta+1)k!},$$

this gives

$$\begin{aligned}\frac{\pi_{n-1}(-1)}{\pi_n(-1)}&=\frac{(2n+\alpha+\beta)(2n+\alpha+\beta-1)}{2n(n+\alpha+\beta)}(-1)\frac{\Gamma(n+\beta)\Gamma(\beta+1)n!}{\Gamma(\beta+1)(n-1)!\Gamma(n+\beta+1)}\\&=-\frac{(2n+\alpha+\beta)(2n+\alpha+\beta-1)}{2n(n+\alpha+\beta)}\,\frac{n}{n+\beta}\\&=-\frac{(2n+\alpha+\beta)(2n+\alpha+\beta-1)}{2(n+\beta)(n+\alpha+\beta)}.\end{aligned}$$

By the hint we have

$$\beta_n=\frac{4n(n+\alpha)(n+\beta)(n+\alpha+\beta)}{(2n+\alpha+\beta)^2(2n+\alpha+\beta+1)(2n+\alpha+\beta-1)},\quad n\geq 1,$$

so that, finally, by (4.32),

$$\begin{aligned}\alpha_n^R&=-1+\frac{4n(n+\alpha)(n+\beta)(n+\alpha+\beta)\cdot(2n+\alpha+\beta)(2n+\alpha+\beta-1)}{(2n+\alpha+\beta)^2(2n+\alpha+\beta+1)(2n+\alpha+\beta-1)\cdot 2(n+\beta)(n+\alpha+\beta)}\\&=-1+\frac{2n(n+\alpha)}{(2n+\alpha+\beta)(2n+\alpha+\beta+1)},\end{aligned}$$

as claimed.

(b) We now have

$$L_k^{(\alpha)}(t)=\frac{(-1)^k}{k!}\,\pi_k^{(\alpha)}(t),$$

so that

$$\frac{L_{n-1}^{(\alpha)}(t)}{L_n^{(\alpha)}(t)}=-n\,\frac{\pi_{n-1}^{(\alpha)}(t)}{\pi_n^{(\alpha)}(t)},$$

that is,

$$\frac{\pi_{n-1}^{(\alpha)}(t)}{\pi_n^{(\alpha)}(t)}=-\frac{1}{n}\,\frac{L_{n-1}^{(\alpha)}(t)}{L_n^{(\alpha)}(t)}.$$

Using, as given in the exercise,

$$L_k^{(\alpha)}(0) = \binom{k+\alpha}{k} = \frac{\Gamma(k+\alpha+1)}{\Gamma(\alpha+1)k!},$$

we get

$$\frac{\pi_{n-1}^{(\alpha)}(0)}{\pi_n^{(\alpha)}(0)} = -\frac{1}{n}\,\frac{\Gamma(n+\alpha)\cdot\Gamma(\alpha+1)n!}{\Gamma(\alpha+1)(n-1)!\cdot\Gamma(n+\alpha+1)} = -\frac{1}{n+\alpha}.$$

By [24, p. 29], we have $\beta_n = n(n+\alpha)$ for $n \geq 1$, so that, by (4.32),

$$\alpha_n^R = 0 - \beta_n\,\frac{\pi_{n-1}^{(\alpha)}(0)}{\pi_n^{(\alpha)}(0)} = n,$$

as claimed.

Exercise 4.15. (Ex. 9 of [27, §4].) The $(n+2)$-point *Gauss-Lobatto formula* is given by (cf. [24, §§1.4.2 and 3.1.1.3])

$$\int_a^b f(t)\mathrm{d}\lambda(t) = \lambda_0^L f(\tau_0^L) + \sum_{\nu=1}^n \lambda_\nu^L f(\tau_\nu^L) + \lambda_{n+1}^L f(\tau_{n+1}^L) + R_n^{a,b}(f),$$

$$-\infty < a < b < \infty,$$

where the remainder term vanishes whenever f is a polynomial of degree $d = 2n+1$. The internal nodes τ_ν^L can be characterized as zeros of the orthogonal polynomial $\pi_n(\cdot;(t-a)(b-t)\mathrm{d}\lambda)$ and the Jacobi–Lobatto matrix as

$$J_{n+2}^L(\mathrm{d}\lambda) = \begin{bmatrix} J_{n+1}(\mathrm{d}\lambda) & \sqrt{\beta_{n+1}^L}\,e_{n+1} \\ \sqrt{\beta_{n+1}^L}\,e_{n+1}^{\mathrm{T}} & \alpha_{n+1}^L \end{bmatrix}, \quad e_{n+1}^{\mathrm{T}} = [0,0,\dots,1] \in \mathbb{R}^{n+1}, \tag{4.35}$$

where α_{n+1}^L, β_{n+1}^L solve the 2×2 linear system

$$\begin{bmatrix} \pi_{n+1}(a) & \pi_n(a) \\ \pi_{n+1}(b) & \pi_n(b) \end{bmatrix} \begin{bmatrix} \alpha_{n+1}^L \\ \beta_{n+1}^L \end{bmatrix} = \begin{bmatrix} a\pi_{n+1}(a) \\ b\pi_{n+1}(b) \end{bmatrix}, \quad \pi_m(\cdot) = \pi_m(\cdot;\mathrm{d}\lambda). \tag{4.36}$$

The quantities α_{n+1}^L, β_{n+1}^L are explicitly known for the Jacobi measure $\mathrm{d}\lambda^{(\alpha,\beta)}(t) = (1-t)^\alpha(1+t)^\beta \mathrm{d}t$ on $[-1,1]$, $\alpha > -1$, $\beta > -1$ (see [24, Example 3.8]; the second formula, as reproduced in [27, Eq. (4.16)], contains a misprint: the square is missing in the second factor of the denominator)

$$\begin{aligned} \alpha_{n+1}^L &= \frac{\alpha-\beta}{2n+\alpha+\beta+2}, \\ \beta_{n+1}^L &= 4\,\frac{(n+\alpha+1)(n+\beta+1)(n+\alpha+\beta+1)}{(2n+\alpha+\beta+1)(2n+\alpha+\beta+2)^2}. \end{aligned} \tag{4.37}$$

Prove (4.37). {Hint: With notation as in the hint to Ex. 4.14(a), use $P_k(1) = \binom{k+\alpha}{k}$ in addition to the information provided there.}

Answer. The α^L_{n+1}, β^L_{n+1} are the solution of the 2×2 system (4.36), where $\pi_k(t) = \pi_k^{(\alpha,\beta)}(t)$ are the monic Jacobi polynomials and $a=-1$, $b=1$. From the information given in Ex. 4.14(a) and the present Ex. 4.15, we have

$$\pi_k(\pm 1)=\gamma_k P_k(\pm 1),\quad \gamma_k=\frac{2^k}{\binom{2k+\alpha+\beta}{k}};$$
$$P_k(-1)=(-1)^k\binom{k+\beta}{k},\quad P_k(+1)=\binom{k+\alpha}{k}.$$

The determinant Δ_n of the 2×2 system is

$$\Delta_n=\gamma_n\gamma_{n+1}D_n,$$

where

$$D_n=\begin{vmatrix} P_{n+1}(-1) & P_n(-1)\\ P_{n+1}(1) & P_n(1)\end{vmatrix}=(-1)^{n+1}\begin{vmatrix}\binom{n+\beta+1}{n+1} & -\binom{n+\beta}{n}\\ \binom{n+\alpha+1}{n+1} & \binom{n+\alpha}{n}\end{vmatrix}.$$

Since

$$\binom{n+\beta+1}{n+1}=\frac{n+\beta+1}{n+1}\binom{n+\beta}{n},\quad \binom{n+\alpha+1}{n+1}=\frac{n+\alpha+1}{n+1}\binom{n+\alpha}{n},$$

we get

$$\begin{aligned}D_n&=\frac{(-1)^{n+1}}{n+1}\binom{n+\alpha}{n}\binom{b+\beta}{n}\begin{vmatrix} n+\beta+1 & -1\\ n+\alpha+1 & 1\end{vmatrix}\\ &=\frac{(-1)^{n+1}}{n+1}(2n+\alpha+\beta+2)\binom{n+\alpha}{n}\binom{n+\beta}{n}.\end{aligned}$$

Writing the solution of the 2×2 system in the form

$$\begin{aligned}\alpha^L_{n+1}&=\frac{1}{\Delta_n}\begin{vmatrix} -\pi_{n+1}(-1) & \pi_n(-1)\\ \pi_{n+1}(1) & \pi_n(1)\end{vmatrix}=\frac{1}{D_n}\begin{vmatrix} -P_{n+1}(-1) & P_n(-1)\\ P_{n+1}(1) & P_n(1)\end{vmatrix}\\ &=-\frac{1}{D_n}[P_{n+1}(-1)P_n(1)+P_{n+1}(1)P_n(-1)],\end{aligned}$$

$$\beta^L_{n+1}=\frac{1}{\Delta_n}\begin{vmatrix} \pi_{n+1}(-1) & -\pi_{n+1}(-1)\\ \pi_{n+1}(1) & \pi_{n+1}(1)\end{vmatrix}=\frac{2\gamma_{n+1}}{\gamma_n D_n}P_{n+1}(-1)P_{n+1}(1),$$

and inserting the explicit expressions for the quantities appearing on the right-hand side yields (4.37).

Exercise 4.16. (Ex. 10 of [27, §4].) The (left-hand) *generalized Gauss–Radau formula* is

$$\int_a^\infty f(t)\,d\lambda(t)=\sum_{\rho=0}^{r-1}\lambda_0^{(\rho)}f^{(\rho)}(a)+\sum_{\nu=1}^{n}\lambda_\nu^R f(\tau_\nu^R)+R^R_{n,r}(f),$$

where $r>1$ is the multiplicity of the endpoint $\tau_0^R=a$, and $R_{n,r}^R(f)=0$ for $f\in\mathbb{P}_{2n-1+r}$. Let $d\lambda^{[r]}(t)=(t-a)^r d\lambda(t)$ and $\tau_\nu^{[r]}$, $\lambda_\nu^{[r]}$, $\nu=1,2,\dots,n$, be the nodes and weights of the n-point Gauss formula for $d\lambda^{[r]}$.

(a) *Show that*

$$\tau_\nu^R=\tau_\nu^{[r]},\quad \lambda_\nu^R=\frac{\lambda_\nu^{[r]}}{(\tau_\nu^R-a)^r},\quad \nu=1,2,\dots,n.$$

(b) *Show that not only the internal weights λ_ν^R are all positive (why?), but also the boundary weights λ_0, λ_0' if $r=2$.*

Answer

(a) Let $g\in\mathbb{P}_{2n-1}$. Then, since $g(t)(t-a)^r\in\mathbb{P}_{2n-1+r}$, the generalized Gauss–Radau formula applied to $f(t)=g(t)(t-a)^r$ has zero remainder term, that is,

$$\int_a^\infty g(t)(t-a)^r d\lambda(t)=\sum_{\nu=1}^n \lambda_\nu^R(\tau_\nu^R-a)^r g(\tau_\nu^R),\quad g\in\mathbb{P}_{2n-1}.$$

Since the left-hand side is $\int_a^\infty g(t)d\lambda^{[r]}(t)$, the right-hand side must be the Gauss quadrature sum for $d\lambda^{[r]}$, and therefore $\tau_\nu^R=\tau_\nu^{[r]}$ and $\lambda_\nu^R(\tau_\nu^R-a)^r=\lambda_\nu^{[r]}$ as claimed.

(b) The positivity of λ_ν^R follows from the result of (a), since $\lambda_\nu^{[r]}>0$ as a Gaussian weight, and $\tau_\nu^R>a$. If $r=2$ and

$$\int_a^\infty f(t)d\lambda(t)=\lambda_0 f(a)+\lambda_0' f'(a)+\sum_{\nu=1}^n \lambda_\nu^R f(\tau_\nu^R),\quad f\in\mathbb{P}_{2n+1},$$

letting in turn $f(t)=\pi_n^2(t)$, $f(t)=(t-a)\pi_n^2(t)$, where $\pi_n(t)=\prod_{\nu=1}^n(t-\tau_\nu^{[r]})$, we get

$$\pi_n^2(a)\lambda_0+2\pi_n(a)\pi_n'(a)\lambda_0'=\int_a^\infty \pi_n^2(a)d\lambda(t),\quad \pi_n^2(a)\lambda_0'=\int_a^\infty (t-a)\pi_n^2(t)d\lambda(t).$$

Since the right-hand integrals are positive, there follows $\lambda_0'>0$ and $\lambda_0>0$, the latter since $\pi_n(a)>0$ and $\pi_n'(a)<0$ if n is even, and $\pi_n(a)<0$ and $\pi_n'(a)>0$ if n is odd, that is, $2\pi_n(a)\pi_n'(a)<0$ in either case.

Exercise 4.17. (Ex. 11 of [27, §4].) The *generalized Gauss–Lobatto formula* is

$$\int_a^b f(t)d\lambda(t)=\sum_{\rho=0}^{r-1}\lambda_0^{(\rho)}f^{(\rho)}(a)+\sum_{\nu=1}^n\lambda_\nu^L f(\tau_\nu^L)+\sum_{\rho=0}^{r-1}(-1)^\rho\lambda_{n+1}^{(\rho)}f^{(\rho)}(b)+R_{n,r}^L(f),$$

where $r>1$ is the multiplicity of the endpoints $\tau_0^L=a$, $\tau_{n+1}^L=b$, and $R_{n,r}^L(f)=0$ for $\mathbb{P}_{2n-1+2r}$. Let $d\lambda^{[r]}(t)=[(t-a)(b-t)]^r d\lambda(t)$ and $\tau_\nu^{[r]}$, $\lambda_\nu^{[r]}$, $\nu=1,2,\dots,n$, be the nodes and weights of the n-point Gauss formula for $d\lambda^{[r]}$.

(a) *Show that*

$$\tau_\nu^L=\tau_\nu^{[r]},\quad \lambda_\nu^L=\frac{\lambda_\nu^{[r]}}{[(\tau_\nu^L-a)(b-\tau_\nu^L)]^r},\quad \nu=1,2,\dots,n.$$

(b) *Show that not only the internal weights λ_ν^L are all positive (why?), but also the boundary weights λ_0, λ_0' and λ_{n+1}, λ_{n+1}' if $r=2$.*

(c) *Show that* $\lambda_0^{(\rho)} = \lambda_{n+1}^{(\rho)}$, $\rho = 0, 1, \ldots, r-1$, *if the measure* $\mathrm{d}\lambda$ *is symmetric in the sense of* Ex. 2.12.

Answer

(a) The proof is similar to the one in Ex. 4.16. Let $g \in \mathbb{P}_{2n-1}$. Then, since $g(t)[(t-a)(b-t)]^r \in \mathbb{P}_{2n-1+2r}$, the generalized Gauss–Lobatto formula applied to $f(t) = g(t)[(t-a)(b-t)]^r$ has zero remainder term, that is,

$$\int_a^b g(t)[(t-a)(b-t)]^r \mathrm{d}\lambda(t) = \sum_{\nu=1}^{n} \lambda_\nu^L [(\tau_\nu^L - a)(b - \tau_\nu^L)]^r g(\tau_\nu^L), \quad g \in \mathbb{R}_{2n-1}.$$

Since the left-hand side is $\int_a^b g(t)\mathrm{d}\lambda^{[r]}(t)$, the right-hand side must be the Gauss quadrature sum for $\mathrm{d}\lambda^{[r]}$, and therefore $\tau_\nu^L = \tau_\nu^{[r]}$, $\lambda_\nu^L[(\tau_\nu^L - a)(b-\tau_\nu^L)]^r = \lambda_\nu^{[r]}$, as claimed.

(b) The positivity of λ_ν^L follows from the result of (a), since $\lambda_\nu^{[r]} > 0$ as a Gaussian weight, and $a < \tau_\nu^L = \tau_\nu^{[r]} < b$. If $r = 2$ and

$$\int_a^b f(t)\mathrm{d}\lambda(t) = \lambda_0 f(a) + \lambda_0' f'(a) + \sum_{\nu=1}^{n} \lambda_\nu^L f(\tau_\nu^L) + \lambda_{n+1} f(b) - \lambda_{n+1}' f'(b), \quad f \in \mathbb{P}_{2n+3},$$

letting in turn $f(t) = \pi_n^2(t)(b-t)^2$ and $f(t) = (t-a)\pi_n^2(t)(b-t)^2$, where $\pi_n(t) = \prod_{\nu=1}^{n}(t - \tau_\nu^{[r]})$, we get

$$\pi_n^2(a)(b-a)^2\lambda_0 + 2[\pi_n(a)\pi_n'(a)(b-a)^2 - \pi_n^2(a)(b-a)]\lambda_0' = \int_a^b \pi_n^2(t)(b-t)^2 \mathrm{d}\lambda(t),$$

$$\pi_n^2(a)(b-a)^2\lambda_0' = \int_a^b (t-a)\pi_n^2(t)(b-t)^2\mathrm{d}\lambda(t),$$

Evidently, $\lambda_0' > 0$, and also $\lambda_0 > 0$ since the coefficient of λ_0' in the first equation is negative. An analogous argument applies for the λ_{n+1}' and λ_{n+1} if we in turn let $f(t) = \pi_n^2(t)(t-a)^2$ and $f(t) = (t-a)^2\pi_n^2(t)(b-t)$.

(c) If $\mathrm{d}\lambda(t) = w(t)\mathrm{d}t$ with w an even function on the symmetric interval $[-a, a]$, then

$$\int_{-a}^{a} f(t)w(t)\mathrm{d}t = \int_{-a}^{a} f(-t)w(t)\mathrm{d}t,$$

as follows readily by the substitution of variables $t \mapsto -\tau$. The generalized Gauss–Lobatto formula applied to the integral on the left must give the same result as when applied to the integral on the right. In particular, the boundary terms must be the same. Those for the integral on the left are

$$\lambda_0^{(\rho)} f^{(\rho)}(-a) \quad \text{and} \quad (-1)^\rho \lambda_{n+1}^{(\rho)} f^{(\rho)}(a),$$

and those for the integral on the right

$$(-1)^\rho \lambda_0^{(\rho)} f^{(\rho)}(a) \quad \text{and} \quad (-1)^\rho \lambda_{n+1}^{(\rho)} (-1)^\rho f^{(\rho)}(-a) = \lambda_{n+1}^{(\rho)} f^{(\rho)}(-a).$$

Comparison yields $\lambda_0^{(\rho)} = \lambda_{n+1}^{(\rho)}$.

With regard to the interior terms, comparison of them, if in both they are ordered with the nodes being increasing, yields $\tau_\nu^L = \tau_{n+1-\nu}^l$, $\lambda_\nu^L = \lambda_{n+1-\nu}^L$.

Exercise 4.18.* (Ex. 12* of [27, §4].) Generalized Gauss-Radau quadrature.

(a) *Write a MATLAB routine* `gradau.m` *for generating the generalized Gauss-Radau quadrature rule of* Ex. 4.16 *for a measure* $d\lambda$ *on* $[a,\infty]$, *having a fixed node* a *of multiplicity* r, $r>1$. {Hint: To compute the boundary weights, set up an (upper triangular) system of linear equations by applying the formula in turn with $\pi_n^2(t)$, $(t-a)\pi_n^2(t),\ldots,(t-a)^{r-1}\pi_n^2(t)$, where $\pi_n(t)=\prod_{\nu=1}^{n}(t-\tau_\nu^R)$.}

(b) *Check your routine against the known formulae with* $r=2$ *for the Legendre and Chebyshev measures* (see [24, Examples 3.10 and 3.11]). *Devise and implement a second check based on moments that works for arbitrary* $r\geq 2$ *and also for* $r=1$.

(c) *Use your routine to explore positivity of the boundary weights and see whether you can come up with any conjectures.*

Answer

(a) The (somewhat intricate) details of the suggested procedure for calculating the boundary weights can be found in [25], to which we refer the reader. The paper also makes reference to a MATLAB routine `gradau.m`, which implements the procedure and which is listed in Appendix A.2 and reproduced below. It calls on the `OPQ` routine `mod_linfac.m` implementing a modification algorithm for linear factors.

```
function xw=gradau(N,ab,r,endl)
%GRADAU Generalized Gauss-Radau quadrature formula.

%
% internal nodes and weights
%
ab1=ab;
for s=r:-1:1
  ab0=ab1;
  ab1=mod_linfac(N+s-1,ab0,endl);
end
xw0=gauss(N,ab1);
xw(r+1:N+r,:)=xw0;
xw(r+1:N+r,2)=xw(r+1:N+r,2)./((xw(r+1:N+r,1)-endl).^r);
xw(1:r,1)=endl*ones(r,1);
%
% external weights
%
rs=zeros(r-1,1);
A=zeros(r); b=zeros(r,1); x=zeros(r,1);
for rho=1:r-1
  rs(rho)=sum((xw0(:,1)-endl).^(-rho));
end
for i=r:-1:1
  A(i,i)=factorial(i-1)*prod((endl-xw0(:,1)).^2);
  if i<r
    for j=i+1:r
      A(i,j)=-2*sum(A(i+1:j,j).*rs(1:j-i))/(j-i);
    end
  end
end
```

```
%condA=cond(A)
ng=N+floor((r+1)/2);
xwg=gauss(ng,ab);
for i=1:r
  s=0;
  for ig=1:ng
    t=xwg(ig,1); w=xwg(ig,2);
    s=s+w*((t-endl)^(i-1)*prod((t-xw0(:,1)).^2));
  end
  b(i)=s;
end
x(r)=b(r)/A(r,r);
for i=r-1:-1:1
  x(i)=(b(i)-sum(A(i,i+1:r).*(x(i+1:r))'))/A(i,i);
end
xw(1:r,2)=x;
```

(b) The first test is carried out by the script `test_gradau_req2.m`, the second by the script `test_gradau_rgt0.m`.

```
%TEST_GRADAU_REQ2 tests gradau.m against known Gauss-Radau
%   formulae with double endpoint for the Legendre and
%   Chebyshev measures.

f1='%3.0f %10.2e %10.2e %10.2e %10.2e\n';
disp('          maximum errors in weights for r=2')
disp(' N   errinleg   errexleg   errincheb  errexcheb')
r=2; endl=-1;
for N=[10 20 40 80 160]
%
% Legendre measure
%
  ab=r_jacobi(N+r);
  xw=gradau(N,ab,r,endl);
  abex=r_jacobi(N,0,2);
  xwex=gauss(N,abex);
  xwex(:,2)=xwex(:,2)./((1+xwex(:,1)).^2);
  errinleg=max(abs((xw(3:N+2,2)-xwex(:,2))./xwex(:,2)));
  den=(N+1)^2*(N+2)^2;
  w1=8*(2*N^2+6*N+3)/(3*den);
  w2=8/den;
  err1=abs((xw(1,2)-w1)/w1);
  err2=abs((xw(2,2)-w2)/w2);
  errexleg=max([err1 err2]);
%
% Chebyshev measure
%
  ab=r_jacobi(N+r,-.5);
  xw=gradau(N,ab,r,endl);
  abex=r_jacobi(N,-.5,1.5);
  xwex=gauss(N,abex);
```

```
    xwex(:,2)=xwex(:,2)./((1+xwex(:,1)).^2);
    errincheb=max(abs((xw(3:N+2,2)-xwex(:,2))./xwex(:,2)));
    den=(N+1)*(2*N+1)*(2*N+3);
    w1=3*pi*(6*N^2+12*N+5)/(5*den);
    w2=3*pi/den;
    err1=abs((xw(1,2)-w1)/w1);
    err2=abs((xw(2,2)-w2)/w2);
    errexcheb=max([err1 err2]);
    fprintf(f1,N,errinleg,errexleg,errincheb,errexcheb)
  end

  %TEST_GRADAU_RGTO tests the accuracy to which the
  %   generalized Gauss-Radau formula computes the
  %   moments up to order 2N-1+r.

  f1='%2.0f %12.2e %12.2e %5.0f %7.0f\n';
  disp('     maximum errors in moments for r up to 10')
  disp(' N     maxerrleg    maxerrcheb rmaxleg rmaxcheb')
  endl=-1;
  for N=[1:5 10:5:40]
  %
  % Legendre measure
  %
    maxerrleg=0;
    for r=1:10
      mom=zeros(2*N+r,1); errmom=zeros(2*N+r,1);
      ab=r_jacobi(N+r);
      xw=gradau(N,ab,r,endl);
      for k=1:2:2*N+r
        mom(k)=2/k;
      end
      for k=1:2*N+r
        s=0;
        for rho=1:r
          s=s+xw(rho,2)*fpow(xw(rho,1),k-1,rho-1);
        end
        s=s+sum(xw(r+1:r+N,2).*xw(r+1:r+N,1).^(k-1));
        errmom(k)=abs(mom(k)-s);
      end
      merr=max(errmom);
      if merr>maxerrleg
        maxerrleg=merr;
        rmaxleg=r;
      end
    end
  %
  % Chebyshev measure
  %
    maxerrcheb=0;
    for r=2:10
```

```
        mom=zeros(2*N+r,1); errmom=zeros(2*N+r,1);
        ab=r_jacobi(N+r,-.5);
        xw=gradau(N,ab,r,endl);
        mom(1)=pi;
        for k=3:2:2*N+r
          mom(k)=(pi/(k-1))*factorial(k-2)/(2^(k-3)...
            *(factorial((k-3)/2))^2);
        end
        for k=1:2*N+r
          s=0;
          for rho=1:r
            s=s+xw(rho,2)*fpow(xw(rho,1),k-1,rho-1);
          end
          s=s+sum(xw(r+1:r+N,2).*xw(r+1:r+N,1).^(k-1));
          errmom(k)=abs(mom(k)-s);
        end
        merr=max(errmom);
        if merr>maxerrcheb
           maxerrcheb=merr;
           rmaxcheb=r;
        end
      end
      fprintf(f1,N,maxerrleg,maxerrcheb,rmaxleg,rmaxcheb)
    end
```

The first test routine, `test_gradau_req2.m`, checking against the known formulae for Legendre and Chebyshev weight functions in the case $r = 2$, produced results shown below, where the first of each pair of errors relates to the interior weights, and the second to the boundary weights. Clearly, the test is successful.

```
>> test_gradau_req2
          maximum errors in weights for r=2
 N   errinleg    errexleg    errincheb   errexcheb
 10  9.62e-015   9.56e-015   1.05e-014   1.36e-014
 20  9.28e-014   1.58e-014   6.08e-014   2.85e-014
 40  3.03e-013   1.44e-013   1.39e-013   5.35e-014
 80  4.51e-012   3.81e-013   1.40e-012   3.00e-013
160  6.06e-011   3.26e-012   8.26e-012   6.99e-013
>>
```

The second test routine, `test_gradau_rgt0.m`, calls on the routine `fpow.m` calculating the derivatives of the power function. The results, shown below, where the last two columns list the values of r for which the maximum errors were observed, are also quite satisfactory.

```
>> test_gradau_rgt0
     maximum errors in moments for r up to 10
 N     maxerrleg    maxerrcheb rmaxleg rmaxcheb
 1     1.73e-013     5.43e-014    10       10
 2     9.20e-014     1.44e-013    10       10
 3     3.79e-014     2.86e-014    10       10
```

```
    4    1.77e-014    1.38e-014    10      10
    5    6.00e-015    1.38e-014    10      10

   10    5.55e-015    2.22e-014     6      10
   15    1.21e-013    1.25e-013    10       7
   20    3.60e-013    3.68e-013    10       8
   25    3.25e-012    3.97e-012    10      10
   30    1.87e-011    2.18e-011    10      10
   35    4.96e-011    5.86e-011    10       9
   40    1.07e-010    3.90e-011    10       9
>>
```

Finally, the script test_gradau_radau.m checks gradau.m for $r = 1$ against radau.m,

```
%TEST_GRADAU_RADAU tests generalized Gauss-Radau for r=1
%   against Radau.
%

f1='%3.0f %10.2e %10.2e %10.2e %10.2e %10.2e %10.2e\n';
errxleg=zeros(100,1); errwleg=zeros(100,1);
errxcheb=zeros(100,1); errwcheb=zeros(100,1);
errxlag=zeros(100,1); errwlag=zeros(100,1);
disp('                              maximum errors for r=1')
disp('          Legendre                      Chebyshev                    Laguerre')
disp(' N    errx         errw         errx         errw         errx         errw')
for N=[1:5 10:5:40 60:20:100]
  r=1;
%
% Legendre measure
%
  endl=-1;
  ab=r_jacobi(N+r);
  xw=gradau(N,ab,r,endl);
  xw0=radau(N,ab,endl);
  errxleg(N)=max(abs(xw(:,1)-xw0(:,1)));
  errwleg(N)=max(abs((xw(:,2)-xw0(:,2))./xw(:,2)));
%
% Chebyshev measure
%
  ab=r_jacobi(N+r,-.5);
  xw=gradau(N,ab,r,endl);
  xw0=radau(N,ab,endl);
  errxcheb(N)=max(abs(xw(:,1)-xw0(:,1)));
  errwcheb(N)=max(abs((xw(:,2)-xw0(:,2))./xw(:,2)));
%
% Laguerre measure
%
  endl=0;
  ab=r_laguerre(N+r);
  xw=gradau(N,ab,r,endl);
  xw0=radau(N,ab,endl);
```

```
  errxlag(N)=max(abs(xw(:,1)-xw0(:,1)));
  errwlag(N)=max(abs((xw(:,2)-xw0(:,2))./xw(:,2)));
end
for N=[1:5 10:5:20 40:20:100]
  fprintf(f1,N,errxleg(N),errwleg(N),errxcheb(N),errwcheb(N),...
    errxlag(N),errwlag(N))
end
```

with results, shown below, that are also quite successful.[3]

```
>> test_gradau_radau
                        maximum errors for r=1
         Legendre                Chebyshev                Laguerre
 N      errx      errw        errx      errw        errx      errw
  1  5.55e-017  5.55e-016  1.11e-016  6.36e-016  0.00e+000  2.22e-016
  2  1.11e-016  6.50e-016  1.11e-016  1.59e-015  3.85e-016  1.83e-015
  3  3.33e-016  2.19e-015  2.22e-016  4.33e-015  2.66e-015  1.29e-015
  4  2.22e-016  1.62e-015  5.00e-016  2.23e-015  3.55e-015  3.89e-015
  5  2.22e-016  6.49e-015  2.22e-016  4.28e-015  3.55e-015  3.50e-015
 10  4.44e-016  1.09e-014  4.58e-016  2.34e-014  1.42e-014  5.95e-015
 15  5.55e-016  1.97e-014  6.66e-016  2.22e-014  2.13e-014  6.62e-014
 20  8.88e-016  1.88e-013  7.77e-016  4.38e-014  4.26e-014  9.98e-014
 40  7.77e-016  3.95e-013  6.66e-016  2.64e-013  1.42e-013  4.68e-013
 60  7.77e-016  9.17e-013  8.88e-016  1.08e-012  1.71e-013  1.24e-012
 80  1.22e-015  8.70e-013  9.99e-016  1.84e-012  4.55e-013  4.09e-013
100  1.55e-015  3.32e-012  9.71e-016  1.90e-012  5.68e-013  5.30e-013
>>
```

(c) Exploration is no longer needed, since in the meantime positivity of generalized Gauss–Radau formulae has been proved; see [44].

Exercise 4.19.* (Ex. 13* of [27, §4].) Generalized Gauss–Lobatto quadrature.

(a) *Write a MATLAB routine* `globatto.m` *for generating the generalized Gauss–Lobatto rule of* Ex. 4.17 *for a measure* $d\lambda$ *on* $[a,b]$, *having fixed nodes at a and b of multiplicity* r, $r>1$. *For simplicity, start with the case* $r\geq 2$ *even; then indicate the changes necessary to deal with odd values of* r. {Hint: Similar to the hint in Ex. 4.18(a).}

(b) *Check your routine against the known formulae with* $r=2$ *for the Legendre and Chebyshev measures* (see [24, Examples 3.13 and 3.14]). *Devise and implement a second check based on moments that works for arbitrary* $r\geq 2$ *and also for* $r=1$.

(c) *Explore the positivity of the boundary weights* $\lambda_0^{(\rho)}$ *and the quantities* $\lambda_{n+1}^{(\rho)}$ *in the quadrature formula.*

Answer

(a) For details, see the reference given in the answer to Ex. 4.18(a). The routine generating the generalized Gauss–Lobatto formula is `globatto.m` (cf. Appendix A.2),

```
function xw=globatto(N,ab,r,endl,endr)
%GLOBATTO Generalized Gauss-Lobatto quadrature formula.
```

[3]In MATLAB Release R2014a, the last four entries in the last column are of the order 10^3–10^4, whereas they are as shown here when using MATLAB Release R2011a. See Demo 1.17 for an explanation.

```
if r<1, error('r too small'), end
if N<3, error('N too small'), end
odd=2*floor(r/2)~=r;
%
% internal nodes and weights
%
ab1=ab;
for s=2*r:-1:r+1
  ab0=ab1;
  ab1=mod_linfac(N+s-1,ab0,endl);
end
for s=r:-1:1
  ab0=ab1;
  ab1=mod_linfac(N+s-1,ab0,endr);
end
ab1(1,2)=abs(ab1(1,2)); xw0=gauss(N,ab1);
xw(2*r+1:N+2*r,:)=xw0;
xw(2*r+1:N+2*r,2)=xw(2*r+1:N+2*r,2)./(((xw(2*r+1:N+2*r,1)...
  -endl).*(endr-xw(2*r+1:N+2*r,1))).^r);
xw(1:r,1)=endl*ones(r,1);
xw(r+1:2*r,1)=endr*ones(r,1);
%
% external weights
%
r0=floor((r+1)/2); rs=zeros(r-1,1); tau=zeros(N+r0,1);
tau(1:N)=xw0(:,1);
for e=1:2
  if e==1
    a=endl; b=endr; tau(N+1:N+r0)=b;
  else
    a=endr; b=endl; tau(N+1:N+r0)=b;
  end
  A=zeros(r); rhs=zeros(r,1); x=zeros(r,1);
  for rho=1:r-1
    rs(rho)=sum((tau-a).^(-rho));
    if odd, rs(rho)=rs(rho)-.5*(b-a)^(-rho); end
  end
  for i=r:-1:1
    A(i,i)=factorial(i-1)*prod((tau-a).^2);
    if odd, A(i,i)=A(i,i)/(b-a); end
    if i<r
      for j=i+1:r
        A(i,j)=-2*sum(A(i+1:j,j).*rs(1:j-i))/(j-i);
      end
    end
  end
  ng=N+r;
  xwg=gauss(ng,ab);
  for i=1:r
    s=0;
```

```
      for ig=1:ng
        t=xwg(ig,1); w=xwg(ig,2);
        p=prod((tau-t).^2);
        if odd, p=p/(b-t); end
        s=s+w*((t-a)^(i-1))*p;
      end
      rhs(i)=s;
    end
    x(r)=rhs(r)/A(r,r);
    for i=r-1:-1:1
      x(i)=(rhs(i)-sum(A(i,i+1:r).*(x(i+1:r))'))/A(i,i);
    end
    xw(1+(e-1)*r:e*r,2)=x;
  end
```

(b) The two tests are test_globatto_req2.m and test_globatto_rgt0.m.

```
%TEST_GLOBATTO_REQ2 tests globatto.m against known Gauss-
%   Lobatto formulae with double endpoints for the Legendre
%   and Chebyshev measures.
%

f1='%3.0f %10.2e %10.2e %10.2e %10.2e\n';
disp('            maximum errors in weights for r=2')
disp(' N    errinleg   errexleg   errincheb  errexcheb')
r=2; endl=-1; endr=1;
for N=[10 20 40 80 160]
%
% Legendre measure
%
  ab=r_jacobi(N+2*r);
  xw=globatto(N,ab,r,endl,endr);
  abex=r_jacobi(N,2);
  xwex=gauss(N,abex);
  xwex(:,2)=xwex(:,2)./((1-xwex(:,1).^2).^2);
  errinleg=max(abs((xw(5:N+4,2)-xwex(:,2))./xwex(:,2)));
  den=(N+1)*(N+2)*(N+3)*(N+4);
  w1=8*(2*N^2+10*N+9)/(3*den);
  w2=8/den;
  err1=abs((xw(1,2)-w1)/w1);
  err2=abs((xw(2,2)-w2)/w2);
  errexleg=max([err1 err2]);
%
% Chebyshev measure
%
  ab=r_jacobi(N+2*r,-.5);
  xw=globatto(N,ab,r,endl,endr);
  abex=r_jacobi(N,1.5);
  xwex=gauss(N,abex);
  xwex(:,2)=xwex(:,2)./((1-xwex(:,1).^2).^2);
  errincheb=max(abs((xw(5:N+4,2)-xwex(:,2))./xwex(:,2)));
```

```
    den=(N+1)*(N+2)*(N+3);
    w1=3*pi*(3*N^2+12*N+10)/(10*den);
    w2=3*pi/(4*den);
    err1=abs((xw(1,2)-w1)/w1);
    err2=abs((xw(2,2)-w2)/w2);
    errexcheb=max([err1 err2]);
    fprintf(f1,N,errinleg,errexleg,errincheb,errexcheb)
end

%TEST_GLOBATTO_RGTO tests the accuracy to which the
%   generalized Gauss-Lobatto formula computes the moments
%   up to order 2N-1+2r.

f1='%1.0f %12.2e %12.2e %6.0f %7.0f\n';
disp('     maximum errors in moments for r up to 10')
disp('N     maxerrleg   maxerrcheb rmaxleg rmaxcheb')
endl=-1; endr=1;
for N=3:7
%
% Legendre measure
%
  maxerrleg=0;
  for r=1:10
    mom=zeros(2*N+2*r,1); errmom=zeros(2*N+2*r,1);
    ab=r_jacobi(N+2*r);
    xw=globatto(N,ab,r,endl,endr);
    for k=1:2:2*N+2*r
      mom(k)=2/k;
    end
    for k=1:2*N+2*r
      s=0;
      for rho=1:r
        s=s+xw(rho,2)*fpow(xw(rho,1),k-1,rho-1);
        s=s+xw(r+rho,2)*fpow(xw(r+rho,1),k-1,rho-1);
      end
      s=s+sum(xw(2*r+1:2*r+N,2).*xw(2*r+1:2*r+N,1).^(k-1));
      errmom(k)=abs(mom(k)-s);
    end
    merr=max(errmom);
    if merr>maxerrleg
       maxerrleg=merr;
       rmaxleg=r;
    end
  end
%
% Chebyshev measure
%
  maxerrcheb=0;
  for r=1:10
    mom=zeros(2*N+2*r,1); errmom=zeros(2*N+2*r,1);
```

```
        ab=r_jacobi(N+2*r,-.5);
        xw=globatto(N,ab,r,endl,endr);
        mom(1)=pi;
        for k=3:2:2*N+2*r
          mom(k)=(pi/(k-1))*factorial(k-2)/(2^(k-3)...
            *(factorial((k-3)/2))^2);
        end
        for k=1:2*N+2*r
          s=0;
          for rho=1:r
            s=s+xw(rho,2)*fpow(xw(rho,1),k-1,rho-1);
            s=s+xw(r+rho,2)*fpow(xw(r+rho,1),k-1,rho-1);
          end
          s=s+sum(xw(2*r+1:2*r+N,2).*xw(2*r+1:2*r+N,1).^(k-1));
          errmom(k)=abs(mom(k)-s);
        end
        merr=max(errmom);
        if merr>maxerrcheb
           maxerrcheb=merr;
           rmaxcheb=r;
        end
      end
      fprintf(f1,N,maxerrleg,maxerrcheb,rmaxleg,rmaxcheb)
    end
```

The output of the first test, in the same format as in the answer to Ex. 4.18(b), is

```
>> test_globatto_req2
              maximum errors in weights for r=2
  N   errinleg    errexleg    errincheb   errexcheb
 10   7.66e-015   4.23e-015   6.65e-015   7.11e-015
 20   7.99e-014   3.68e-014   3.48e-014   3.12e-014
 40   5.21e-013   6.33e-014   3.91e-014   1.97e-013
 80   4.99e-013   2.32e-013   2.99e-013   3.85e-013
160   4.96e-012   1.13e-013   1.02e-012   5.53e-012
>>
```

while the output of the second is

```
>> test_globatto_rgt0
      maximum errors in moments for r up to 10
N     maxerrleg     maxerrcheb   rmaxleg  rmaxcheb
3     8.99e-015      8.10e-015      10        10
4     2.75e-015      1.15e-014      10        10
5     5.34e-015      5.33e-015      10         9
6     6.44e-015      7.11e-015      10         9
7     3.68e-015      8.41e-015       9        10
>>
```

both equally satisfactory.

Finally, the output of `test_globatto_lobatto.m`, checking the generalized Gauss–Lobatto rule, with $r = 1$, against the ordinary Gauss–Lobatto rule, is

```
>> test_globatto_lobatto
                      maximum errors for r=1
                 Legendre                    Chebyshev
  N       errx          errw          errx          errw
   3   1.11e-16   8.16e-16   2.22e-16   1.78e-15
   4   2.22e-16   1.03e-15   3.33e-16   1.77e-15
   5   2.37e-16   3.21e-15   5.00e-16   1.27e-15
  10   4.44e-16   6.36e-15   2.78e-16   9.52e-15
  15   8.88e-16   9.16e-15   5.55e-16   1.19e-14
  20   6.66e-16   1.89e-14   5.55e-16   4.10e-14
  40   8.88e-16   7.27e-14   8.33e-16   8.51e-14
  60   7.77e-16   2.60e-13   7.77e-16   1.42e-13
  80   1.22e-15   2.39e-13   9.99e-16   5.39e-13
 100   1.33e-15   2.64e-13   9.99e-16   1.97e-12
>>
```

(c) There is no need for exploration, since positivity has been proved in the reference cited in the answer to part (c) of Ex. 4.18.

4.4 ▪ Gauss–Kronrod quadrature

Exercise 4.20. (Ex. 14 of [27, §4].) *Show that the monic Stieltjes polynomial* ω_{n+1}^K *in* (4.3) *and* (4.4) *exists uniquely.*

Answer. The orthogonality in (4.4) can be expressed as

$$\int_{\mathbb{R}} \omega_{n+1}^K(t)\pi_\nu(t)\pi_n(t)\mathrm{d}\lambda(t)=0, \quad \nu=0,1,2,\ldots,n,$$

where $\pi_k(t)=\pi_k(t;\mathrm{d}\lambda)$. Since ω_{n+1}^K is assumed monic, it can be written in the form

$$\omega_{n+1}^K(t)=\sum_{\mu=0}^{n} c_\mu\pi_\mu(t)+\pi_{n+1}(t),$$

so that the orthogonality conditions above become

$$\int_{\mathbb{R}}\left(\sum_{\mu=0}^{n} c_\mu\pi_\mu(t)\right)\pi_\nu(t)\pi_n(t)\mathrm{d}\lambda(t)=-\int_{\mathbb{R}}\pi_{n+1}(t)\pi_\nu(t)\pi_n(t)\mathrm{d}\lambda(t),$$

which is a system of $n+1$ linear algebraic equations (for the coefficients c_μ) whose matrix is

$$A=[a_{\nu\mu}], \quad a_{\nu\mu}=\int_{\mathbb{R}}\pi_\nu(t)\pi_\mu(t)\pi_n(t)\mathrm{d}\lambda(t), \quad \nu,\mu=0,1,\ldots,n.$$

By orthogonality, $a_{\nu\mu}=0$ if $\nu+\mu<n$. Thus, nonzero elements can only occur when $\nu+\mu\geq n$, i.e., in the right lower triangle of the matrix. On the upward diagonal $\nu+\mu=n$, we have $\pi_\nu(t)\pi_\mu(t)=\pi_n(t)+$lower-degree terms, hence $a_{\nu\mu}=\int_{\mathbb{R}}\pi_n^2(t)\mathrm{d}\lambda(t)>0$. Therefore, the system is nonsingular and has a unique solution.

4.5 ▪ Gauss–Turán quadrature

Exercise 4.21. (Ex. 15 of [27].) (a) *Let* $\mathrm{d}\lambda$ *be a positive measure. Use approximation theory to show that the minimum of* $\int_{\mathbb{R}} |\pi(t)|^p \mathrm{d}\lambda(t)$, $1 < p < \infty$, *extended over all monic polynomials* π *of degree* n *is uniquely determined.*

(b) *Show that the minimizer of the extremal problem in* (a), *when* $p = 2s+2$, $s \geq 0$ *an integer, is the* s*-orthogonal polynomial* $\pi = \pi_{n,s}$. {Hint: Differentiate the integral in (a) partially with respect to the variable coefficients of π.}

Answer

(a) The problem

$$\int_{\mathbb{R}} |\pi(t)|^p \mathrm{d}\lambda(t) = \min \quad \text{over all } \pi \in \mathbb{P}_n^0,$$

where $\mathbb{P}_n^0$ is the class of monic polynomials of degree n, is equivalent to

$$\int_{\mathbb{R}} |t^n - p(t)|^p \mathrm{d}\lambda(t) = \min \quad \text{over all } p \in \mathbb{P}_{n-1}.$$

In the language of approximation theory, we are thus trying to approximate the power t^n best by a polynomial of degree $\leq n-1$ in the norm $\|u\| = \left(\int_{\mathbb{R}} |u|^p \mathrm{d}\lambda\right)^{1/p}$. It is known from approximation theory (see, e.g., [6, Ch. VII]) that this problem has a unique solution.

(b) In the case $p = 2s+2$, $s \geq 0$ an integer, following the hint, we consider

$$\int_{\mathbb{R}} [t^n - p(t)]^{2s+2} \mathrm{d}\lambda(t)$$

as a function of the coefficients c_k of $p(t) = c_{n-1}t^{n-1} + c_{n-2}t^{n-2} + \cdots + c_0$ and set the partial derivatives with respect to c_k equal to zero, giving

$$\int_{\mathbb{R}} [t^n - p(t)]^{2s+1} t^k \mathrm{d}\lambda(t) = 0, \quad k = 0, 1, \ldots, n-1.$$

This identifies $t^n - p(t) = \pi_{n,s}(t)$ as the s-orthogonal polynomial of degree n.

Exercise 4.22. (Ex. 16 of [27, §4].) A *Gauss–Turán quadrature formula*

$$\int_{\mathbb{R}} f(t)\mathrm{d}\lambda(t) = \sum_{\nu=1}^{n} [\lambda_\nu f(\tau_\nu) + \lambda'_\nu f'(\tau_\nu) + \cdots + \lambda_\nu^{(r-1)} f^{(r-1)}(\tau_\nu)] + R_n(f) \tag{4.38}$$

has degree of exactness $rn - 1 + k$, $k > 0$, if and only if it is interpolatory and

$$\int_{\mathbb{R}} \omega_n^r(t) p(t) \mathrm{d}\lambda(t) = 0 \quad \text{for all } p \in \mathbb{P}_{k-1}, \tag{4.39}$$

where $\omega_n(t) = \prod_{\nu=1}^n (t - \tau_\nu)$ is the node polynomial of (4.38) (cf. [24, §3.1.3]).

(a) *Show that* r *in* (4.39) *has to be odd.*

(b) *Show that in* (4.39) *with* $r = 2s+1$, $s \geq 0$, *one cannot have* $k > n$.

Answer

(a) If r were even, then

$$\int_{\mathbb{R}} \omega_n^r(t)\mathrm{d}\lambda(t) > 0, \quad r \text{ even},$$

so that ω_n^r cannot be orthogonal to a constant function, let alone to $\mathbb{P}_{k-1}$, $k > 1$.

(b) If we had

$$\int_{\mathbb{R}} \omega_n^{2s+1}(t)p(t)\mathrm{d}\lambda(t) = 0 \quad \text{for } p \in \mathbb{P}_{k-1} \text{ with } k > n,$$

then any $p \in \mathbb{P}_n$ would be admissible ($k = n+1$), in particular $p(t) = \omega_n(t)$. This leads to $\int_{\mathbb{R}} \omega_n^{2s+2}(t)\mathrm{d}\lambda(t) = 0$, which is not possible. Thus, necessarily, $k \le n$.

4.6 ▪ Polynomial/rational quadrature formulae of Gaussian type

Exercise 4.23. (Ex. 17 of [27, §4].) Generalized Fermi–Dirac integral

$$F_k(\eta, \theta) = \int_0^\infty \frac{t^k\sqrt{1+\theta t/2}}{e^{-\eta+t}+1}\mathrm{d}t.$$

Here, $\eta \in \mathbb{R}$, $\theta \ge 0$, and k is the Boltzmann constant ($= \frac{1}{2}, \frac{3}{2}$, or $\frac{5}{2}$). The integrand has poles at $t = \eta + \mu i\pi$, $\mu = \pm1, \pm3, \pm5, \ldots$. Let $2 \le m(\text{even}) \le 2n$ and

$$\omega_m(t) = \prod_{\mu=1}^{m}(1 + \zeta_\mu t),$$

where ζ_μ are the negative reciprocals of the first $m/2$ pairs of conjugate complex poles.

Show that

$$\omega_m(t) = \prod_{\nu=1}^{m/2}[(1+\xi_\nu t)^2 + \eta_\nu^2 t^2], \quad 2 \le m(\mathit{even}) \le 2n, \tag{4.40}$$

where

$$\xi_\nu = \frac{-\eta}{\eta^2 + (2\nu-1)^2\pi^2}, \quad \eta_\nu = \frac{(2\nu-1)\pi}{\eta^2+(2\nu-1)^2\pi^2}. \tag{4.41}$$

Answer. The negative reciprocals of the first $m/2$ pairs of conjugate complex poles are

$$\zeta_\nu = -\frac{1}{\eta + (2\nu-1)i\pi}, \quad \overline{\zeta}_\nu = -\frac{1}{\eta-(2\nu-1)i\pi}, \quad \nu = 1, 2, \ldots, m/2.$$

Therefore,

$$\omega_m(t) = \prod_{\nu=1}^{m/2}(1+\zeta_\nu t)(1+\overline{\zeta}_\nu t) = \prod_{\nu=1}^{m/2}|1+\zeta_\nu t|^2,$$

where

$$|1+\zeta_\nu t|^2 = \left|1 - \frac{t}{\eta+(2\nu-1)i\pi}\right|^2 = \left|1 - t\,\frac{\eta-(2\nu-1)i\pi}{\eta^2+(2\nu-1)^2\pi^2}\right|^2.$$

If we let

$$\xi_\nu = \frac{-\eta}{\eta^2+(2\nu-1)^2\pi^2}, \quad \eta_\nu = \frac{(2\nu-1)\pi}{\eta^2+(2\nu-1)^2\pi^2},$$

then

$$|1+\zeta_\nu t|^2 = |1+\xi_\nu t + i\eta_\nu t|^2 = (1+\xi_\nu t)^2 + \eta_\nu^2 t^2,$$

hence

$$\omega_m(t) = \prod_{\nu=1}^{m/2} [(1+\xi_\nu t)^2 + \eta_\nu^2 t^2].$$

(There is a misprint in [27, Eq. (4.33)]: η_ν should read η_ν^2.)

4.7 ▪ Quadrature estimates of matrix functionals

Exercise 4.24. (Ex. 23 of [27, §4].) Lanczos Algorithm.
Let $A \in \mathbb{R}^{N\times N}$ be a symmetric positive definite matrix of order N.

initialization:

$$\boldsymbol{h}_0 \text{ prescribed with } \|\boldsymbol{h}_0\| = 1, \quad \boldsymbol{h}_{-1} = 0.$$

continuation: for $j = 0, 1, \ldots, N-1$ do

$$\begin{aligned}
\alpha_j &= \boldsymbol{h}_j^{\mathrm{T}} A \boldsymbol{h}_j, \\
\tilde{\boldsymbol{h}}_{j+1} &= (A - \alpha_j I)\boldsymbol{h}_j - \gamma_j \boldsymbol{h}_{j-1}, \\
\gamma_{j+1} &= \|\tilde{\boldsymbol{h}}_{j+1}\|, \\
\boldsymbol{h}_{j+1} &= \tilde{\boldsymbol{h}}_{j+1}/\gamma_{j+1}.
\end{aligned}$$

The constant γ_0 is arbitrary, since it multiplies $\boldsymbol{h}_{-1} = 0$. We define it here to be $\gamma_0 = 1$.

(a) *Prove that the Lanczos vectors* $\boldsymbol{h}_0, \boldsymbol{h}_1, \ldots, \boldsymbol{h}_{N-1}$ *are mutually orthonormal.*

(b) *Show that the vectors* $\{\boldsymbol{h}_j\}_{j=0}^n$, $n < N$, *form an orthonormal basis of the Krylov space*

$$\mathcal{K}_n(A, \boldsymbol{h}_0) = \operatorname{span}(\boldsymbol{h}_0, A\boldsymbol{h}_0, \ldots, A^n \boldsymbol{h}_0).$$

(c) *Prove that*

$$\boldsymbol{h}_j = p_j(A)\boldsymbol{h}_0, \quad j = 0, 1, \ldots, N,$$

where p_j *is a polynomial of degree* j *satisfying the three-term recurrence relation*

$$\begin{gathered}
\gamma_{j+1} p_{j+1}(\lambda) = (\lambda - \alpha_j) p_j(\lambda) - \gamma_j p_{j-1}(\lambda), \quad j = 0, 1, \ldots, N-1, \\
p_0(\lambda) = 1, \quad p_{-1}(\lambda) = 0.
\end{gathered}$$

{Hint: Use mathematical induction.}

Answer

(a) All b_j, $j = 0, 1, \ldots, N-1$, are normalized, b_0 by assumption, and b_j, $j > 0$, by virtue of the last two lines of the Lanczos algorithm. We first show that b_1 is orthogonal to b_0. It suffices to show this for $\tilde{b}_1$. We have

$$b_0^T \tilde{b}_1 = b_0^T (A - \alpha_0 I) b_0 = b_0^T A b_0 - \alpha_0 b_0^T b_0 = b_0^T A b_0 - \alpha_0 = 0,$$

by the definition of α_0 in the first line of the continuation part of the algorithm. We proceed by induction. Suppose $b_0, b_1, \ldots, b_j$ are mutually orthogonal for some $j < N-1$. We show that $\tilde{b}_{j+1}$ is orthogonal to b_j and to all b_i with $0 \le i < j$. The first claim is proved similarly as above,

$$b_j^T \tilde{b}_{j+1} = b_j^T (A - \alpha_j I) b_j - \gamma_j b_j^T b_{j-1} = (b_j^T A b_j - \alpha_j) - \gamma_j b_j^T b_{j-1} = 0,$$

by the definition of α_j and the induction assumption. Furthermore, for $i = j-1$ we have that

$$\begin{aligned} b_i^T \tilde{b}_{j+1} &= b_{j-1}^T \tilde{b}_{j+1} = b_{j-1}^T [(A - \alpha_j I) b_j - \gamma_j b_{j-1}] \\ &= b_{j-1}^T A b_j - \alpha_j b_{j-1}^T b_j - \gamma_j b_{j-1}^T b_{j-1} \\ &= b_{j-1}^T A b_j - \gamma_j \end{aligned}$$

by the induction assumption. Using $u^T A v = (Au)^T v$ by the symmetry of A, we obtain

$$b_i^T \tilde{b}_{j+1} = (A b_{j-1})^T b_j - \gamma_j = \tilde{b}_j^T b_j + \alpha_{j-1} b_{j-1}^T b_j + \gamma_{j-1} b_{j-2}^T b_j - \gamma_j = \tilde{b}_j^T b_j - \gamma_j = 0,$$

using again the induction assumption and $\tilde{b}_j^T b_j = \gamma_j b_j^T b_j = \gamma_j$ from the last line of the algorithm. Furthermore, when $i < j-1$,

$$b_i^T \tilde{b}_{j+1} = b_i^T A b_j - \alpha_j b_i^T b_j - \gamma_j b_i^T b_{j-1} = (A b_i)^T b_j,$$

by the induction assumption. Since Ab_i is a linear combination of b_{i+1}, b_i, and b_{i-1}, where $i+1 \le j-1$, the last expression in the above chain of equalities is zero, again by the induction assumption. Thus, $b_0, b_1, \ldots, b_j, b_{j+1}$ are also mutually orthogonal, and, by induction, so are all the Lanczos vectors $b_0, b_1, \ldots, b_{N-1}$. (Note that $b_N = 0$, which is the only vector that can be orthogonal to all Lanczos vectors.)

(b) We have that Ab_0 is a linear combination of b_0 and b_1, $A^2 b_0 = A(Ab_0)$ is a linear combination of b_0, b_1, b_2, etc., that is, $\mathcal{K}_n(A, b_0) = \operatorname{span}(b_0, Ab_0, \ldots, A^n b_0) = \operatorname{span}(b_0, b_1, \ldots, b_n)$. Since by (a), $b_0, b_1, \ldots, b_n$ are mutually orthonormal, they form an orthonormal basis of $\mathcal{K}_n(A, b_0)$.

(c) Assume

$$b_j = p_j(A) b_0, \quad j = 0, 1, \ldots, n \ (n < N), \tag{4.42}$$

which is true for $n = 0$. Then

$$\gamma_{n+1} b_{n+1} = \tilde{b}_{n+1} = [(A - \alpha_n I) p_n(A) - \gamma_n p_{n-1}(A)] b_0 = \gamma_{n+1} p_{n+1}(A) b_0,$$

by the definition of the polynomials p_j. Thus,

$$b_{n+1} = p_{n+1}(A) b_0,$$

establishing the truth of (4.42) for $j = n+1$.

Exercise 4.25. (Ex. 24 of [27, §4].) *Let A be as in* Ex. 4.24 *and assume that $\boldsymbol{h}_0=\sum_{k=1}^N \rho_k \boldsymbol{v}_k$, $\sum_{k=1}^N \rho_k^2=1$, where $\boldsymbol{v}_k$ are normalized eigenvectors of A corresponding to the eigenvalues $\lambda_1,\lambda_2,\dots,\lambda_N$. Prove that the polynomial p_k of* Ex. 4.24(c) *is then equal to the orthonormal polynomial $\tilde{\pi}_k(\,\cdot\,;\mathrm{d}\rho_N)$, where $\mathrm{d}\rho_N$ is the discrete measure having as abscissae the eigenvalues $\lambda_1,\lambda_2,\dots,\lambda_N$ and as weights $\rho_1^2,\rho_2^2,\dots,\rho_N^2$.* {Hint: Use the spectral resolution of A and Exs. 4.24(a) and (c).}

Answer. Let

$$AV=V\Lambda,\quad \Lambda=\mathrm{diag}(\lambda_1,\lambda_2,\dots,\lambda_N),\quad V=[\boldsymbol{v}_1,\boldsymbol{v}_2,\dots,\boldsymbol{v}_N], \tag{4.43}$$

be the spectral resolution of A. Then

$$p_n(\Lambda)=p_n(V^{\mathrm{T}}AV)=V^{\mathrm{T}}p_n(A)V;$$

hence, by Ex. 4.24(c),

$$\boldsymbol{h}_n=p_n(A)\boldsymbol{h}_0=V p_n(\Lambda)V^{\mathrm{T}}\boldsymbol{h}_0.$$

Orthonormality $\boldsymbol{h}_n^{\mathrm{T}}\boldsymbol{h}_m=\delta_{nm}$ of the Lanczos vectors $\boldsymbol{h}_j$ then yields

$$\boldsymbol{h}_0^{\mathrm{T}}V p_n(\Lambda)V^{\mathrm{T}}V p_m(\Lambda)V^{\mathrm{T}}\boldsymbol{h}_0=\boldsymbol{h}_0^{\mathrm{T}}V p_n(\Lambda)p_m(\Lambda)V^{\mathrm{T}}\boldsymbol{h}_0=\delta_{nm}. \tag{4.44}$$

Since, by assumption,

$$\boldsymbol{h}_0=\sum_{k=1}^N \rho_k\boldsymbol{v}_k,$$

we have, by the orthonormality of the vectors $\boldsymbol{v}_k$,

$$V^{\mathrm{T}}\boldsymbol{h}_0=\sum_{k=1}^N \rho_k\boldsymbol{e}_k,$$

where $\boldsymbol{e}_k$ is the kth coordinate vector, and therefore, by (4.44),

$$\begin{aligned}&\sum_{k,\ell=1}^N \rho_k\boldsymbol{e}_k^{\mathrm{T}}\mathrm{diag}(p_n(\lambda_1)p_m(\lambda_1),\dots,p_n(\lambda_N)p_m(\lambda_N))\rho_\ell\boldsymbol{e}_\ell\\ &=\sum_{k,\ell=1}^N \rho_k\rho_\ell\delta_{k\ell}p_n(\lambda_\ell)p_m(\lambda_\ell)=\sum_{k=1}^N \rho_k^2 p_n(\lambda_k)p_m(\lambda_k)=\delta_{nm},\end{aligned}$$

as claimed.

Exercise 4.26. (a) *Write a MATLAB routine* `lanczos_algorithm.m` *implementing the algorithm of* Ex. 4.24.

(b) *Write a driver script* `test_lanczos.m` *that tests the routine of* (a) *on the $N\times N$ matrix $A=\boldsymbol{L}^{\mathrm{T}}\boldsymbol{L}$, where $\boldsymbol{L}$ is the lower triangular matrix with all 1's on the diagonal and below. With $V^{\mathrm{T}}AV=\Lambda$ being the spectral resolution of A, and $V=[\boldsymbol{v}_1,\boldsymbol{v}_2,\dots,\boldsymbol{v}_N]$, select $\boldsymbol{h}_0=(1/\sqrt{N})\sum_{k=1}^N \boldsymbol{v}_k$, i.e., $\rho_k=1/\sqrt{N}$ for $k=1,2,\dots,N$* (cf. Ex. 4.25). *For N=*[5 10 20 40], *verify the orthogonality of the Lanczos vectors $\boldsymbol{h}_0,\boldsymbol{h}_1,\dots,\boldsymbol{h}_{N-1}$ and the claims $\alpha_k=\alpha_k(\mathrm{d}\rho_N)$, $\gamma_k=\sqrt{\beta_k(\mathrm{d}\rho_N)}$, $k=0,1,\dots,N-1$, where $\mathrm{d}\rho_N(t)=\sum_{k=1}^N \rho_k^2\delta(t-\lambda_k)$.* (Note that $\gamma_N=0$ in theory.)

(c) *Determine explicitly the matrix* $\mathbf{A}$ *of* (b) *and its inverse. Show how the characteristic polynomial of* $\mathbf{A}^{-1}$ *can be computed recursively, and hence how the eigenvalues of* $\mathbf{A}$ *can be computed. Use part of the routine* `test_lanczos.m` *to check your answers.*

Answer

(a) The routine in question is (cf. Appendix A.2)

```
function [H,a,g]=lanczos_algorithm(N,A,h0)
%LANCZOS_ALGORITHM  Lanczos algorithm.

H=zeros(N); Ht=zeros(N); a=zeros(N,1); g=zeros(N+1,1);
g(1)=1;
for j=0:N-1
  if j==0
    a(1)=h0'*A*h0;
    Ht(:,1)=(A-a(1)*eye(N))*h0;
    g(2)=norm(Ht(:,1));
    H(:,1)=(1/g(2))*Ht(:,1);
  elseif j==1
    a(j+1)=H(:,1)'*A*H(:,1);
    Ht(:,2)=(A-a(2)*eye(N))*H(:,1)-g(2)*h0;
    g(3)=norm(Ht(:,2));
    H(:,2)=(1/g(3))*Ht(:,2);
  else
    a(j+1)=H(:,j)'*A*H(:,j);
    Ht(:,j+1)=(A-a(j+1)*eye(N))*H(:,j)-g(j+1)*H(:,j-1);
    g(j+2)=norm(Ht(:,j+1));
    H(:,j+1)=(1/g(j+2))*Ht(:,j+1);
  end
end
```

(b) The driver script in question is

```
%TEST_LANCZOS

f0='%21.14e %24.14e\n';
N=5;
L=zeros(N);
for i=1:N
  for j=1:i
    L(i,j)=1;
  end
end
A=L'*L; r=ones(N,1); r=(1/norm(r))*r;
[V,D]=eig(A);
h0=V*r;
[H,a,g]=lanczos_algorithm(N,A,h0);
omax=0;
for i=1:N-1
  om=max(abs(h0'*H(:,i)));
  if om>omax, omax=om; end
```

```
    for j=1:N-1
      if j~=i
        om=max(abs(H(:,i)'*H(:,j)));
        if om>omax, omax=om; end
      end
    end
  end
  omax
  xw=[diag(D) (1/N)*ones(N,1)];
  ab=lanczos(N,xw); ab(:,2)=sqrt(ab(:,2));
  fprintf('\n')
  disp('           a                         g')
  for k=1:N
    fprintf(f0,a(k),g(k))
  end
  fprintf('\n')
  disp('         alpha                    sqrt(beta)')
  for k=1:N
    fprintf(f0,ab(k,1),ab(k,2))
  end
```

We show the output of the test routine only for $N = 5$.

```
>> test_lanczos

omax =

  4.5451e-012

           a                            g
3.00000000000000e+000      1.00000000000000e+000
9.90000000000000e+000      4.69041575982343e+000
1.21111111111111e+000      9.94987437106619e-001
5.42735042735042e-001      2.41581025113159e-001
3.46153846153846e-001      7.77922918226823e-002

         alpha                     sqrt(beta)
3.00000000000000e+000      1.00000000000000e+000
9.90000000000000e+000      4.69041575982343e+000
1.21111111111111e+000      9.94987437106620e-001
5.42735042735042e-001      2.41581025113159e-001
3.46153846153845e-001      7.77922918226816e-002

>>
```

The output of the test, where omax is $\max |\boldsymbol{h}_i^{\mathrm{T}}\boldsymbol{h}_j|$, $i \neq j$, certainly validates the correctness of the Lanczos algorithm routine. A slight deterioration of accuracy, however, is noticeable toward the end of the routine, both in the orthogonality of the Lanczos vectors and in the agreement between a and alpha, resp., g and sqrt(beta). This becomes more pronounced as N increases. When $N=10$, for example, the accuracy begins to deteriorate at $n=7$ and is completely lost by the time n reaches N.

This prompted us to write symbolic routines slanczos_algorithm.m and stest_lanczos.m to determine the deterioration of accuracy when $N>10$, say $N=20$ and $N=40$. When $N=20$, comparing 52-digit with 64-digit results revealed that the loss of accuracy is significantly more serious in the vectors a and g than in the two columns of the array ab=[alpha,beta]. In the former case, the last components of a and g suffer a loss of 34, resp., 31, digits, whereas in the corresponding components of ab there is no loss of digits whatsoever. When $N=40$, comparing 132-digit with 144-digit results shows a loss of 114, resp., 112, digits in the last components of a and g, and again no loss of digits in the corresponding components of ab.

(c) One easily computes

$$A=L^{\mathrm{T}}L=\begin{bmatrix} 1 & 1 & 1 & \cdots & 1 & 1 \\ & 1 & 1 & \cdots & 1 & 1 \\ & & 1 & \cdots & 1 & 1 \\ & & & \ddots & \vdots & \vdots \\ & & & & 1 & 1 \\ 0 & & & & & 1 \end{bmatrix}\begin{bmatrix} 1 & & & & & 0 \\ 1 & 1 & & & & \\ 1 & 1 & 1 & & & \\ \vdots & \vdots & \vdots & \ddots & & \\ 1 & 1 & 1 & \cdots & 1 & \\ 1 & 1 & 1 & \cdots & 1 & 1 \end{bmatrix}$$

$$=\begin{bmatrix} N & N-1 & N-2 & \cdots & 2 & 1 \\ N-1 & N-1 & N-2 & \cdots & 2 & 1 \\ N-2 & N-2 & N-2 & \cdots & 2 & 1 \\ \cdots & \cdots & \cdots & \cdots & \cdots & \cdots \\ 2 & 2 & 2 & \cdots & 2 & 1 \\ 1 & 1 & 1 & \cdots & 1 & 1 \end{bmatrix},$$

that is,

$$A_{i,k}=N-i+1,\ \ i\geq k;\quad A_{i,k}=N-k+1,\ \ i<k.$$

To compute the inverse of A, we first note that L^{-1} is a lower bidiagonal matrix with 1's on the diagonal and -1's on the subdiagonal. Therefore,

$$A^{-1}=L^{-1}L^{-\mathrm{T}}=\begin{bmatrix} 1 & & & & 0 \\ -1 & 1 & & & \\ & -1 & 1 & & \\ & & \ddots & \ddots & \\ & & & -1 & 1 \end{bmatrix}\begin{bmatrix} 1 & -1 & & & \\ & 1 & -1 & & \\ & & \ddots & \ddots & \\ & & & 1 & -1 \\ 0 & & & & 1 \end{bmatrix}$$

$$=\begin{bmatrix} 1 & -1 & & & 0 \\ -1 & 2 & -1 & & \\ & \ddots & \ddots & \ddots & \\ & & -1 & 2 & -1 \\ 0 & & & -1 & 2 \end{bmatrix}.$$

Thus,

$$D_N(\lambda) = \det(A^{-1} - \lambda I) = \begin{vmatrix} 1-\lambda & -1 & & & 0 \\ -1 & 2-\lambda & -1 & & \\ & \ddots & \ddots & \ddots & \\ & & -1 & 2-\lambda & -1 \\ 0 & & & -1 & 2-\lambda \end{vmatrix} \tag{4.45}$$

$$= (1-\lambda)d_{N-1}(\lambda) - d_{N-2}(\lambda), \quad N \geq 3,$$

$$D_2(\lambda) = \lambda^2 - 3\lambda + 1, \quad D_1(\lambda) = 1 - \lambda,$$

where $d_k(\lambda)$ is the $k \times k$ determinant with $2-\lambda$ on the diagonal, -1 on the two side diagonals, and zero elsewhere. Expanding the determinant $d_{k+2}(\lambda)$ along the first row, one gets

$$\begin{aligned} d_{k+2}(\lambda) &= (2-\lambda)d_{k+1}(\lambda) - d_k(\lambda), \quad k = 1,2,3,\ldots, \\ d_1(\lambda) &= 2-\lambda, \; d_2(\lambda) = (\lambda-1)(\lambda-3). \end{aligned} \tag{4.46}$$

Thus the characteristic polynomial of A^{-1} can be computed from (4.45), using (4.46) to compute $d_1(\lambda), d_2(\lambda), \ldots$. This is implemented in the routine `eigA.m`, which produces the vector `D(N)` containing the coefficients of the Nth-degree characteristic polynomial of A^{-1} in the order of increasing powers. The MATLAB command `r=roots(D(N,:))` then generates the reciprocals of the eigenvalues of A^{-1}, i.e., the eigenvalues of A. Comparison with the eigenvalues produced by the MATLAB routine `eig.m` yields maximum discrepancies as shown below.

N	Discrepancy
5	1.42(–14)
10	1.44(–11)
15	1.19(–07)
20	1.15(–04)
25	3.16(–02)

Evidently, the conditioning of the eigenvalues as roots of the characteristic polynomial worsens with increasing N. Unfortunately, the routine `roots.m` does not accept symbolic input in order to verify this statement.

Exercise 4.27. *Let A be as in* Ex. 4.24 *and f a function sufficiently regular so that $f(A)$ is meaningful* (for example, analytic in a domain containing all eigenvalues of A). *Show that the matrix functional $\boldsymbol{u}^{\mathrm{T}} f(A)\boldsymbol{u}$, $\|u\| = 1$, can be expressed as*

$$\boldsymbol{u}^{\mathrm{T}} f(A)\boldsymbol{u} = \int_{\mathbb{R}_+} f(t)\mathrm{d}\rho_N(t), \tag{4.47}$$

with $\mathrm{d}\rho_N$ the discrete measure of Ex. 4.25 *obtained when Lanczos's algorithm is applied with $\boldsymbol{h}_0 = \boldsymbol{u}$.*

Answer. Since we have $\boldsymbol{u} = \boldsymbol{h}_0 = \sum_{k=1}^N \rho_k \boldsymbol{v}_k$ in Ex. 4.25, that is, $\boldsymbol{u} = V\boldsymbol{\rho}$, where $V = [\boldsymbol{v}_1, \boldsymbol{v}_2, \ldots, \boldsymbol{v}_N]$, $\boldsymbol{\rho} = [\rho_1, \rho_2, \ldots, \rho_N]$, there holds, using (4.43),

$$\begin{aligned}\boldsymbol{u}^{\mathrm{T}} f(A)\boldsymbol{u} &= \boldsymbol{\rho}^{\mathrm{T}} V^{\mathrm{T}} f(V\Lambda V^{\mathrm{T}})V\boldsymbol{\rho}\\ &= \boldsymbol{\rho}^{\mathrm{T}} V^{\mathrm{T}} V f(\Lambda) V^{\mathrm{T}} V\boldsymbol{\rho}\\ &= \boldsymbol{\rho}^{\mathrm{T}} f(\Lambda)\boldsymbol{\rho} = \textstyle\sum_{k=1}^N \rho_k^2 f(\lambda_k),\end{aligned}$$

which is precisely (4.47).

Exercise 4.28. (Ex. 25 of [27, §4], revised.) (a) *Use the Lanczos algorithm routine of* Ex. 4.26(a) *to compute lower bounds for the first and last diagonal elements of A^{-1}, where A is the Nth-order matrix of* Ex. 4.26(a). *Comment on problems of numerical stability and resolve them by appropriate variable-precision arithmetic.* {Hint: Express the two diagonal elements as a matrix functional $\boldsymbol{u}^{\mathrm{T}} f(A)\boldsymbol{u}$ and apply n-point, $n = 2, 3, \ldots, N$, Gauss quadrature to the right-hand side of (4.47).}

(b) *Determine the 2×2 Jacobi matrix generated in* (a) *by the Lanczos algorithm explicitly, i.e., determine α_0, α_1, γ_1 explicitly as functions of N.*

(c) *Show, and implement, how Gauss–Radau and Gauss–Lobatto quadratures can be used in Lanczos's algorithm to produce other lower and upper bounds. Obtain suitable intervals $[a, b]$ of integration by computing the eigenvalues of A, using the MATLAB routine* `eig.m`.

Answer

(a) Evidently, the diagonal elements of A^{-1} are

$$(A^{-1})_{ii} = \boldsymbol{e}_i^{\mathrm{T}} A^{-1} \boldsymbol{e}_i,$$

so, with $f(t) = t^{-1}$, we have, by (4.47),

$$(A^{-1})_{ii} = \int_{\mathbb{R}_+} t^{-1} \mathrm{d}\rho_N(t). \tag{4.48}$$

Since the constants α_j and $\gamma_j = \sqrt{\beta_j}$ produced by the Lanczos algorithm, by Ex. 4.24(c), are nothing but the elements on the diagonal, resp., side diagonals, of the Jacobi matrix $J_N(\mathrm{d}\rho_N)$ for the measure $\mathrm{d}\rho_N$, the n-point Gauss quadrature rule for (4.48) is readily obtained by the `OPQ` routine `gauss.m`. Since $f^{(2n)}(t) = (2n)!t^{-(2n+1)} > 0$ on $\mathbb{R}_+$, the remainder term in n-point Gaussian quadrature is positive, so that (4.48), using Ex. 4.4 and the fact that $\beta_0 = 1$ in this application, yields the lower bound

$$(A^{-1})_{ii} > \boldsymbol{e}_1^{\mathrm{T}} J_n^{-1} \boldsymbol{e}_1, \quad \boldsymbol{e}_1^{\mathrm{T}} = [1, 0, \ldots, 0] \in \mathbb{R}^n.$$

The exact answers, by Ex. 4.26(c), are

$$(A^{-1})_{11} = 1, \quad (A^{-1})_{ii} = 2 \ \text{ for } i > 1.$$

The procedure above is implemented in the script `recdiag_Gauss.m` for $N = 5, 10, 20$.

```
%RECDIAG_GAUSS Lower and upper bounds for the first and
%   last diagonal elements of the Nth-order matrix A^{-1},
%   using n-point Gauss quadrature, n=2,3,...,N.
```

```
f0='%2.0f %18.11f %19.11f\n';
fprintf('\n')
disp(' n   lb for a_{11}^{-1}  lb for a_{NN}^{-1}')
for N=[5 10 20]
  L=zeros(N);
  for i=1:N
    for j=1:i
      L(i,j)=1;
    end
  end
  A=L'*L;
  I=eye(N);
  for r=1:2
    if r==1
%
% first diagonal element
%
      h0=I(:,1);
    else
%
% last diagonal element
%
      h0=I(:,N);
    end
    [H,a,g]=lanczos_algorithm(N,A,h0);
    J=diag(a)+diag(g(2:N),1)+diag(g(2:N),-1);
    bG=zeros(N-1,1);
    for n=2:N
      Jn=J(1:n,1:n);
      In=eye(n);
      if r==1
        bG1(n-1)=In(1,:)*inv(Jn)*In(:,1);
      else
        bG2(n-1)=In(1,:)*inv(Jn)*In(:,1);
      end
    end
  end
  for k=1:N-1
    fprintf(f0,k+1,bG1(k),bG2(k))
  end
  fprintf('\n')
end
```

The results produced are shown below.

```
>> recdiag_Gauss

 n   lb for a_{11}^{-1}  lb for a_{NN}^{-1}
 2      0.74545454545       1.53333333333
 3      0.97202797203       1.92929292929
 4      0.99930069930       1.99766899767
```

```
   5      1.00000000000      2.00000000000

   2      0.47012987013      1.28421052632
   3      0.79841897233      1.67464114833
   4      0.95162055336      1.90805075931
   5      0.99296066253      1.98443935927
   6      0.99941743414      1.99851803422
   7      0.99997533491      1.99992845682
   8      0.99999956026      1.99999855760
   9      0.99999965678      1.99999998797
  10      0.99999999793      1.99999999238

   2      0.26515679443      1.14615384615
   3      0.53781003455      1.40994371482
   4      0.76421613357      1.67066625944
   5      0.90280186357      1.85203846438
   6      0.96778576050      1.94674434090
   7      0.99147270131      1.98474062339
   8      0.99821230454      1.99654765234
   9      0.99961968674      1.99938923444
  10      0.99970823617      1.99940429829
  11      0.99996290753      1.99991726846
  12      0.99999641451      1.99999143651
  13      0.99999972285      1.99999934158
  14      0.99999975474      1.99999981180
  15      0.99999998709      1.99999996491
  16      0.99999999143      1.99999999863
  17      0.99999999953      1.99999999864
  18      0.99999999994      1.99999999996
  19      0.99999999999      1.99999999997
  20      1.00000000000      2.00000000000

>>
```

The results, qualitatively, are as expected. Notice that the bounds are sharp, i.e., have zero error, when $n = N$. In this case, indeed, the functional $\int_{\mathbb{R}_+} f(t)\mathrm{d}\rho_N(t)$ by the very definition is an N-point quadrature sum and hence trivially Gaussian. This is confirmed when $N = 5$ and $N = 20$, but only approximately when $N = 10$. In fact, all results shown are subject to possibly considerable rounding errors, which are peaking somewhere near the middle of the range $2 \le n \le N$.

Running the same procedure in 50-digit arithmetic (using the script `srecdiag_Gauss.m` and the SOPQ routine `slanczos_algorithm.m`; cf. Appendix B.2) yields the following results rounded to 24 decimal digits after the decimal point.

```
>> srecdiag_Gauss

bG1 =
  0.745454545454545454545455
  0.972027972027972027972028
  0.999300699300699300699301
```

```
                        1.0
bG2 =
 1.5333333333333333333333
 1.9292929292929292929292929
 1.997668997668997668997 67
                       2.0

bG1 =
 0.4701298701298701298701 3
 0.7984189723320158102766 8
 0.9516205533596837944664 03
 0.9929606625258799171842 65
 0.9994174341400728207324 91
 0.9999753349131885670068 19
 0.9999995603371334860429 02
 0.9999999979063673023144 9
                        1.0
bG2 =
 1.2842105263157894736842 1
 1.6746411483253588516746 4
 1.9080507593093405450384 9
 1.9844393592677345537757 4
 1.9985180342159747194072 1
 1.9999284568242194692127 6
 1.9999985575972624892986 4
 1.9999999922866163769481 2
                       2.0

bG1 =
 0.2651567944250871080139 37
 0.5378100345485484453153 2
 0.7642161335667957286246 27
 0.9028018635714184519596 2
 0.9677857604979558297923 31
 0.9914727013082824255332 64
 0.9982123063539376154157 79
 0.9997063074724326082468 78
 0.9999626887896012387972 44
 0.9999963953576394417143 1
 0.9999997407554938995627 43
 0.9999999864976819739355 6
 0.9999999995090066172340 2
 0.9999999999881454626636 84
 0.9999999999998232863999 56
 0.9999999999999985481276 99
 0.9999999999999999946246 86
 0.9999999999999999999943 42
                        1.0
bG2 =
 1.1461538461538461538461 5
```

```
1.4099437148217636022514 1
1.6706662594354029407914 8
1.8520384643840216110802 3
1.9467443409041779912221 4
1.9847406233937685509542 6
1.9965476523515313463697 4
1.9993896268657843680164 5
1.9999167672998796865477
1.9999913897206772089532 1
1.9999993384192158759421 7
1.9999999632649485282781 2
1.9999999985787033656774 3
1.9999999999635564965558 3
1.9999999999994240710605 8
1.9999999999999949919222 7
1.9999999999999999804065 8
1.9999999999999999999782 4
                       2.0
>>
```

It can be seen that for $N = 5$ the less accurate double-precision results are correct to all digits shown, but for $N = 10$ and $N = 20$ there are substantial discrepancies, the largest involving the last four digits, occurring at $n = 9$ when $N = 10$, and involving the last eight digits, occurring at $n = 10$ when $N = 20$.

(b) *First diagonal element of* A^{-1}. Here $\boldsymbol{b}_0 = \boldsymbol{e}_1$, where $\boldsymbol{e}_1^{\mathrm{T}} = [1, 0, \ldots, 0] \in \mathbb{R}^N$, and (for the explicit form of $\boldsymbol{A}$, see Ex. 4.26(c))

$$\alpha_0 = \boldsymbol{b}_0^{\mathrm{T}} \boldsymbol{A} \boldsymbol{b}_0 = a_{11} = N.$$

To compute γ_1, we need

$$\tilde{\boldsymbol{b}}_1 = (\boldsymbol{A} - \alpha_0 \boldsymbol{I})\boldsymbol{b}_0 = (\boldsymbol{A} - N\boldsymbol{I})\boldsymbol{e}_1 = [0, N-1, N-2, \ldots, 1]^{\mathrm{T}}$$

to get

$$\gamma_1 = \|\tilde{\boldsymbol{b}}_1\| = \sqrt{\sum_{j=1}^{N-1} j^2}.$$

Since

$$\sum_{j=1}^{n} j^2 = \frac{n}{6}(n+1)(2n+1),$$

we obtain

$$\gamma_1^2 = \frac{N}{6}(N-1)(2N-1).$$

Now

$$\boldsymbol{b}_1 = \frac{1}{\gamma_1} \tilde{\boldsymbol{b}}_1$$

and

$$\alpha_1 = \boldsymbol{b}_1^{\mathrm{T}} \boldsymbol{A} \boldsymbol{b}_1 = \frac{1}{\gamma_1^2} [0, N-1, N-2, \ldots, 1] \boldsymbol{A} [0, N-1, N-2, \ldots, 1]^{\mathrm{T}}.$$

Using the explicit formula in Ex. 26(c) for A and doing a bit of (tedious) algebra, one finds

$$\begin{aligned}\gamma_1^2\alpha_1 &= \frac{1}{6}\sum_{k=1}^{N-1}(N-k)^2(N-k+1)(2N-2k+1)+\sum_{k=2}^{N-1}(N-k)^2\sum_{\ell=1}^{k-1}(N-\ell)\\ &= \frac{1}{6}\sum_{k=1}^{N-1}(N-k)^2(N-k+1)(2N-2k+1)+\frac{1}{2}\sum_{k=2}^{N-1}(N-k)^2(k-1)(2N-k).\end{aligned}$$

To evaluate the right-hand side, we use Maple to get (in factored form)

$$\gamma_1^2\alpha_1 = \frac{N}{30}(N-1)(2N-1)(2N^2-2N+1),$$

and hence

$$\alpha_1 = \frac{1}{5}(2N^2-2N+1).$$

Last diagonal element of A^{-1}. Here $\boldsymbol{h}_0 = \boldsymbol{e}_N$, where $\boldsymbol{e}_N^{\mathrm{T}} = [0,0,\ldots,0,1]\in\mathbb{R}^N$ and

$$\alpha_0 = \boldsymbol{h}_0^{\mathrm{T}} A\boldsymbol{h}_0 = a_{NN} = 1.$$

Furthermore,

$$\tilde{\boldsymbol{h}}_1 = (A-I)\boldsymbol{e}_N = [1,1,\ldots,1,0]^{\mathrm{T}},$$

so that

$$\gamma_1 = \|\tilde{\boldsymbol{h}}_1\| = \sqrt{N-1}, \quad \boldsymbol{h}_1 = \frac{1}{\gamma_1}\tilde{\boldsymbol{h}}_1.$$

Therefore,

$$\alpha_1 = \boldsymbol{h}_1^{\mathrm{T}} A\boldsymbol{h}_1 = \frac{1}{\gamma_1^2}[1,1,\ldots,1,0]A[1,1,\ldots,1,0]^{\mathrm{T}}.$$

Simple algebra and Maple yield

$$\begin{aligned}\gamma_1^2\alpha_1 &= \sum_{k=1}^{N-1}\left(\frac{1}{2}(N-k)(N-k+3)+(N-k+1)(k-1)\right)\\ &= \frac{1}{6}(N-1)(2N^2+5N-6).\end{aligned}$$

Thus,

$$\alpha_1 = \frac{1}{6}(2N^2+5N-6).$$

(c) The procedure is the same as in (a) except for the choice of quadrature, which is now either Gauss–Radau or Gauss–Lobatto on $[a,b]$, $a>0$. (We cannot take $a=0$ since the quadrature rule involves $f(a)$ and is applied to $f(t)=t^{-1}$.)

We begin with the case of Gauss–Radau quadrature. We need the n-point formula for $n=2,3,\ldots,N$, which can be obtained from the nth-order Jacobi matrix of $d\lambda_N$ (which is produced as part of the Lanczos algorithm) by replacing the (n,n) element by $a-\beta_{n-1}\pi_{n-2}(a)/\pi_{n-1}(a)$, where $\pi_k(\cdot)=\pi_k(\cdot;d\lambda_N)$ and $\beta_{n-1}=\gamma_{n-1}^2$ (cf. (4.32), which holds for the $(n+1)$-point formula). We first choose a to be the smallest eigenvalue of A. This gives rise to the script `recdiag_Radau.m`.

```
%RECDIAG_RADAU Lower and upper bounds for the first and
%   last diagonal elements of the Nth-order matrix A^{-1},
%   using n-point Gauss-Radau quadrature, n=2,3,...,N.

f0='%2.0f %18.11f %19.11f\n';
fprintf('\n')
disp(' n   lb for a_{11}^{-1}  lb for a_{NN}^{-1}')
for N=[5 10 20]
  L=zeros(N);
  for i=1:N
    for j=1:i
      L(i,j)=1;
    end
  end
  A=L'*L;
  I=eye(N);
  bG1=zeros(N-1,1); bg2=zeros(N-1,1);
  for r=1:2
    if r==1
%
% first diagonal element
%
      h0=I(:,1);
    else
%
% last diagonal element
%
      h0=I(:,N);
    end
    [H,a,g]=lanczos_algorithm(N,A,h0);
    J=diag(a)+diag(g(2:N),1)+diag(g(2:N),-1);
    t=min(eigvalA(N));
    p(1)=1; p(2)=t-a(1);
    for k=2:N-1
      p(k+1)=(t-a(k))*p(k)-g(k)^2*p(k-1);
    end
    for n=2:N
      Jn=J(1:n,1:n); Jn(n,n)=t-g(n)^2*p(n-1)/p(n);
      In=eye(n);
      if r==1
        bG1(n-1)=In(1,:)*inv(Jn)*In(:,1);
      else
        bG2(n-1)=In(1,:)*inv(Jn)*In(:,1);
      end
    end
  end
  for k=1:N-1
    fprintf(f0,k+1,bG1(k),bG2(k))
  end
  fprintf('\n')
```

```
end
```

Run with the same values of N as in (a), it produces the following results.

```
>> recdiag_Radau

 n    lb for a_{11}^{-1}   lb for a_{NN}^{-1}
 2        2.14762994062        3.26925052251
 3        1.16691976970        2.30053310171
 4        1.00667576657        2.01751591633
 5        1.00000000000        2.00000000000

 2        2.94009883528        3.68877759850
 3        1.78520357031        2.91202828240
 4        1.20131455330        2.30199640783
 5        1.03259582513        2.06060769153
 6        1.00317796430        2.00708148023
 7        1.00017017171        2.00044355829
 8        1.00000424174        2.00001271325
 9        1.00000004493        2.00000011571
10        1.00000003354        2.00000009737

 2        3.43523736892        3.86374056000
 3        2.54672294956        3.39093141385
 4        1.78050144510        2.82501905632
 5        1.31841907125        2.39074465341
 6        1.10576729479        2.14816076962
 7        1.02855844366        2.04493856486
 8        1.00622252437        2.01085301602
 9        1.00108250883        2.00206999272
10        1.00104517582        2.00041554301
11        1.00014710893        2.00030616551
12        1.00001558060        2.00003506689
13        1.00000125040        2.00000302710
14        1.00000056172        2.00000019722
15        1.00000007087        2.00000018075
16        1.00000000306        2.00000000833
17        1.00000000299        2.00000000270
18        1.00000000009        2.00000000026
19        1.00000000007        2.00000000001
20        1.00000000000        2.00000000000

>>
```

As can be seen, upper bounds are obtained. This is because the remainder terms of Gauss–Radau formulae involve odd-order derivatives of f multiplied by positive constants. The accuracy is similar to that in (b) (as determined by comparing with 112-digit results using `srecdiag_Radau.m`, which produces results correct to 24 digits after the decimal point). The magnitude of the error, as before, bulges somewhere near the middle of the range $2 \le n \le N$, affecting the last 4 digits if $N = 10$, and the last 8–9 digits if $N = 20$.

If we use the Gauss–Radau formula with the fixed node at the right endpoint b, the sign of the remainder term is reversed, and lower bounds will be produced. The only change in the program is redefining the variable t, for example, making it the maximum eigenvalue of A. The respective results (again not fully accurate) are shown below.

```
>> recdiag_Radau

 n    lb for a_{11}^{-1}  lb for a_{NN}^{-1}
 2      0.73137011279       1.50056853105
 3      0.97106827722       1.92664848471
 4      0.99928328013       1.99760940640
 5      1.00000024172       2.00000002631

 2      0.45322898894       1.25309030365
 3      0.79442826458       1.66713807682
 4      0.95092455373       1.90660172317
 5      0.99288128202       1.98425292034
 6      1.00120618482       1.99864055917
 7      0.99942279320       1.99853074956
 8      0.99997552884       1.99992903240
 9      0.99999956335       1.99999856760
10      0.99999999792       1.99999999233

 2      0.25257949380       1.12558487042
 3      0.53289664813       1.40242031438
 4      0.76239360057       1.66779666709
 5      0.90220756769       1.85105788322
 6      0.94194465812       1.94596441315
 7      0.96794703892       1.94718792090
 8      0.99150889806       1.98480880350
 9      0.99821891197       1.99656094101
10      0.99970726833       1.99939169068
11      0.99996279832       1.99991701824
12      0.99999640443       1.99999141314
13      0.99999659299       1.99999927624
14      0.99999972560       1.99999934326
15      0.99999975553       1.99999982489
16      0.99999998712       1.99999996499
17      0.99999999953       1.99999999864
18      0.99999999994       1.99999999996
19      0.99999999999       1.99999999997
20      0.99999999999       2.00000000000

>>
```

Based on higher-precision results, it is found (not unexpectedly) that the Gauss–Radau lower bounds are generally somewhat smaller than the corresponding Gauss lower bounds, but practically equal as n approaches N.

If the left endpoint a of the interval $[a,b]$ is chosen to be less than the smallest eigenvalue of A, for example, half its value, one still gets upper bounds, but of lesser quality. Similarly, if b is chosen larger than the maximum eigenvalue, for example, three halves of it, one still gets lower bounds, but of lesser quality.

Proceeding now to Gauss–Lobatto quadrature, the modification of the $(n \times n)$ Jacobi matrix involves the three elements in positions (n,n), $(n-1,n)$, and $(n,n-1)$ (cf. (4.35), (4.36), which hold for the $(n+2)$-point formula). This is implemented in the script `recdiag_Lobatto.m` and its symbolic counterpart `srecdiag_Lobatto.m`.

```
%RECDIAG_LOBATTO  Upper bounds for the first and last
%   diagonal elements of the Nth-order matrix A^{-1},
%   using n-point Gauss-Lobatto quadrature, n=2,3,...,N.

f0='%2.0f %18.11f %19.11f\n';
fprintf('\n')
disp(' n   lb for a_{11}^{-1}  lb for a_{NN}^{-1}')
for N=[5 10 20]
  L=zeros(N);
  for i=1:N
    for j=1:i
      L(i,j)=1;
    end
  end
  A=L'*L;
  I=eye(N);
  bG1=zeros(N-1,1); bg2=zeros(N-1,1);
  for r=1:2
    if r==1
%
% first diagonal element
%
      h0=I(:,1);
    else
%
% last diagonal element
%
      h0=I(:,N);
    end
    [H,a,g]=lanczos_algorithm(N,A,h0);
    J=diag(a)+diag(g(2:N),1)+diag(g(2:N),-1);
    ta=min(eigvalA(N)); tb=max(eigvalA(N));
    pa(1)=1; pa(2)=ta-a(1); pb(1)=1; pb(2)=tb-a(1);
    for k=2:N-1
      pa(k+1)=(ta-a(k))*pa(k)-g(k)^2*pa(k-1);
      pb(k+1)=(tb-a(k))*pb(k)-g(k)^2*pb(k-1);
    end
    for n=2:N
      P=[pa(n) pa(n-1);pb(n) pb(n-1)];
      p=[ta*pa(n);tb*pb(n)]; q=inv(P)*p;
      Jn=J(1:n,1:n); Jn(n,n)=q(1);
      Jn(n-1,n)=sqrt(q(2)); Jn(n,n-1)=Jn(n-1,n);
      In=eye(n);
      if r==1
        bG1(n-1)=In(1,:)*inv(Jn)*In(:,1);
```

```
        else
          bG2(n-1)=In(1,:)*inv(Jn)*In(:,1);
        end
      end
    end
    for k=1:N-1
      fprintf(f0,k+1,bG1(k),bG2(k))
    end
    fprintf('\n')
  end
```

It should produce upper bounds since the remainder term involves an even-order derivative of $f(t)=t^{-1}$, multiplied by a negative constant. This is confirmed by the following output of `recdiag_Lobatto.m`.

```
>> recdiag_Lobatto

  n    lb for a_{11}^{-1}  lb for a_{NN}^{-1}
  2      2.27184700970       3.46518629669
  3      1.17531369920       2.31729682725
  4      1.00688845833       2.01809799972
  5      0.99999999456       1.99999999941

  2      3.05979865923       3.84611542913
  3      1.80869472790       2.94300545385
  4      1.20515051016       2.30819992785
  5      1.03304602732       2.06149085761
  6      1.00319201856       2.00715905168
  7      1.00320192783       2.01268961004
  8      1.00016866954       2.00043954055
  9      1.00000420982       2.00001261605
 10      1.00000003377       2.00000011492

  2      3.51570843993       3.95909320493
  3      2.57237075084       3.41697328115
  4      1.78888352731       2.83446766270
  5      1.32090062885       2.39394268683
  6      1.10703055297       2.14910590948
  7      1.10448647818       2.09236556148
  8      1.02841776694       2.04470786581
  9      1.00619662344       2.01080626230
 10      1.00107810249       2.00206229761
 11      1.00014757906       2.00030717830
 12      1.00001562521       2.00003516948
 13      1.00000144438       2.00000303530
 14      1.00000124704       2.00000273885
 15      1.00000052890       2.00000019619
 16      1.00000007070       2.00000017994
 17      1.00000000300       2.00000000831
 18      1.00000000009       2.00000000026
 19      1.00000000007       2.00000000001
```

```
20        1.00000000000          2.00000000000

>>
```

They are, to be sure, inaccurate as before, but also of lesser quality. Accurate 24-digit results are obtained by running `srecdiag_Lobatto.m` in 100-digit arithmetic.

Chapter 5

Answers to Exercises on Approximation

This chapter deals with four kinds of approximation: evaluation of special functions, polynomial least squares approximation, moment-preserving approximation, and the summation of slowly convergent series.

Special functions can often be evaluated by numerical quadrature, especially Gaussian quadrature, if suitable integral representations are available. Examples of this are given in Section 5.1

For a system of discrete orthogonal polynomials (not necessarily monic) $\pi_i(\cdot;d\lambda_N)$, $i = 0, 1, \ldots, N-1$, relative to the N-point discrete measure $d\lambda_N = \sum_{k=1}^{N} w_k \delta(t-t_k)$, where δ is the Dirac delta function, w_k positive weights, and t_k mutually distinct real numbers, the *Fourier coefficients* of a function f relative to the orthogonal system $\pi_0, \pi_1, \ldots, \pi_{N-1}$ are defined by

$$\hat{c}_i = \hat{c}_i(f) = \frac{(f, \pi_i)}{\|\pi_i\|^2}, \quad i = 0, 1, \ldots, N-1, \tag{5.1}$$

where $(u, v) = (u, v)_{d\lambda_N}$, $\|u\|^2 = \|u\|^2_{d\lambda_N}$. The polynomial

$$\hat{p}(t) = \sum_{i=0}^{n} \hat{c}_i(f)\pi_i(t; d\lambda_N), \quad 0 \le n < N,$$

then solves (Ex. 5.7(a)) the *least squares problem*: Find $\hat{p} \in \mathbb{P}_n$ such that

$$E_n^2(\hat{p}) \le E_n^2(p) \quad \text{for all } p \in \mathbb{P}_n, \tag{5.2}$$

where

$$E_n^2(p) = \|p - f\|^2_{d\lambda_N}.$$

The least squares error achieved is (Ex. 5.7(b))

$$\min_{p \in \mathbb{P}_n} E_n^2(p) = \|f\|^2 - \sum_{i=0}^{n} \frac{(f, \pi_i)^2}{\|\pi_i\|^2}. \tag{5.3}$$

Adding constraints,

$$p(s_j) = f_j, \quad j = 1, 2, \ldots, m, \quad m \le n,$$

gives rise to a *constrained least squares problem*, which, however, can be reduced to an unconstrained problem for a new function f^* and a new measure $d\lambda_N^*$ (Exs. 5.11–5.12).

For approximating data that contains not only function values, but also values of derivatives, the use of discrete *Sobolev orthogonal polynomials* is called for. An example of this is discussed in Ex. 5.14.

In the applied sciences, moments of a function often have real physical meaning and therefore should be preserved if the function is being approximated. This gives rise to the problem of *moment-preserving approximation*, specifically moment-preserving approximation on $\mathbb{R}_+$ by spline functions. The existence of such approximations, and a method based on Gaussian quadrature to compute them, is the subject of Ex. 5.16* of Section 5.3, and examples are given in Exs. 5.17–5.18.

Section 5.4 deals with infinite series whose kth term can be expressed in terms of the Laplace transform $(\mathcal{L}f)(s)$ of some function f evaluated at $s = k$. The summation of the series can then be effected by evaluating a definite integral, typically by Gaussian quadrature relative to a nonclassical weight function. This is particularly useful if the series converges slowly, and Gaussian quadrature fast, and is demonstrated by a number of examples including one involving the spiral of Theodorus in Section 5.5.

5.1 ▪ Special functions

Exercise 5.1. Repeated integrals of the coerror function.

These are

$$f_n(x) = \mathrm{i}^n \operatorname{erfc} x, \quad n = -1, 0, 1, 2, \ldots, \quad x \in \mathbb{R},$$

where

$$f_{-1}(x) = \frac{2}{\sqrt{\pi}} e^{-x^2},$$

$$f_n(x) = \int_x^\infty f_{n-1}(t)\mathrm{d}t, \quad n = 0, 1, 2, \ldots.$$

(a) *Represent* $f_n(x)$, $n \geq 0$, *as a single integral extended from* 0 *to* ∞. *What is the appropriate Gaussian quadrature rule for evaluating this integral? Comment on the merits of this quadrature procedure.*

(b) *When* x *is large negative, what precaution should be taken to avoid unnecessary overflow?*

(c) *Write a MATLAB routine,* `inerfc.m`, *that evaluates* $f_n(x)$ *for* $|x| \leq 10$, $0 \leq n \leq 50$, *using numerical quadrature as suggested in* (a). {Hint: It has been shown in [33] that $f_n(x)$ neither overflows nor underflows in MATLAB double-precision arithmetic when (n, x) is in a domain considerably larger than the one specified in this exercise. Applied to the present domain, it has also been determined that the appropriate Gauss quadrature rule yields at least 12 correct decimal digits if the number N of quadrature points is taken to be $N = \lceil 24 + 2.388x \rceil$ when $x \geq 0$, and $N = \lceil 10 + 10.5|x| + (.3223 + .00747|x|)n \rceil$ when $x < 0$.}

(d) *Test the accuracy of the quadrature routine written in* (c). *Use the routine to reproduce Table* 7.4 *of* [1].

Answer

(a) Any repeated integral, as is well known from calculus, can be expressed as a single integral, in our case

$$f_n(x) = \frac{2}{\sqrt{\pi}} \int_x^\infty \frac{(t-x)^n}{n!} e^{-t^2} \mathrm{d}t, \quad n \geq 0.$$

Make the change of variables, $t - x = \tau$, to write

$$f_n(x) = \frac{2}{\sqrt{\pi}} \int_0^\infty \frac{\tau^n}{n!} e^{-(\tau+x)^2} \mathrm{d}\tau,$$

that is,

$$f_n(x) = \frac{2}{\sqrt{\pi}} \frac{e^{-x^2}}{n!} \int_0^\infty t^n e^{-2xt} e^{-t^2} \mathrm{d}t. \tag{5.4}$$

To the integral on the right we can apply the half-range Gauss–Hermite quadrature rule,

$$\int_0^\infty t^n e^{-2xt} e^{-t^2} \mathrm{d}t = \sum_{\nu=1}^N \lambda_\nu^G [\tau_\nu^G]^n e^{-2x\tau_\nu^G} + R_N,$$

where τ_ν^G, λ_ν^G are the Gauss nodes and weights for the weight function $w(t) = e^{-t^2}$ on $[0, \infty]$. The first 200 recurrence coefficients for the required orthogonal polynomials are available in the file `ab_hrhermite.txt` (cf. Ex. 2.17 and Appendix A.5) to 32 decimal digits and can be downloaded by the routine `loadvpa.m`.

Among the merits of this evaluation procedure is a relatively high degree of stability, all terms in the quadrature sum being positive, and also the fact, not shared by methods using a recurrence relation, that the function $f_n(x)$ can be evaluated for a single value of n.

(b) When $x < 0$ is large in absolute value, there is a danger of the integrand in (5.4) overflowing. To diminish the chance of this happening, move the factor e^{-x^2} inside the integral, that is, write

$$f_n(x) = \frac{2}{\sqrt{\pi} n!} \int_0^\infty t^n e^{-(x^2+2xt)} e^{-t^2} \mathrm{d}t.$$

(c) The routine is

```
function y=inerfc(n,x)
%INERFC Repeated integral of the coerror function.
%   Y=INERFC(N,X) evaluates the N times repeated integral
%   of the coerror function to at least 12 correct decimal
%   digits in the domain |X|<=10, -1<=N<=50.

global ab
if abs(x)>10 | n>50 | n<-1
  error('x or n out of range');
end
if n==-1
  y=(2/sqrt(pi))*exp(-x^2); return
end
```

```
if x>=0
  N=ceil(24+2.388*x);
else
  N=ceil(10+10.5*abs(x)+(.3223+.00747*abs(x))*n);
end
uv=gauss(N,ab); u=uv(:,1); v=uv(:,2);
if x>=0
  y0=exp(-x^2)*sum(v.*u.^n.*exp(-2*x*u));
else
  y0=sum(v.*u.^n.*exp(-x^2-2*x*u));
end
y=2*y0/(sqrt(pi)*factorial(n));
```

The required array `ab` has to be declared `global` and downloaded in the driver routine by the commands

```
ab32=loadvpa('ab_hrhermite.txt',3,32,200,2);
ab=double(ab32);
```

(d) To check the accuracy of the routine `inerfc.m`, we copied it into another one, `inerfc_test.m`, in which the estimates for N were increased by 5. We then determined the maximum relative discrepancy between the results produced by the two routines as $n = 0:50$, $x = -.2:-.2:-10$, and $x = 0:.2:10$. The script `test_inerfc.m` implementing this produced the maximum discrepancy of 1.2810×10^{-13} for negative x, and 2.1189×10^{-14} for positive x. The former was attained at $n = 50$, $x = -10$, the latter at $n = 50$, $x = .2$. This confirms the accuracy of 12 decimal digits striven for.[4]

The maximum discrepancy is even less, 4.2475×10^{-15}, when run to reproduce Table 7.4 of [1], where $n \le 11$ and $0 \le x \le 5$; cf. the script `Table7_4.m`. The entries in that table, by the way, were found to be correct to all six digits given, except for occasional end figure errors of one unit.

Exercise 5.2. *Complex error function,* also known as *Faddeeva function* or *plasma dispersion function,*

$$w(z) = e^{-z^2}\operatorname{erfc}(-\mathrm{i}z), \quad z \in \mathbb{C},$$

where erfc is the complementary error function.

It suffices to consider $w(z)$ for z in the first quadrant $\mathbb{Q}_1$ of the complex plane, since for the other quadrants $\mathbb{Q}_2$, $\mathbb{Q}_3$, $\mathbb{Q}_4$ one has, resp.,

$$\begin{aligned} w(-x+\mathrm{i}y) &= \overline{w(x+\mathrm{i}y)}, \\ w(-x-\mathrm{i}y) &= 2e^{-(x+\mathrm{i}y)^2} - w(x+\mathrm{i}y), \\ w(x-\mathrm{i}y) &= 2e^{-(x-\mathrm{i}y)^2} - \overline{w(x+\mathrm{i}y)}. \end{aligned}$$

(a) *Assuming* $z = x + \mathrm{i}y$, $x \ge 0$, $y \ge 0$, *investigate the numerical use of the two integral representations*

$$w(z) = \frac{2}{\sqrt{\pi}} \int_0^\infty e^{2\mathrm{i}zt} e^{-t^2} \mathrm{d}t, \quad z \in \mathbb{C}, \tag{5.5}$$

[4]In MATLAB Release R2014a, the maximum discrepancies are found, disappointingly, to be 2.100×10^0 for $x < 0$, and 1.657×10^{-10} for $x > 0$, attained, resp., at $[n,x] = [2,-6.8]$ and $[n,x] = [50,.6]$. The reason for this poor performance is believed to be the failure of the MATLAB routine `eig.m` to deliver eigenvectors with sufficient relative accuracy; cf. Demo 1.17.

and

$$w(z) = -\frac{\mathrm{i}}{\pi}\int_{-\infty}^{\infty}\frac{e^{-t^2}}{t-z}\,dt, \quad \mathrm{Im}\, z > 0. \tag{5.6}$$

(See Eq. (5.4) with $n=0$, $x=-\mathrm{i}z$ for the former, and [13, Eq. (2.3)] for the latter.)

(b) *Write routines that evaluate* $w(z)$ *to an accuracy of about* 16, *resp.,* 32, *significant decimal digits. Check the accuracies achieved.*

Answer

(a) Use Gauss quadrature relative to the half-range Hermite weight function for the integral in (5.5), and Gauss–Hermite quadrature for the one in (5.6). The former requires more quadrature points, the latter fewer quadrature points, as $r=|z|$ is increased. It is thus natural to find a value $r=r_0$ such that the work for computing (5.5) by half-range Hermite quadrature is comparable to the work involved in computing (5.6) by Gauss–Hermite quadrature. A suitable choice of r_0 was found empirically, using the scripts `sN_erfc.m` and `sN_erfcas.m`, to be $r_0 = 6.05$ for 16-digit accuracy and $r_0 = 8.29$ for 30-digit accuracy. The numbers n of quadrature points then needed in (5.5) and (5.6) are shown in Table 5.1, in the left half for 16-digit accuracy and in the right half for 30-digit accuracy.

Table 5.1. *Number n of quadrature points needed in the sector* $S_q = \{z = re^{\mathrm{i}\theta} : q\pi/20 \le \theta \le (q+1)\pi/20\}$ *to approximate* (5.5) *and* (5.6) *to* 16 *digits (left half), resp.,* 30 *digits (right half)*

q	(5.5)	(5.6)	q	(5.5)	(5.6)	q	(5.5)	(5.6)	q	(5.5)	(5.6)
0	39	21	5	29	15	0	71	37	5	53	27
1	36	19	6	28	15	1	66	33	6	51	27
2	33	17	7	27	15	2	61	31	7	50	26
3	31	16	8	27	15	3	58	29	8	48	26
4	30	16	9	26	14	4	55	28	9	48	26

(b) The routines in question require a certain number n of nodes and weights of Gauss quadrature rules for the half-range Hermite and the Hermite measure. They can be obtained by first generating the respective array `ab` of the first n recurrence coefficients, and then calling the routine `gauss(n,ab)`, resp., its symbolic `dig`-digit analogue `sgauss(dig,n,ab)`. For the half-range Hermite measure, the (first 200) recurrence coefficients are available in the file `ab_hrhermite.txt` (cf. Appendix A.5) to 32 decimal digits, and for the Hermite measure can be obtained by means of `ab=r_hermite(n)` for double-precision accuracy, and `ab=sr_ hermite(dig,n)` for `dig`-digit accuracy.

The routines themselves are

```
function y=cplxerr(z)
%CPLXERR Complex error function.
%   Y=CPLXERR(Z) evaluates the complex error function W(Z) for Z
%   in the first quadrant of the complex plane to an accuracy of
%   at least 14 decimal digits. The global variables xghrh, wghrh
%   are to hold the first forty 16-digit nodes and weights of the
%   half-range Hermite quadrature rule, and the global variables
%   xgh, wgh the first twenty-one 16-digit nodes and weights of the
%   Hermite quadrature rule.

global xghrh wghrh xgh wgh
```

```
if angle(z)<0 | angle(z)>pi/2
  error('phase of z out of range')
end
if abs(z)<=6.05
  s=sum(wghrh.*exp(2*i*z*xghrh));
  y=2*s/sqrt(pi);
else
  s=sum(wgh./(xgh-z));
  y=-i*s/pi;
end
```

and

```
function y=scplxerr(z)
%SCPLXERR Symbolic analogue of cplxerr.m.
%   Y=SCPLXERR(Z) evaluates the complex error function W(Z) for Z
%   in the first quadrant of the complex plane to an accuracy of
%   at least 29 decimal digits. The global variables xghrh, wghrh
%   are to hold the first seventy-one 32-digit nodes and weights of
%   the half-range Hermite quadrature rule, and the global variables
%   xgh, wgh the first thirty-seven 32-digit nodes and weights of
%   the Hermite quadrature rule.

global xghrh wghrh xgh wgh
if angle(z)<0 | angle(z)>pi/2
  error('phase of z out of range')
end
digits=32; sz=vpa(z); spi=vpa(pi);
if abs(z)<=8.29
  s=sum(wghrh.*exp(2*i*sz*xghrh));
  y=2*s/sqrt(spi);
else
  s=sum(wgh./(xgh-sz));
  y=-i*s/spi;
end
```

The script[5] test_cplxerr.m,

```
%TEST_CPLXERR
%
global xghrh wghrh xgh wgh
f0='%3.0f %16.12f %16.12f %12.4e %5.1f\n';
dig=32;
abhrh=loadvpa('ab_hrhermite.txt',3,dig,200,2);
abhrh=double(abhrh);
xw=gauss(40,abhrh); xghrh=xw(:,1); wghrh=xw(:,2);
abh=r_hermite(21);
xw=gauss(21,abh); xgh=xw(:,1); wgh=xw(:,2);
disp('  q       Re(w)              Imag(w)            err       rerr')
```

[5]This script, contrary to MATLAB Release R2014a, does not work in MATLAB Release R2011a, which does not allow symbolic arguments z in the function `erfc` z.

```
errmmax=0; ic=0; ie=0;
for q0=0:10:90
  for q=q0:q0+10
    errmax=0;
    for r=0:.1:20
      ic=ic+1;
      z=r*exp(q*(pi/200)*i); w=cplxerr(z);
      sz=vpa(z,32); sw=exp(-sz^2)*erfc(-i*sz);
      err=abs(double(w-sw));
      if err<.5e-16, ie=ie+1; end
      if err>errmax, errmax=err; rerrmax=r; end
    end
    fprintf(f0,q,real(w),imag(w),errmax,rerrmax)
    if errmax>errmmax, errmmax=errmax; end
  end
  fprintf('\n');
end
errmmax
[ic ie]
```

and its 32-digit analogue `test_scplxerr.m` determine the maximum errors in each of the sectors $S_q = \{z = re^{i\theta} : 0 \le r \le 20,\ q\pi/20 \le \theta \le (q+1)\pi/20\}$, $q = 0,1,\dots,9$. They are displayed in Table 5.2, in the left half for the routine `cplxerr.m`, and in the right half for the routine `scplxerr.m`. The error for the former is $< .5 \times 10^{-16}$ in 69% of all cases tested, and for the latter $< .5 \times 10^{-32}$ in 90% of all cases.

Table 5.2. *Maximum errors in the sector S_q for the results produced by* `cplxerr.m` *(in the left half) and by* `scplxerr.m` *(in the right half)*

q	maxerr	q	maxerr	q	maxerr	q	maxerr
0	1.22(−15)	5	8.39(−16)	0	1.25(−30)	5	3.90(−33)
1	1.10(−15)	6	9.47(−16)	1	1.89(−32)	6	3.25(−33)
2	1.00(−15)	7	9.07(−15)	2	1.08(−32)	7	2.92(−33)
3	8.42(−16)	8	9.49(−16)	3	7.01(−33)	8	2.92(−33)
4	8.85(−16)	9	8.42(−16)	4	5.02(−33)	9	2.92(−33)

Exercise 5.3. Variable-precision incomplete gamma functions.

There is a lower and upper incomplete gamma function, defined, resp., by

$$\gamma(a,x) = \int_0^x t^{a-1}e^{-t}\,dt, \quad \Gamma(a,x) = \int_x^\infty t^{a-1}e^{-t}\,dt. \tag{5.7}$$

They sum up to $\Gamma(a)$. Thus,

$$\gamma(a,x) = \Gamma(a) - \Gamma(a,x). \tag{5.8}$$

Alternatively,

$$\gamma(a,x) = \Gamma(a)x^a\gamma^*(a,x), \tag{5.9}$$

where $\gamma^*(a,x) = x^{-a}\gamma(a,x)/\Gamma(a)$ is Tricomi's form of the lower incomplete gamma function, an entire function in both variables.

The Symbolic Math Toolbox of MATLAB presently has a routine only for $\Gamma(a,x)$, presumably because $\gamma(a,x)$ can be obtained from it by (5.8). In principle, this is not unreasonable for symbolic work; yet, when (5.8) is subject to a great deal of cancellation, then (5.8) may require a lot of additional precision to obtain $\gamma(a,x)$ to a given, typically high accuracy. It is desirable, therefore, to have also a symbolic routine for $\gamma(a,x)$.

(a) *Write a variable-precision MATLAB routine for evaluating* $\gamma^*(a,x)$ *for real* a, x.

(b) *Write a variable-precision MATLAB routine for evaluating* $\gamma(a,x)$ *for* $a>0$, $x>0$, *without incurring any cancellation error. Analyze the condition for avoiding cancellation.*

(c) *Write a variable-precision MATLAB routine for evaluating* $\Gamma(a,x)$ *for* $a>0$, $x>0$, *using the procedure provided in the MATLAB Symbolic Toolbox.*

Answer

(a) One way to compute $\gamma^*(a,x)$ is by means of its power series (cf. [53, Eq. 8.7.1]),

$$\gamma^*(a,x)=e^{-x}\sum_{n=0}^{\infty}\frac{x^n}{\Gamma(a+n+1)}. \tag{5.10}$$

The routine for it is `sgammastar.m`,

```
function y=sgammastar(dig,a,x)
%SGAMMASTAR Tricomi's form of the lower incomplete gamma function.
%   Y=SGAMMASTAR(DIG,A,X) computes to an accuracy of DIG decimal
%   digits the function X^{-A}*gamma(A,X)/gamma(A) for real A and
%   X, where gamma(A,X) is the lower incomplete gamma function.

digits(dig)
eps0=.5*10^(-dig);
sa=vpa(a); sx=vpa(x);
if rem(double(a),1)==0 & double(a)<0
  y=sx^(-sa);
  return
end
t=vpa(1)/gamma(sa+1); y0=vpa(0); y=t; k=0;
while double(abs((y-y0)/y))>eps0
  k=k+1; y0=y; t=sx*t/(sa+k); y=y0+t;
end
y=exp(-sx)*y;
```

If $a>0$, $x>0$, the evaluation of $\gamma^*(a,x)$ by its series in (5.10) is very stable, since all terms are positive. However, the series should be avoided if $|x|$ is very large.

(b) Any sort of cancellation can be avoided if (5.8) is used only when

$$\Gamma(a)-\Gamma(a,x)>1/2, \tag{5.11}$$

and (5.9) is used otherwise. The routine required, therefore, is

```
function y=sgam_lowinc(dig,a,x)
%SGAM_LOWINC Lower incomplete gamma function.
%   Y=SGAM_LOWINC(DIG,A,X) computes to an accuracy of DIG decimal
%   digits the function int_0^X t^{A-1}*EXP(-t)dt, A>0, X>0.

digits(dig); sa=vpa(a); sx=vpa(x);
```

```
d=gamma(sa)-feval(symengine,'igamma',sa,sx);
if double(d)>1/2
  y=vpa(d,dig);
else
  y=gamma(sa)*sx^sa*sgammastar(dig,sa,sx);
end
```

If we let

$$d(a,x) = \Gamma(a) - \Gamma(a,x) - 1/2,$$

it is of interest to know how large x can be when $d(a,x) \leq 0$, i.e., $\gamma^*(a,x)$ in (5.9) is invoked to compute $\gamma(a,x)$. For this, we determine the curve $x = x(a)$ for which $d(a,x(a)) \equiv 0$, which is done in the script `test_incgamma`, using Newton's method.

```
%TEST_INCGAMMA

f0='%8.2f %16.12f %12.4e %4.0f\n';
disp('     a            zero              err          it')
eps0=.5e-10;
k=0; z=zeros(size((.2:.1:10)',1),1);
for a=.2:.1:10
  k=k+1;
  if a<=1
    x1=.000001;
  elseif a<=1.55
    x1=.1;
  elseif a<=8
    x1=3;
  else
    x1=1.5;
  end
  x0=0; it=0;
  while abs(x1-x0)>eps0
    x0=x1; it=it+1;
    if it>50, error('Newton diverges'); end
    x1=x0-(gamma(a)*(1-gammainc(x0,a,'upper'))-1/2) ...
       *exp(x0)/x0^(a-1);
  end
  z(k)=x1;
  err=abs(gamma(a)*(1-gammainc(x1,a,'upper'))-1/2);
% fprintf(f0,a,x1,err,it)
end
aa=(.2:.1:10)'; [aa z]
plot(aa,z);set(gca,'FontSize',14)
axis([0 10 -.2 2])
```

It produces the graph in Fig. 5.1, below which $d(a,x)$ is negative. Thus, when the series in (5.10) is used, then x is smaller than the maximum shown in Fig. 5.1, that is, $x \leq 1.7484$. It therefore converges very quickly.

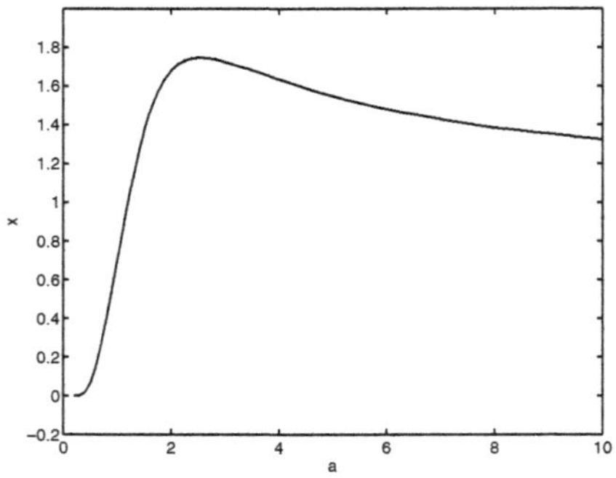

Figure 5.1. *The curve $x = x(a)$*

(c) We simply use the routine from the MATLAB Symbolic Toolbox,

```
function y=sgam_upinc(dig,a,x)
%SGAM_UPINC Upper incomplete gamma function.
%   Y=SGAM_UPINC(DIG,A,X) computes to an accuracy of DIG decimal
%   digits the function int_X^Inf t^{A-1}*EXP(-t)dt, A>0, X>0.

digits(dig); sa=vpa(a); sx=vpa(x);
y=feval(symengine,'igamma',sa,sx);
```

Exercise 5.4. Complex Airy function (see also Ex. 5.6).

(a) *Show that*

$$\mathrm{Ai}(z) = \frac{1}{3\sqrt{\pi}} z^{5/4} e^{-\zeta} \int_0^\infty \frac{w(t)}{t+\zeta} \, \mathrm{d}t, \quad |\arg z| < 2\pi/3, \tag{5.12}$$

where $\zeta = \frac{2}{3} z^{3/2}$ and w is the normalized weight function $w(t) = 2^{-1/2}\pi^{-3/2}t^{-1/2}e^{-t}$ $K_{1/3}(t)$, with $K_{1/3}$ the modified Bessel function of order 1/3 (cf. Eq. (2.73)). {Hint: Use $\int_0^\infty t^{-1/2}e^{-t}K_{1/3}(t)\mathrm{d}t/(t+y) = 2\pi y^{-1/2}e^{y}K_{1/3}(y)$, $|\arg y| < \pi$ (cf. [9, §14.3, Eq. (39) for $\nu = 1/3$, $\alpha = 1$]) in combination with $K_{1/3}(y) = \pi\, 2^{1/3}3^{1/6}y^{-1/3}\mathrm{Ai}((3y/2)^{2/3})$; cf. [53, Eq. 9.6.1].

(b) *Use Gaussian quadrature relative to the weight function w to evaluate* Ai(z) *for z in the domain $|\arg z| < 2\pi/3$. In particular, determine the curves in the complex plane along which n-point Gaussian quadrature, $n = 25, 50, 100, 200$, attains a fixed accuracy ε; choose $\varepsilon = .5 \times 10^{-6}$, $.5 \times 10^{-12}$, and $.5 \times 10^{-24}$.* {Hint: Use the file `ab_Airy.txt` of the first 200 recurrence coefficients for the relevant orthogonal polynomials to 32 digits; cf. Ex. 2.33(c) and Appendix A.5.}

Answer

(a) Following the hint, we have

$$2\pi y^{-1/2}e^{y} \cdot \pi\, 2^{1/3}3^{1/6}y^{-1/3}\mathrm{Ai}\Big((3y/2)^{2/3}\Big) = \int_0^\infty \frac{t^{-1/2}e^{-t}K_{1/3}(t)}{t+y}\,\mathrm{d}t,$$

that is,

$$\mathrm{Ai}\Big((3y/2)^{2/3}\Big) = \frac{1}{\pi^2} 2^{-4/3}3^{-1/6}y^{5/6}e^{-y} \int_0^\infty \frac{t^{-1/2}e^{-t}K_{1/3}(t)}{t+y}\,\mathrm{d}t.$$

Letting $(3y/2)^{2/3} = z$, hence $y = (2/3)z^{3/2} = \zeta$, this yields

$$\mathrm{Ai}(z) = \frac{1}{3\pi^2} 2^{-1/2} z^{5/4} e^{-\zeta} \int_0^\infty \frac{t^{-1/2} e^{-t} K_{1/3}(t)}{t+\zeta} \, dt,$$

or, since $t^{-1/2} e^{-t} K_{1/3}(t) = \pi^{3/2} 2^{1/2} w(t)$,

$$\mathrm{Ai}(z) = \frac{1}{3\sqrt{\pi}} z^{5/4} e^{-\zeta} \int_0^\infty \frac{w(t)}{t+\zeta} \, dt,$$

as claimed.

(b) Since $\mathrm{Ai}(\overline{z}) = \overline{\mathrm{Ai}(z)}$, it suffices to consider the upper half-plane, $\mathrm{Im}\, z \geq 0$. Given θ, $0 \leq \theta < 2\pi/3$, we let z move along the ray $z = re^{i\theta}$, $0 \leq r < \infty$, and use a bisection method to determine $r(\theta)$ such that the relative error of Gauss quadrature is less than ε when $r < r(\theta)$. In the case $\varepsilon = .5 \times 10^{-24}$, a variable-precision bisection routine is required. This is implemented in the script run_complex_Airy.m and its symbolic analogue srun_complex_Airy.m,

```
%RUN_COMPLEX_AIRY
%
global ab0 eps0 th n
ab0=loadvpa('ab_Airy.txt',4,32,200,2)
ab0=double(ab0)
for n=[25 50 100 200]
  eps0=.5e-6;
%  eps0=.5e-12;
  tol=.5e-4;
  k=0; R=zeros(50,1); Th=zeros(50,1);
  for kappa=0:.02:.98
    k=k+1;
    r=.1; err=1; fe=ferr(err,eps0);
    while fe>0
      r=r+.1; th=kappa*(2*pi/3); z=r*exp(i*th);
      y=Ai(r);
      err=abs((y-airy(z))/airy(z));
      fe=ferr(err,eps0);
    end
    a=r-.1; b=r;
    R(k)=bisec(@ferr,a,b,tol); Th(k)=th;
  end
  k=(1:50)';
  axis('equal')
  axis([-2.5 1 0 4.3]) %for eps0=.5e-6
%  axis([-4 2 0 7]) %for eps0=.5e-12
  plot(real(R(k).*exp(i*Th(k))),imag(R(k).*exp(i*Th(k)))); ...
  set(gca,'FontSize',14)
  axis(axis)
  hold on
  z0=4.6*exp(i*2*pi/3); %for eps0=.5e-6
%  z0=7.4*exp(i*2*pi/3); %for eps0=.5e-12
  plot([0 real(z0)],[0 imag(z0)],'--')
```

```
end

function y=ferr(err,eps0)
if err>eps0, y=1; else y=0; end
```

where `bisec.m` is a bisection routine,

```
function r=bisec(fun,a,b,tol)
%BISEC Bisection method.
%   R=BISEC(FUN,A,B,TOL) computes to an absolute accuracy TOL a
%   solution R of FUN(X)=0 by the method of bisection, assuming
%   A<B and FUN(A)*FUN(B)<0. In the calling script the name of
%   the function FUN has to be prefixed by the symbol @.

global eps0 th
ntol=floor(1+log((b-a)/tol)/log(2));
s=1;
z=a*exp(i*th); y=Ai(a); err=abs((y-airy(z))/airy(z));
if fun(err,eps0)>0, s=-s; end
for n=1:ntol
  r=(a+b)/2;
  z=r*exp(i*th); y=Ai(r); err=abs((y-airy(z))/airy(z));
  if s*fun(err,eps0)<0
    a=r;
  else
    b=r;
  end
end
```

and `Ai.m` the function approximating Ai along the ray $re^{i\theta}$ by Gaussian quadrature,

```
function y=Ai(r)
global ab0 th n
z=r*exp(i*th);
zeta=(2/3)*z^(3/2); u=real(zeta); v=imag(zeta);
xw=gauss(n,ab0); x=xw(:,1); w=xw(:,2);
I=sum(w.*(x+u-i*v)./((x+u).^2+v^2));
y=(1/(3*sqrt(pi)))*z^(5/4)*exp(-2*z^(3/2)/3)*I;
```

The results are displayed in Fig. 5.2, from left to right for $\varepsilon = .5\times 10^{-6}$, $.5\times 10^{-12}$, and $.5\times 10^{-24}$. Note that in the domain away (in the northeastward direction) from the curve corresponding to n-point Gaussian quadrature, the latter will provide answers whose relative error is less than ε. It is seen that Gauss quadrature provides good approximations in the whole domain where (5.12) is defined, except for a relatively narrow band off the boundary (dashed line). Near that boundary, as one would expect, the number of Gauss quadrature points required for any given accuracy becomes prohibitively large.

Exercise 5.5. Complex inhomogeneous Airy functions (Scorer functions).

These are the solutions $y = -\mathrm{Gi}(z)$ and $y = \mathrm{Hi}(z)$ of the inhomogeneous Airy differential equation $y'' - zy = 1/\pi$, having initial values $\mathrm{Gi}(0) = \frac{1}{2}\mathrm{Hi}(0) = 1/\left(3^{7/6}\Gamma(2/3)\right)$, $\mathrm{Gi}'(0) = \frac{1}{2}\mathrm{Hi}'(0) = 1/\left(3^{5/6}\Gamma(1/3)\right)$. They are entire functions possessing the integral rep-

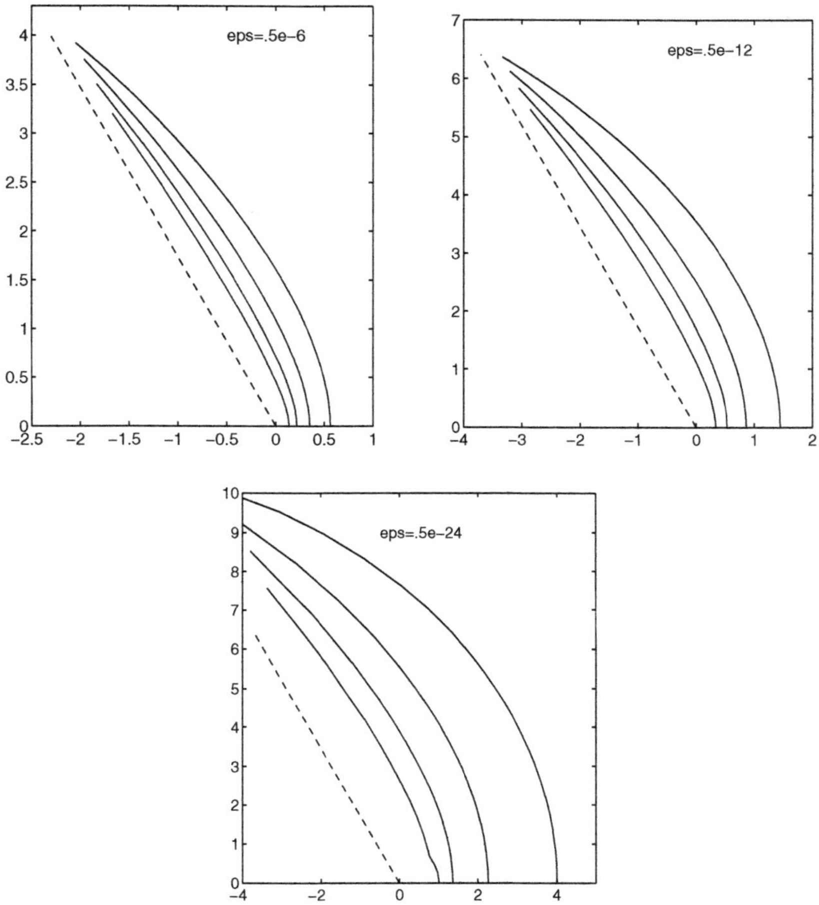

Figure 5.2. *Gauss quadrature approximation of the complex Airy function; lines of constant accuracy ε of n-point quadrature, $n = 25, 50, 100, 200$ (from right to left)*

resentations (cf. [53, Eq. 9.12.21])

$$\mathrm{Gi}(z) = -\frac{1}{\pi}\int_0^\infty e^{-zt/2}\cos(\sqrt{3}zt/2 + 2\pi/3)e^{-t^3/3}\mathrm{d}t \tag{5.13}$$

and (cf. [53, Eq. 9.12.20])

$$\mathrm{Hi}(z) = \frac{1}{\pi}\int_0^\infty e^{zt}e^{-t^3/3}\mathrm{d}t. \tag{5.14}$$

Explore the use of Gaussian quadrature relative to the weight function $w(t) = e^{-t^3/3}$ on $[0,\infty]$ (cf. Ex. 2.25) *to evaluate these functions in the complex plane.*

(a) *In each sector $S_q = \{z = re^{i\theta} : 0 \le r \le R,\ q\pi/10 \le \theta \le (q+1)\pi/10\}$, $q = 0, 1, \ldots, 9$, of the complex plane, where $R = 20$, determine the number of Gauss quadrature points needed to obtain* Gi(z) *to full double precision, i.e., to a relative accuracy of* $.5 \times 10^{-16}$. *Also determine the amount of cancellation occurring among the terms of the quadrature sums.*

(b) *Write a routine* `cGi.m` *for computing* Gi *to machine precision in the domain $|z| \le R$, $R = 20$, and test the accuracy of the routine, using a 32-digit routine* `scGi.m`. *Use* `cGi.m` *to recompute the 7-digit table in* [57, p. 112] *to 16 digits.*

(c) *Same as* (a), *but for the function* Hi.
(d) *Same as* (b), *but for the function* Hi.

Answer

(a) The first 200 recurrence coefficients for the weight function $w(t) = e^{-t^3/3}$ are available in the file `ab_hrf3mod.txt` to 32-digit accuracy (cf. Ex. 2.25(c) and Appendix A.5). The basic routine for exploration is `scorerGi.m`,

```
function [y,can]=scorerGi(z,n)
%SCORERGI The inhomogeneous Airy function (Scorer function) GI.
%    [Y,CAN]=SCORERGI(Z,N) approximates the function Y=GI(Z) for
%    complex Z by N-point Gaussian quadrature. The output
%    variable CAN measures the amount of cancellation occurring
%    within the quadrature sum. The global variables xg, wg
%    are the N Gaussian double-precision nodes and weights
%    for the weight function W(T)=EXP(-T^3/3), generated, and
%    declared global, in the calling program.

global xg wg
s=wg.*exp(-z*xg/2).*cos(z*xg*sqrt(3)/2+2*pi/3);
gmax=max(abs(s));
y=sum(s);
can=gmax/abs(y);
y=-y/pi;
```

and its symbolic analogue `sscorerGi.m`, which evaluate the n-point Gauss quadrature approximation of $\mathrm{Gi}(z)$ and also determine the amount can of cancellation occurring in the evaluation of the Gauss quadrature sum. The degree of cancellation is defined to be the absolutely largest term in the quadrature sum divided by the absolute value of the sum.

The number of quadrature points required for an accuracy level `eps0`$= .5 \times 10^{-16}$ and $.5 \times 10^{-24}$, as well as 9-digit values of Gi, are determined by the script `sN_scorerGi.m`.

```
%SN_SCORERGI

f0='%3.0f %5.1f %15.8e %15.8e %3.0f\n';
global xg wg
dig=32; digits(dig);
ab=loadvpa('ab_hrf3mod.txt',3,dig,200,2);
nest=[30;40;50;50;50;40;40;40;50;50];
%nest=[40;50;60;60;60;55;55;55;55;60];
eps0=.5e-16;
%eps0=.5e-24;
s=0;
disp(' q    r        real Gi             imag Gi        n')
for q0=0:10:90
  s=s+1;
  for q=q0:q0+10
    for r=20
      z=vpa(r)*exp(vpa(q)*vpa(pi/100)*i);
      y1=vpa(1);
```

```
        err=1; n=nest(s);
        while err>eps0
          n=n+1; if n>200, disp('n exceeds 200'); end
          y0=y1;
          xw=sgauss(dig,n,ab); xg=xw(:,1); wg=xw(:,2);
          y1=sscorerGi(dig,z,n);
          err=double(abs(y1-y0)/(1+abs(y1)));
        end
        fprintf(f0,floor(q),r,real(double(y1)),imag(double(y1)),n)
      end
    end
    fprintf('\n')
  end
```

For each sector S_q, the maximum value of n, shown in the left half of Table 5.3, is used in the script `run_sscorerGi.m` to determine the maximum degree of cancellation `canmax` along rays $z = re^{i\theta}, 0 \leq r \leq R(=20)$ and eleven equally spaced phases θ in the sector S_q, $q = 0, 1, \ldots, 9$, including the boundary phases.

Table 5.3. *Maximum number n of Gauss quadrature points needed in sector* S_q *when* eps0 $= .5 \times 10^{-16}$ *(left half) and* eps0 $= .5 \times 10^{-24}$ *(right half)*

q	n_{max}	q	n_{max}	q	n_{max}	q	n_{nmax}
0	40	5	53	0	49	5	62
1	47	6	50	1	56	6	59
2	56	7	52	2	64	7	61
3	59	8	56	3	66	8	64
4	57	9	59	4	66	9	68

```
%RUN_SSCORERGI

f0='%8.0f %12.4e %4.0f\n';
global xg wg
disp('       q     canmax    rcanmax')
N=[40;47;56;59;57;53;50;52;56;59];
dig=32; digits(dig);
ab=loadvpa('ab_hrf3mod.txt',3,dig,200,2);
s=0;
for q0=0:10:90
  s=s+1; n=N(s);
  xw=sgauss(dig,n,ab); xg=xw(:,1); wg=xw(:,2);
  for q=q0:q0+10
    canmax=0;
    for r=0:.1:20
      z=vpa(r)*exp(vpa(q)*vpa(pi/100)*i);
      [y,can]=sscorerGi(dig,z,n);
      if can>canmax
        canmax=can; rcanmax=r;
      end
    end
```

```
    fprintf(f0,s-1,canmax,rcanmax)
  end
  fprintf('\n')
end
```

The results are shown in Table 5.4. The maxima are always attained at $r = R$. Cancellation is seen to be worst in the sector S_3 with phases $54^\circ \le \theta \le 72^\circ$ and in the sector S_9 with phases $162^\circ \le \theta \le 180^\circ$, where it amounts to a loss of as much as eight decimal digits. Note that the number of quadrature points needed (as shown in Table 5.3) is largest where cancellation is worst.

Table 5.4. *Maximum degree of cancellation in Gauss quadrature in the sector* S_q

q	canmax	q	canmax
0	4.56(−1)	5	6.42(1)
1	1.04(1)	6	5.01(−1)
2	7.17(6)	7	8.34(0)
3	2.64(8)	8	1.16(4)
4	2.48(5)	9	7.83(8)

(b) The routine requested is

```
function y=cGi(z)
%CGI The inhomogeneous Airy function (Scorer function) GI.
%   Y=CGI(Z) evaluates the Scorer function Y=GI(Z) to 16 decimal
%   places for complex Z with |Z|<=20. The global variables xg,
%   wg are the nodes and weights to 32 decimal digits of the
%   59-point Gaussian quadrature rule for the weight function
%   W(T)=EXP(-T^3/3) on [0,Inf].

global xg wg
digits(32); sz=vpa(z); s3=vpa(3); spi=vpa(pi);
sy=-sum(wg.*exp(-sz*xg/2).*cos(sz*xg*sqrt(s3)/2+2*spi/s3))/spi;
y=double(sy);
```

To enhance efficiency, the generation of the 59-point Gaussian quadrature rule is done outside the routine and the result transmitted to the routine via the global variables `xg`, `wg`. The routine `sGi.m` is the same, except for the last command, which is omitted, and the global variables `xg`, `wg` being the nodes and weights of the 68-point Gaussian quadrature rule.

The script `test_cGi.m` tests the accuracy achieved by the routine `cGi.m`.

```
%TEST_CGI

f0='%8.0f %12.4e %5.1f\n';
global xg wg sxg swg
sab=loadvpa('ab_hrf3mod.txt',3,32,200,2);
xw=sgauss(32,59,sab); xg=xw(:,1); wg=xw(:,2);
sxw=sgauss(32,68,sab); sxg=sxw(:,1); swg=sxw(:,2);
disp('        q    errmax       r')
for q0=0:10:90
```

```
    for q=q0:q0+9
      for r=0:.1:20
        z=r*exp(q*(pi/100)*i);
        errmax=0;
        y=cGi(z); sy=scGi(z);
        error=double(abs(sy-y))/(1+abs(double(sy)));
        if error>=errmax
          errmax=error; rerrmax=r;
        end
      end
      fprintf(f0,floor(q/10),errmax,rerrmax)
    end
    fprintf('\n')
  end
```

It uses the maximum value $n=59$ from Table 5.3 for the routine `cGi.m`, and $n=68$ for the routine `scGi.m`. The accuracy is tested sector by sector. The relative error is found to be consistently smaller in absolute value than 9.86×10^{-17} and is one or two decimal orders smaller in sectors where cancellation is insignificant. Along each ray $z=re^{i\theta}$, $0\le r\le 20$, the error is always largest when $r=20$.

The script `table_Gi.m` generates the table in [57] to 16 digits.

```
%TABLE_GI

f0='%8.1f %21.16f\n';
global xg wg
disp('      x             Gi(x)')
ab=loadvpa('ab_hrf3mod.txt',3,32,200,2);
xw=sgauss(32,59,ab); xg=xw(:,1); wg=xw(:,2);
for x=0:1/10:10
  y=cGi(x);
  fprintf(f0,x,y)
  if abs(rem(x,1)-2/5)<1/100|abs(rem(x,1)-9/10)<1/100
    fprintf('\n')
  end
end
```

All seven digits given in this reference are successfully reproduced, except in the case $x=6.2$, where the terminal digit seems to be off by one unit (and is believed to be 2).

(c) All routines and scripts developed in part (a) have analogues for this part of the exercise, now using Gauss quadrature on the integral of (5.14). They are named in the same way, except for "`Gi`" being replaced by "`Hi`". The results regarding the number of Gauss quadrature points needed and the degrees of cancellation occurring are very similar, but side-reversed, owing to the different sign of z in the integrand of (5.14). They are summarized in Table 5.5, the number n of Gauss quadrature points for eps0 $=.5\times 10^{-24}$ in the last third of the table.

(d) The routine requested is

```
function y=cHi(z)
%CHI The inhomogeneous Airy function (Scorer function) HI.
%  Y=CHI(Z) evaluates the Scorer function Y=HI(Z) to 16
%   decimal places for complex Z with |Z| <= 20. The global
```

```
%    variables xg, wg are the nodes and weights to 32
%    digits of the 59-point Gaussian quadrature rule
%    for the weight function W(T)=exp(-T^3/3).

global xg wg
digits(32); sz=vpa(z); spi=vpa(pi);
sy=sum(wg.*exp(sz*xg))/spi;
y=double(sy);
```

Table 5.5. *The number n of Gauss quadrature points needed and the degree of cancellation in Gauss quadrature of* (5.14)

q	n	canmax	q	n	canmax	q	n	q	n
0	51	1.63(0)	5	45	2.00(0)	0	60	5	54
1	54	7.41(2)	6	38	5.71(−1)	1	63	6	48
2	59	6.65(6)	7	34	4.42(−1)	2	67	7	43
3	59	3.60(8)	8	31	4.26(−1)	3	68	8	40
4	53	7.89(4)	9	30	3.98(−1)	4	62	9	39

with `scHi.m` being the same, except for the last command, which is omitted, and the global variables `xg`, `wg` being the nodes and weights of the 68-point Gauss quadrature rule.

The relative error found by the script `test_cHi.m`, testing the routine `cHi.m`, is very similar to the one for `cGi.m`, being consistently smaller in absolute value than 9.93×10^{-17} and always largest when $r=20$. On large portions of the complex domain, including the left half-plane, the error is smaller by one or two decimal orders.

The table of $\mathrm{Hi}(-x)$ in [57] was successfully reproduced by the script `table_Hi.m`, except for two end-figure errors of one unit when $x=6.0$ and $x=7.6$, where the correct digits are believed to be 7, resp., 8.

Exercise 5.6. Complex Airy function, revisited.

The complex Airy function was already considered in Ex. 5.4, but only in the subdomain $|\arg z| < 2\pi/3$ of the complex plane. This seems unnatural since $\mathrm{Ai}(z)$ is an entire function, i.e., free of any singularities in $\mathbb{C}\backslash\infty$. The problem caused by the singular lines $|\arg z| = 2\pi/3$ in Ex. 5.4 is entirely self-inflicted by our choice (5.12) of integral representation. In Ex. 5.5, we dealt with related entire functions, the inhomogeneous Airy functions $\mathrm{Gi}(z)$ and $\mathrm{Hi}(z)$, and were able to evaluate them accurately on the whole circular domain $|z|\le 20$. This suggests expressing Ai in terms of Gi and Hi.

(a) *Prove the two identities*

$$\mathrm{Ai}(z) = -\mathrm{i}\,\mathrm{Gi}(z) + e^{\pi\mathrm{i}/6}\,\mathrm{Hi}\left(ze^{2\pi\mathrm{i}/3}\right) \tag{5.15}$$

and

$$\mathrm{Ai}(z) = \mathrm{i}\,\mathrm{Gi}(z) + e^{-\pi\mathrm{i}/6}\,\mathrm{Hi}\left(ze^{-2\pi\mathrm{i}/3}\right). \tag{5.16}$$

{Hint: Show that both sides satisfy the same differential equation and have the same initial values at the origin.}

(b) *Determine which of the two formulae* (5.15), (5.16) *provides more stability, numerically.*

(c) *Prepare a routine* `cAi.m` *evaluating the complex Airy function* Ai(z) *to* 16 *digits for* $|z| \le 20$, *using the more stable formula of* (a).

(d) *Use asymptotics to compute* Ai(z) *for* $|z| > 20$ *to comparable accuracy. Amend the routine* `cAi.m` *accordingly.* {Hint: Use a connection formula for Ai near the negative real axis.} *Also provide and test a symbolic analogue of* `cAi.m` *evaluating* Ai(z) *to an accuracy of* 24 *digits.*

(e) *Prepare routines* `cBi.m` *and* `scBi.m` *evaluating the complex Airy function* Bi(z) *to* 16, *resp.*, 24, *digits, using* $\mathrm{Bi}(z) = \mathrm{Gi}(z) + \mathrm{Hi}(z)$, *and test their accuracy.*

Answer

(a) We prove only the first identity; the proof of the second is similar.

The differential equation satisfied by the left-hand side, Ai(z), is the homogeneous Airy equation,

$$\frac{d^2 w}{dz^2} = zw,$$

while $-$Gi(z) and Hi(z) both satisfy the inhomogeneous Airy equation (cf. [53, §§9.2 and 9.12]),

$$\frac{d^2 w}{dz^2} = zw + \frac{1}{\pi}.$$

Denote the right-hand side of (5.15) by $y(z)$. Then

$$\begin{aligned}
y''(z) &= -\mathrm{i}\,\mathrm{Gi}''(z) + e^{\pi \mathrm{i}/6} e^{4\pi \mathrm{i}/3}\,\mathrm{Hi}''\left(z e^{2\pi \mathrm{i}/3}\right) \\
&= -\mathrm{i}\left[z\,\mathrm{Gi}(z) - \frac{1}{\pi}\right] + e^{\pi \mathrm{i}/6} e^{4\pi \mathrm{i}/3}\left[z e^{2\pi \mathrm{i}/3}\,\mathrm{Hi}\left(z e^{2\pi \mathrm{i}/3}\right) + \frac{1}{\pi}\right] \\
&= z\left[-\mathrm{i}\,\mathrm{Gi}(z) + e^{\pi \mathrm{i}/6}\,\mathrm{Hi}\left(z e^{2\pi \mathrm{i}/3}\right)\right] + \frac{1}{\pi}\left[\mathrm{i} + e^{3\pi \mathrm{i}/2}\right] \\
&= z y(z) + \frac{1}{\pi}[\mathrm{i} - \mathrm{i}] = z y(z),
\end{aligned}$$

so that y also satisfies the homogeneous Airy equation. The initial values of Ai are [53, Eqs. 9.2.3–9.2.4]

$$\mathrm{Ai}(0) = \frac{1}{3^{2/3}\Gamma(2/3)}, \quad \mathrm{Ai}'(0) = \frac{-1}{3^{1/3}\Gamma(1/3)}.$$

We need to show that y has the same initial values.

Since (cf. Ex. 5.5) $\mathrm{Gi}(0) = 1/\left(3^{7/6}\Gamma(2/3)\right)$, $\mathrm{Gi}'(0) = 1/\left(3^{5/6}\Gamma(1/3)\right)$ and $\mathrm{Hi}(0) = 2\mathrm{Gi}(0)$, $\mathrm{Hi}'(0) = 2\mathrm{Gi}'(0)$, we have

$$\begin{aligned}
y(0) &= -\mathrm{i}\,\mathrm{Gi}(0) + e^{\pi \mathrm{i}/6}\,\mathrm{Hi}(0) = \frac{1}{3^{7/6}\Gamma(2/3)}[-\mathrm{i} + 2(\cos(\pi/6) + \mathrm{i}\sin(\pi/6))] \\
&= \frac{1}{3^{7/6}\Gamma(2/3)}\sqrt{3} = \frac{1}{3^{2/3}\Gamma(2/3)}, \\
y'(0) &= -\mathrm{i}\,\mathrm{Gi}'(0) + e^{\pi \mathrm{i}/6} e^{2\pi \mathrm{i}/3}\,\mathrm{Hi}'(0) = \frac{1}{3^{5/6}\Gamma(1/3)}[-\mathrm{i} + 2(\cos(5\pi/6) \\
&\quad + \mathrm{i}\sin(5\pi/6))] = \frac{1}{3^{5/6}\Gamma(1/3)}[-\sqrt{3}] = \frac{-1}{3^{1/3}\Gamma(1/3)},
\end{aligned}$$

as was to be shown.

(b) The script `test_cAi.m`, using the routine `cAi_expl`,

```
%TEST_CAI

f0='%8.0f %12.4e %5.1f\n';
global xg wg
sab=loadvpa('ab_hrf3mod.txt',3,32,200,2);
xw=sgauss(32,59,sab); xg=xw(:,1); wg=xw(:,2);
disp('       q     errmax       r')
for q0=0:10:90
  for q=q0:q0+10
    errmax=0;
    for r=0:.1:20
      z=r*exp(q*(pi/100)*i);
      sz=vpa(z,24);
      ai=cAi_expl(z); sai=airy(sz);
%      ai=cAi(z);
      error=double(abs(ai-sai))/(1+abs(double(sai)));
      if error>errmax
        errmax=error; rerrmax=r;
      end
    end
    fprintf(f0,q,errmax,rerrmax)
  end
  fprintf('\n')
end

function y=cAi_expl(z)
%CAI_EXPL Exploratory complex Airy function.
%   Y=CAI_EXPL(Z) evaluates the complex Airy function AI(Z)
%   in terms of the complex inhomogeneous Airy functions
%   GI(Z) and HI(Z) in two ways, one being commented out.

global xg wg
z1=z*exp(2*pi*i/3);
y=-i*cGi(z)+exp(pi*i/6)*cHi(z1);
%z1=z*exp(-2*pi*i/3);
%y=i*Gi(z)+exp(-pi*i/6)*Hi(z1);
```

checks both formulae (5.15), (5.16) against the 24-digit MATLAB routine `airy.m` (which requires MATLAB Release R2012a or later) in the ten sectors

$$S_q = \{z = re^{i\theta} : 0 \le r \le 20,\ q\pi/10 \le \theta \le (q+1)\pi/10\}, \quad q = 0,1,2,\dots,9, \tag{5.17}$$

of the complex plane. It determines the maximum relative error along eleven equally spaced rays in S_q, including the boundary rays, in steps of $\Delta r = .1$. The results for the two formulae are shown in Table 5.6. It can be seen that the second formula (5.16) is less accurate than (5.15), by 2 decimal orders, in all sectors S_q with $q \ge 4$. The relatively large error 7.26(−15) for $q = 9$, Eq. (5.15) occurs only at, or near, $z = -20$.

(c) To obtain `cAi.m`, it suffices to delete the two commands in `cAi_expl.m` relating to (5.16).

Table 5.6. *Accuracy of $Ai(z)$ in the sector S_q using Eqs.* (5.15) *and* (5.16)

q	(5.15)	(5.16)	q	(5.15)	(5.16)
0	7.43(–17)	8.76(–17)	5	1.95(–16)	4.83(–14)
1	7.04(–17)	1.43(–16)	6	1.95(–16)	4.99(–14)
2	7.09(–17)	3.89(–16)	7	1.93(–16)	5.93(–14)
3	1.06(–15)	5.17(–14)	8	1.96(–16)	6.54(–14)
4	1.82(–16)	5.59(–14)	9	7.26(–15)	5.01(–14)

(d) For $|z| > 20$, we propose to use the Poincaré-type asymptotic expansion [53, Eq. 9.7.5] when $|\arg z| \le 9\pi/10$,

$$\mathrm{Ai}(z) \sim \frac{e^{-\zeta}}{2\sqrt{\pi} z^{1/4}} \sum_{k=0}^{\infty} (-1)^k \frac{u_k}{\zeta^k}, \quad z \to \infty \text{ in } |\arg| < \pi, \tag{5.18}$$

where

$$\zeta = \tfrac{2}{3} z^{3/2}, \quad u_0 = 1, \quad u_k = \frac{(2k+1)(2k+3)(2k+5)\cdots(6k-1)}{216^k\, k!}, \quad k = 1,2,\ldots.$$

It is implemented in the routine

```
function y=Ai_asym(z,eps0)
zeta=2*z^(3/2)/3;
u=1; s=1; k=0; sum1=1; sum0=0;
while abs(sum1-sum0)>eps0*(1+abs(sum1))
  sum0=sum1;
  k=k+1; s=-s; l=1:2:4*k-1;
  u=prod(2*k+l)/((216^k)*factorial(k));
  term=s*u/(zeta^k);
  sum1=sum0+term;
end
y=exp(-zeta)*sum1/(2*sqrt(pi)*z^(1/4));
```

and in the corresponding symbolic routine `sAi_asym.m`, where the input variable `eps0` is used to terminate the series in (5.18) as soon as two consecutive partial sums differ in absolute value by less than `eps0`. When $9\pi/10 < |\arg z| \le \pi$, we use the connection formula [53, Eq. 9.2.14]

$$\mathrm{Ai}(z) = e^{\pi i/3}\mathrm{Ai}(ze^{-2\pi i/3}) + e^{-\pi i/3}\mathrm{Ai}(ze^{2\pi i/3}) \tag{5.19}$$

in combination with `sAi_asym.m`.

The script `test_Aiasym.m`,

```
%TEST_AIASYM

f0='%8.0f %12.4e\n';
eps0=.5e-16;
disp('        q      errmax')
for q0=0:10:80
  for q=q0:q0+10
    errmax=0;
```

```
      for r=20.1:.1:50
        z=r*exp(q*pi*i/100);
        digits=24; sz=vpa(z);
%         ya=Ai_asym(z,eps0); y=airy(sz);
        ya=sAi_asym(z,eps0); y=airy(sz);
%         ya=cAi(z); y=airy(sz);
        err=abs(double(ya-y))/(1+abs(double(y)));
        if err>errmax, errmax=err; end
      end
      fprintf(f0,q,errmax)
    end
    fprintf('\n')
  end
```

checks the accuracy of the asymptotic approximation (5.18) in the nine sectors S_q, $q = 0, 1, \ldots, 8$, of (5.17) for $|z| = 20.1:.1:50$. The resulting maximum errors in S_q are shown in the left half of Table 5.7. The accuracy shown, at least for $q \geq 3$, is not quite optimal, probably because of rounding errors. Replacing the routine `Ai_asym.m` by `sAi_asym.m`, and employing 24-digit arithmetic, yields the more satisfactory maximum errors shown in the right half of Table 5.7. The maximum error in the sector S_9, where (5.19) is used, is shown under $q = 9$ in the right half of Table 5.7. It is obtained as part (q0 = 90 and r = 20.1 : .1 : 50 instead of r = 0 : .1 : 20) of the script `test_cAi`.

Table 5.7. *Accuracy of* $Ai(z)$ *for* $|z| > 20$

q	errmax	q	errmax	q	errmax	q	errmax
0	2.58(−39)	5	7.01(−14)	0	5.28(−42)	5	8.28(−18)
1	7.68(−31)	6	5.69(−14)	1	4.41(−34)	6	8.33(−18)
2	1.17(−19)	7	9.20(−14)	2	8.20(−23)	7	8.32(−18)
3	6.39(−14)	8	6.14(−14)	3	7.75(−18)	8	8.15(−18)
4	7.01(−14)	9		4	8.07(−18)	9	1.06(−16)

The routine for computing Ai(z) for $z \in \mathbb{C}$, incorporating the techniques described, is `cAi.m`,

```
function y=cAi(z)
%CAI Complex Airy function.
%   Y=CAI(Z) evaluates the complex Airy function AI(Z) to about 16
%   decimal digits using a linear combination of the inhomogeneous
%   Airy functions GI(Z) and HI(Z) when |Z|<=20 and asymptotics
%   otherwise. The global variables xg, wg are the nodes and
%   weights to 32 digits of the 59-point Gaussian quadrature rule
%   for the weight function W(T)=EXP(-T^3/3) on [0,Inf].

global xg wg
if abs(z) <= 20
  z1=z*exp(2*pi*i/3);
  y=-i*cGi(z)+exp(pi*i/6)*cHi(z1);
else
  eps0=.5e-16;
  digits=24; sz=vpa(z);
```

```
    c1=exp(vpa(pi*i/3)); c2=exp(vpa(2*pi*i/3));
    if real(z)<0 & abs(imag(z)/real(z))<abs(tan(9*pi/10))
      y=c1*sAi_asym(sz/c2,eps0)+sAi_asym(c2*sz,eps0)/c1;
    else
      y=sAi_asym(z,eps0);
    end
    y=double(y);
end
```

A slightly edited script `test_cAi.m` (calling `cAi.m` in place of `cAi_expl.m` and adjusting the range of the `r`-loop) provides an overall test of the accuracy of `cAi.m` by computing the maximum relative error for $z = re^{i\theta}$ with $r = 0 : .5 : 50$, $\theta = k\pi i/100$, $k = 0, 1, 2, \ldots, 100$. The maximum error is found to be $\leq 1.89 \times 10^{-16}$ for most of the upper half-plane (more precisely for $0 \leq k \leq 95$) and deteriorates as θ approaches π, being $\leq 2.40 \times 10^{-16}, 2.57 \times 10^{-16}, 3.96 \times 10^{-16}, 8.70 \times 10^{-16}, 6.51 \times 10^{-15}$ for, resp., $k = 96, 97, 98, 99, 100$. In about 75% of all the 11,110 cases tested, the error is $\leq .5 \times 10^{-17}$.

The symbolic analogue of `cAi.m` is

```
function y=scAi(z)
%SCAI Symbolic analogue of cAi.m.
%   The global variables sxg, swg are the nodes and weights to 32
%   digits of the 68-point Gauss quadrature rule for the weight
%   function W(T)=EXP(-T^3/3) on [0,Inf].

global sxg swg
digits(32); sz=vpa(z);
if abs(z)<=20
  s3=vpa(3); s5=vpa(5); s6=vpa(6); spi=vpa(pi);
  z1=sz*exp(2*spi*i/s3);
  y=-i*sGi(sz)+exp(spi*i/s6)*sHi(z1);
else
  eps0=.5e-32;
  e1=exp(vpa(pi*i/3)); e2=exp(vpa(2*pi*i/3));
  if real(z)<0 & abs(imag(z)/real(z))<abs(tan(9*pi/10))
    y=e1*sAi_asym(sz/e2,eps0)+sAi_asym(e2*sz,eps0)/e1;
  else
    y=sAi_asym(sz,eps0);
  end
end
```

It is tested in the script `test_scAi.m` against the 32-digit MATLAB routine `airy.m` in the ten sectors S_q of Ex. 5.5(a) with $R = 50$. The maximum relative errors observed are shown in Table 5.8, for $0 \leq |z| \leq 20$ in the left half, and for $20 < |z| \leq 50$ in the right half. They are quite satisfactory, being $< .5 \times 10^{-24}$ in 99.7% of all cases tested.

(e) The relevant routines are

```
function y=cBi(z)
%CBI Complex Airy function BI
%   Y=CBI(Z) evaluates the complex Airy function BI(Z) to about
%   16 decimal digits using BI(Z)=GI(Z)+HI(Z) when |Z|<=20 and
%   an expression in terms of AI otherwise. The global variables
%   xg, wg must be the nodes and weights to 32 decimal digits of
```

```
%   the 59-point Gaussian quadrature rule for the weight function
%   W(T)=exp(-T^3/3) on [0,Inf].

global xg wg
if abs(z)<=20
  y=cGi(z)+cHi(z);
else
  e1=exp(pi*i/6); e2=exp(2*pi*i/3);
  y=cAi(z/e2)/e1+e1*cAi(z*e2);
end
```

Table 5.8. *Accuracy of* `scAi.m`

q	errmax	q	errmax	q	errmax	q	errmax
0	4.65(−33)	5	1.39(−29)	0	8.70(−58)	5	3.03(−33)
1	5.01(−32)	6	4.22(−31)	1	7.32(−50)	6	3.09(−33)
2	1.70(−26)	7	1.69(−30)	2	1.82(−38)	7	3.08(−33)
3	8.80(−24)	8	6.78(−28)	3	2.45(−33)	8	1.40(−26)
4	1.91(−26)	9	7.98(−24)	4	2.76(−33)	9	2.27(−33)

and

```
function y=scBi(z)
%SCBI Symbolic analogue of cBi.m.
%   The global variables sxg, swg must be the nodes and weights to
%   32 decimal digits of the 70-point Gaussian quadrature rule for
%   the weight function W(T)=exp(-T^3/3) on [0,Inf].

global sxg swg
digits=32; sz=vpa(z);
if abs(double(z))<=20
  y=scGi(sz)+scHi(sz);
else
  spi=vpa(pi); s6=vpa(6); s3=vpa(3);
  e1=exp(spi*i/s6); e2=exp(2*spi*i/s3);
  y=scAi(sz/e2)/e1+e1*scAi(sz*e2);
end
```

In the script `test_cBi.m`, we use the second routine `scBi.m` to check the accuracy of the first, `cBi.m`. The maximum errors in the sector S_q (cf. Ex. 5.5(a)) are shown in Table 5.9, in the left half for $|z| \le 20$, and in the right half for $20 < |z| \le 50$. The relatively large errors of $O(10^{-14})$, when $|z| \le 20$, are isolated, but systemic when $|z| > 20$. In the former case,

Table 5.9. *Accuracy of* `cBi.m`

q	errmax	q	errmax	q	errmax	q	errmax
0	1.83(−14)	5	1.89(−16)	0	1.83(−14)	5	1.89(−16)
1	1.89(−16)	6	1.27(−15)	1	1.89(−16)	6	1.35(−13)
2	1.89(−16)	7	1.92(−16)	2	1.89(−16)	7	1.35(−13)
3	1.84(−14)	8	1.97(−16)	3	1.84(−14)	8	1.05(−13)
4	1.82(−14)	9	1.93(−14)	4	1.82(−14)	9	9.36(−14)

the error indeed is $\leq .5\times 10^{-16}$ in 61%, and $\leq 2\times 10^{-16}$ in 99.98% of all cases tested. The accuracy of `sCBi.m` could not be tested, since the variable-precision MATLAB routine for Bi(z) is currently being rewritten to correct an error. (The erroneous performance of the routine was detected while trying to test `scBi.m`.)

5.2 ▪ Polynomial least squares approximation

Exercise 5.7. (a) *Show that*

$$\hat{p}(t)=\sum_{i=0}^{n}\hat{c}_i(f)\pi_i(t;d\lambda_N),$$

where $\hat{c}_i(f)$ *are the Fourier coefficients* (5.1) *of* f, *solves the least squares problem* (5.2).

(b) *Show that the least squares error* $\|\hat{p}-f\|^2$ *is given by* (5.3).

Answer

(a) For any $p\in\mathbb{P}_n$, letting

$$p(t)=\sum_{i=0}^{n}c_i\pi_i(t;d\lambda_N),$$

we have

$$\begin{aligned}E_n^2=\|p-f\|^2&=\left(\sum_{i=0}^{n}c_i\pi_i-f,\sum_{j=0}^{n}c_j\pi_j-f\right)\\&=\sum_{i,j=0}^{n}c_ic_j(\pi_i,\pi_j)-2\sum_{i=0}^{n}c_i(f,\pi_i)+\|f\|^2.\end{aligned}$$

By orthogonality, $(\pi_i,\pi_j)=\|\pi_i\|^2\delta_{ij}$, so that

$$\begin{aligned}E_n^2&=\sum_{i=0}^{n}(c_i^2\|\pi_i\|^2-2c_i(f,\pi_i))+\|f\|^2\\&=\sum_{i=0}^{n}\left(c_i\|\pi_i\|-\frac{(f,\pi_i)}{\|\pi_i\|}\right)^2+\|f\|^2-\sum_{i=0}^{n}\frac{(f,\pi_i)^2}{\|\pi_i\|^2}.\end{aligned}\tag{5.20}$$

Evidently, this is smallest when

$$c_i=\frac{(f,\pi_i)}{\|\pi_i\|^2}=\hat{c}_i(f),$$

which proves the assertion.

(b) From (a) and (5.20), we get

$$\min_{p\in\mathbb{P}_n}E_n^2=\|f\|^2-\sum_{i=0}^{n}\frac{(f,\pi_i)^2}{\|\pi_i\|^2}.$$

Exercise 5.8. (Ex. 1 of [27, §5].) *Show that*

$$\sum_{i=0}^{n}|\hat{c}_i(f)|^2\|\pi_i\|^2\leq\|f\|^2,\quad n<N,$$

with equality holding for $n=N-1$.

Answer. The least squares error is necessarily nonnegative; hence, by Ex. 5.7(b),

$$\|f\|^2 - \sum_{i=0}^{n} \frac{(f,\pi_i)^2}{\|\pi_i\|^2} \geq 0, \quad n < N.$$

Thus, using (5.1),

$$\|f\|^2 - \sum_{i=0}^{n} |\hat{c}_i(f)|^2 \|\pi_i\|^2 \geq 0, \tag{5.21}$$

as claimed.

If $n = N-1$, the least squares polynomial $\hat{p}_{N-1}$ is the interpolation polynomial of degree $N-1$ interpolating f at the N nodes t_k, making the least squares error equal to zero. Therefore, (5.21) must hold with equality sign.

Exercise 5.9. (Ex. 2 of [27, §5].) *Prove the following alternative form for the Fourier coefficients:*

$$\hat{c}_i(f) = \frac{1}{\|\pi_i\|^2}\left(f - \sum_{j=0}^{i-1} \hat{c}_j(f)\pi_j, \pi_i\right), \quad i = 0, 1, \ldots, n. \tag{5.22}$$

Discuss its possible advantages over the original form.

Answer. By the linearity of the inner product and the orthogonality of the π_k one has

$$\left(f - \sum_{j=0}^{i-1} \hat{c}_j(f)\pi_j, \pi_i\right) = (f,\pi_i) - \sum_{j=0}^{i-1} \hat{c}_j(f)(\pi_j,\pi_i) = (f,\pi_i),$$

so that both sides divided by $\|\pi_i\|^2$, together with the definition $\hat{c}_i(f) = (f,\pi_i)/\|\pi_i\|^2$, prove the claim. The modified form (5.22) of the Fourier coefficients may reduce somewhat the cancellation error incurred when forming the inner product (f,π_i) in the original form of the Fourier coefficients.

Exercise 5.10. An equally weighted least squares approximation on $N(=10)$ equally spaced points on $[-1,1]$ (including endpoints) is implemented in the MATLAB program

```
N=10; k=(1:N)'; d=ones(1,N);
xw(k,1)=-1+2*(k-1)/(N-1); xw(:,2)=2/N;
ab=r_hahn(N-1); ab(:,1)=-1+2*ab(:,1)/(N-1);
ab(:,2)=(2/(N-1))^2*ab(:,2); ab(1,2)=2;
[phat,c]=least_squares(N-1,f,xw,ab,d);
```

Explain the details of the program and run it to reproduce the results of [27, Demo#5].

Answer. The relevant discrete orthogonal polynomials are $t_n(x;N)$, the Chebyshev orthogonal polynomials of a discrete variable (cf. [24, Example 1.15, p. 6]). They are orthogonal, however, with regard to $d\lambda_N(x) = \sum_{k=0}^{N-1} \delta(x-k)$, i.e., $x_k = k$, $w_k = 1$, $k = 0, 1, \ldots N-1$. To get the desired discrete measure and corresponding orthogonal polynomials, we must apply a preliminary transformation,

$$x = \frac{N-1}{2}(t+1), \quad -1 \leq t \leq 1,$$

that is,

$$t_k = -1 + \frac{2(k-1)}{N-1}, \quad k = 1,2,\ldots,N.$$

For the constant weights we select $w_k = 2/N$. This accounts for the first two lines of the MATLAB program, the choice d=[1,1,...,1]$\in \mathbb{R}^N$ being made because we use *monic* polynomials throughout.

The monic discrete Chebyshev polynomials $p_k(x)$ satisfy a recurrence relation (cf. [24, Table 1.2, p. 33])

$$p_{k+1}(x) = (x-a_k)p_k(x) - b_k p_{k-1}, \quad k = 0,1,2,\ldots,N-1,$$
$$p_0(x) = 1, \quad p_{-1}(x) = 0$$

(where $p_N(x) = \prod_{k=0}^{N-1}(x-k)$). Its coefficients are generated by the OPQ routine r_hahn.m with input parameters a=b=0 (cf. Appendix A.1).

```
function ab=r_hahn(N,a,b)
%R_HAHN Recurrence coefficients for monic Hahn polynomials.

if nargin<2, a=0; end; if nargin<3, b=a; end
if((N<=0)|(a<-1)|(b<-1)), error('parameter(s) out of range'), end
n=1:N; ab(1,2)=prod(1+(a+b+1)./n);
if a+b==0
  n=(1:N+1)';
  ab(n,1)=((2*n+a+b-1)*N+(b-a)*n+a)./(2*(2*n-1));
  n=(1:N)';
  ab(n+1,2)=.25*((N+1)^2)*(1+a./n).*(1+b./n).*(1-(n./(N+1)).^2)...
    ./(4-(1./n).^2);
elseif a+b+1==0
  n=(1:N+1)';
  ab(n,1)=((2*(n-1).^2+b)*N+(2*b+1)*(n-1).^2)./(4*(n-1).^2-1);
  n=(1:N)';
  ab(n+1,2)=.25*((N+1)^2)*(1+a./n).*(1+b./n).*(1-n./(N+1)).*...
    (1+(n-1)./(N+1))./(4-(1./n).^2);
else
  n=(1:N+1)';
  ab(n,1)=((n+a+b).*(n+a).*(N-n+1)./(2*n+a+b)+(n-1).*(n+b-1).*...
  (N+n+a+b)./(2*n+a+b-2))./(2*n+a+b-1);
  n=(1:N)';
  ab(n+1,2)=((N+1)^2)*(1+a./n).*(1+b./n).*(1+(a+b)./n).*...
    (1-n./(N+1)).*(1+(n+a+b)./(N+1))./(((2+(a+b)./n).^2).*...
    ((2+(a+b)./n).^2-(1./n).^2));
end
```

What we need are the monic polynomials $\pi_k(t) = (2/(N-1))^k p_k((N-1)(t+1)/2)$, which satisfy (put $x = (N-1)(t+1)/2$ in the recurrence relation above and multiply through by $(2/(N-1))^{k+1}$)

$$\pi_{k+1}(t) = \left[\frac{N-1}{2}(t+1) - a_k\right]\frac{2}{N-1}\pi_k(t) - \left(\frac{2}{N-1}\right)^2 b_k \pi_{k-1}(t),$$
$$k = 0,1,\ldots,N-1$$

(where $\pi_N(t)=\prod_{k=0}^{N-1}(t+1-2k/(N-1))$. Simplifying, we get

$$\pi_{k+1}(t)=\left[t-\left(\frac{2}{N-1}a_k-1\right)\right]\pi_k(t)-\left(\frac{2}{N-1}\right)^2 b_k\pi_{k-1}(t), \quad k=0,1,\dots,N-1,$$

so that

$$\alpha_k=\frac{2}{N-1}a_k-1,\ k=0,1,\dots,N-1;\quad \beta_0=2,\ \beta_k=\left(\frac{2}{N-1}\right)^2 b_k, \quad k=1,2,\dots,N-1.$$

This accounts for the third and fourth lines of the MATLAB program.

To produce the results of Demo#5, especially the maximum errors E_n^∞, we still need a routine for evaluating the partial sums $\hat{p}_{n-1}(t)=\sum_{i=0}^{n-1}\hat{c}_i(f)\pi_i(t;d\lambda_N)$, $n=1,2,\dots,N$. For this, we use Clenshaw's algorithm (cf. [24, Algorithm 2.2,p. 79] implemented in the OPQ routine `clenshaw.m` (cf. Appendix A.5).

```
function s=clenshaw(N,x,y0,ym1,ab,c)
%CLENSHAW Clenshaw algorithm.

if N<0, error('N out of range'), end
if size(x,2)>1, x=x'; end
lx=length(x);
if N==0, s=c(1)*y0*ones(lx,1); return, end
if size(ab,1)<N, error('array ab too short'), end
if size(c,2)<N+1, error('array c too short'), end
u=zeros(lx,N+1);
u(:,N+1)=c(N+1); u(:,N)=(x-ab(N,1)).*u(:,N+1)+c(N);
for k=N-1:-1:1
  u(:,k)=(x-ab(k,1)).*u(:,k+1)-ab(k+1,2).*u(:,k+2)+c(k);
end
s=u(:,1)*y0-ab(1,2)*u(:,2)*ym1;
```

Finally, the OPQ routine `least_squares.m` (cf. Appendix A.2) in the last line of the MATLAB program does the main work.

```
function [phat,c]=least_squares(n,f,xw,ab,d)
%LEAST_SQUARES Polynomial least squares approximation.

N=size(xw,1);
if n>N-1, error('n too large'), end
%
% Generate the matrix of orthogonal polynomials
%
p=zeros(N,n+1); p2=zeros(1,n+1); pn=ab(1,2);
p(:,1)=d(1); p2(1)=d(1)^2*pn; pn=ab(2,2)*pn;
p(:,2)=d(2)*(xw(:,1)-ab(1,1)); p2(2)=d(2)^2*pn;
for k=2:n
  pn=ab(k+1,2)*pn;
  p(:,k+1)=d(k+1)*((xw(:,1)-ab(k,1)).*p(:,k)/d(k)-ab(k,2)...
  *p(:,k-1)/d(k-1));
```

```
  p2(k+1)=d(k+1)^2*pn;
end
%
% Compute the matrix of least squares approximants and the array
% of Fourier coefficients
%
phat=zeros(N,n+1); e=f;
c(1)=sum(xw(:,2).*e.*p(:,1))/p2(1); phat(:,1)=c(1)*p(:,1);
for i=2:n+1
  c(i)=sum(xw(:,2).*e.*p(:,i))/p2(i);
  phat(:,i)=phat(:,i-1)+c(i)*p(:,i);
  if i==n+1, return, end
  e=e-c(i)*phat(:,i);
end
```

Altogether, the required driver script for Demo#5 looks as follows.

```
%DEMO5_5

N=10;
f0='%8.0f %10.2e %10.2e\n';
disp('       n    hat{E}_n    E_n^inf')
k=(1:N); E2=zeros(N,1); Einf=zeros(N,1);
xw(k,1)=-1+2*(k-1)/(N-1); xw(:,2)=2/N; d=ones(1,N);
x=xw(:,1);
f=log(2+x);
ab=r_hahn(N-1);
ab(:,1)=-1+2*ab(:,1)/(N-1);
ab(:,2)=(2/(N-1))^2*ab(:,2); ab(1,2)=2;
[phat,c]=least_squares(N-1,f,xw,ab,d);
t=linspace(-1,1)'; p=zeros(100,N);
ft=log(2+t);
for n=1:N
  E2(n)=sqrt(sum(xw(:,2).*((phat(:,n)-f).^2)));
  p(:,n)=clenshaw(n-1,t,1,0,ab,c);
  Einf(n)=max(abs(p(:,n)-ft));
end
for n=[1 4 7 10]
  fprintf(f0,n-1,E2(n),Einf(n))
end
```

It produces the results displayed in the small table of [27, p. 59]:

```
>> Demo5_5
       n    hat{E}_n    E_n^inf
       0    4.88e-01    6.37e-01
       3    2.96e-03    3.49e-03
       6    2.07e-05    7.06e-05
       9    2.45e-15    3.44e-06
>>
```

Exercise 5.11. Constrained least squares problem.

The problem is to minimize $\|p-f\|_{\mathrm{d}\lambda_N}$ over all $p\in\mathbb{P}_n$ satisfying

$$p(s_j)=f_j,\quad j=1,2,\ldots,m,\quad m\le n,$$

where s_j are given real numbers, mutually distinct and distinct from any of the support points t_k of $\mathrm{d}\lambda_N$, and f_j are prescribed data.

Reduce the problem to an unconstrained least squares problem for a new function f^* *and a new measure* $\mathrm{d}\lambda_N^*$.

Answer. The competing polynomial $p\in\mathbb{P}_n$ can be written in the form

$$p(t)=p_m(f;t)+\sigma_m(t)q(t),$$

where $p_m(f;\cdot)$ is the interpolation polynomial of degree $\le m-1$ satisfying the m constraints, and $\sigma_m(t)=\prod_{j=1}^m(t-s_j)$ the constraint polynomial. Here, $q\in\mathbb{P}_{n-m}$ is freely variable. The constrained problem then becomes

$$\text{minimize }\|f-p_m(f;\cdot)-\sigma_m q\|^2_{\mathrm{d}\lambda_N}=\int_{\mathbb{R}}\left[\frac{f(t)-p_m(f;t)}{\sigma_m(t)}-q(t)\right]^2\sigma_m^2(t)\mathrm{d}\lambda_N(t)$$

over all $q\in\mathbb{P}_{n-m}$, which is an unconstrained problem

$$\text{minimize }\|f^*-q\|_{\mathrm{d}\lambda_N^*},\quad q\in\mathbb{P}_{n-m},$$

where

$$f^*(t)=\frac{f(t)-p_m(f;t)}{\sigma_m(t)},\quad \mathrm{d}\lambda_N^*(t)=\sigma_m^2(t)\mathrm{d}\lambda_N(t).$$

With $\hat{q}\in\mathbb{P}_{n-m}$ the solution of this unconstrained problem, the polynomial

$$\hat{p}(t)=p_m(f;t)+\sigma_m(t)\hat{q}(t)$$

is the solution of the constrained problem. The function f^*, incidentally, can be given the form of a divided difference

$$f^*(t)=[s_1,s_2,\ldots,s_m,t]f,\quad t\in\operatorname{supp}\mathrm{d}\lambda_N^*,$$

as follows from the theory of interpolation. It may also be worth noting that the discrete orthogonal polynomials $\pi_k(\cdot;\mathrm{d}\lambda_N^*)$ can be obtained from the polynomials $\pi_k(\cdot;\mathrm{d}\lambda_N)$ by m modifications of the measure $\mathrm{d}\lambda_N$ by the quadratic factors $(t-s_j)^2$ (cf. [24, p. 127]).

Exercise 5.12. (Ex. 3 of [27, §5].) *Discuss the modifications needed in the solution to* Ex. 5.11 *when* ν *of the points* s_j $(0\le\nu\le m)$ *are equal to one of the support points* t_k *of* $\mathrm{d}\lambda_N$.

Answer. As in Ex. 5.11, the quantity to be minimized is

$$E_n^2=\|f-p\|^2_{\mathrm{d}\lambda_N}=\|f-p_m(f;\cdot)-\sigma_m q\|^2_{\mathrm{d}\lambda_N}.$$

The function involved on the far right trivially vanishes at all the points s_j. Therefore those points s_j that are part of the support points of $\mathrm{d}\lambda_N$ may be deleted, that is, $\mathrm{d}\lambda_N$ may be replaced by $\mathrm{d}\lambda_{N-\nu}^*$ whose support is the one of $\mathrm{d}\lambda_N$ with the ν points s_j (those equal to some t_k) deleted and whose jumps at the surviving support points are the original jumps of $d\lambda_N$. The unconstrained least squares procedure, then, is the same as the one in Ex. 5.11, but with $d\lambda^*_{N-\nu}$ instead of $d\lambda^*_N$.

Exercise 5.13. (Ex. 4 of [27, §5].) *Consider $f(t)=J_0(t)$, $0\le t\le j_{0,3}$, in the constrained least squares problem of* Ex. 5.11, *where J_0 is the Bessel function of order zero, $j_{0,3}$ the third positive zero of J_0, and where the constraints are that p, like J_0, should also vanish at the first three positive zeros $j_{0,1}, j_{0,2}, j_{0,3}$ of J_0. What are $p_m(f;\cdot)$, f^*, and σ_m in this case?*

Answer. Here, $m=3$, and the interpolation polynomial $p_3(J_0;t)\equiv 0$. Furthermore,

$$f^*(t)=\frac{J_0(t)}{\sigma_3(t)},\quad \sigma_3(t)=(t-j_{0,1})(t-j_{0,2})(t-j_{0,3}).$$

Exercise 5.14. (a) (Ex. 5 of [27, §5].) *Calculate the first and second derivatives of the complementary error function $f(t)=\frac{2}{\sqrt{\pi}}e^{t^2}\int_t^\infty e^{-u^2}du$.*

(b) *Consider the problem of computing the Sobolev least squares approximation $p\in\mathbb{P}_n$ to the complementary error function of* (a) *on* $[0,2]$, *that is, the solution of*

$$\sum_{\sigma=0}^{s}\sum_{k=1}^{N} w_k^{(\sigma)}[p^{(\sigma)}(t_k)-f_k^{(\sigma)}]^2=\min,$$

where f is the complementary error function and $s=2$, $N=5$, $t_k=2(k-1)/(N-1)=(k-1)/2$, and $w_k^{(\sigma)}=1/N$ for $\sigma=0,1,2$. Write MATLAB programs to reproduce the results of [27, Demo#8] *comparing Sobolev least squares errors with maximum errors and ordinary least squares errors.*

Answer

(a) The product rule of differentiation yields

$$f'(t)=\frac{2}{\sqrt{\pi}}\left\{2te^{t^2}\int_t^\infty e^{-u^2}du-1\right\}=2tf(t)-\frac{2}{\sqrt{\pi}},$$
$$f''(t)=2f(t)+2tf'(t)=2f(t)+2t\left[2tf(t)-\frac{2}{\sqrt{\pi}}\right]=2(1+2t^2)f(t)-\frac{4}{\sqrt{\pi}}t.$$

(b) There are four routines, altogether, that will accomplish the task. First the main routine `least_squares_sob.m`, the analogue for Sobolev orthogonal polynomials of the earlier least squares routine (cf. Appendix A.2).

```
function [phat,c]=least_squares_sob(n,f,xw,B)
%lEAST_SQUARES_SOB Polynomial least squares approximation in
%Sobolev space.

N=size(xw,1); s=size(xw,2)-2;
Ns=N*(s+1); p=zeros(n+1,Ns); p2=zeros(1,n+1);
%
% Generate the matrix of Sobolev orthogonal polynomials
% and their derivatives, along with the array of their norms
%
```

```
sN=1:s+1:Ns; p(1,sN)=1;
p2(1)=sum(xw(:,2));
t=xw(:,1)';
for k=1:n
  for sig=1:s+1
    sigN=sig:s+1:Ns;
    bsum=zeros(1,N);
    for j=1:k
      bsum=bsum+B(j,k)*p(k-j+1,sigN);
    end
    if sig==1
      p(k+1,sigN)=t.*p(k,sigN)-bsum;
    else
      p(k+1,sigN)=t.*p(k,sigN)+(sig-1).*p(k,sigN-1)-bsum;
    end
    p2(k+1)=p2(k+1)+sum(xw(:,1+sig)'.*(p(k+1,sigN).^2));
  end
end
%
% Compute the matrix of least squares approximants and their
% derivatives, along with the array of Fourier coefficients
%
c=zeros(n+1,1);
phat=zeros(n+1,Ns); e=f;
for k=1:n+1
  for sig=1:s+1
    sigN=sig:s+1:Ns;
    c(k)=c(k)+sum(xw(:,1+sig)'.*e(:,sig)'.*p(k,sigN));
  end
  c(k)=c(k)/p2(k);
  if k==1
    for sig=1:s+1
      sigN=sig:s+1:Ns;
      phat(1,sigN)=c(1)*p(1,sigN);
    end
  else
    for sig=1:s+1
      sigN=sig:s+1:Ns;
      phat(k,sigN)=phat(k-1,sigN)+c(k)*p(k,sigN);
    end
  end
  if k==n+1, return, end
  for sig=1:s+1
    sigN=sig:s+1:Ns;
    e(:,sig)=e(:,sig)-c(k)*p(k,sigN)';
  end
end
```

It requires as input the matrix `B` of the recurrence coefficients of Sobolev orthogonal polynomials, which here is generated by the routine `stieltjes_sob.m` (cf. Appendix A.1).

```
function [B,normsq]=stieltjes_sob(n,s,nd,xw,a0,same)
%STIELTJES_SOB Stieltjes algorithm for Sobolev orthogonal polynomials.

if n<1, error('n out of range'), end
if s<0, error('s out of range'), end
if s>=n, s=n-1; end
if any(nd<1), error('nd has nonpositive components'), end
md=max(nd); if size(xw,1)<md, error('columns of xw too short'), end
B=zeros(n);
%
% Initialization
%
p=zeros(n,s+1,md); p(1,1,:)=1; xp=zeros(n,s+1,s+1,md);
for is=1:s+1
  for js=1:s+1
    if(same&is==js | ~same)
      for mu=1:nd(js)
        if is==1
          xp(1,is,js,mu)=xw(mu,js);
        elseif is==2
          xp(1,is,js,mu)=1;
        else
          xp(1,is,js,mu)=0;
        end
      end
    end
  end
end
B(1,1)=a0;
if n==1, return, end
%
% Continuation
%
for k=2:n
  for j=1:k
    for is=1:s+1
      for nu=1:nd(is)
        sm=xp(k-1,is,is,nu);
        for l=1:k-1
          sm=sm-B(l,k-1)*p(k-l,is,nu);
        end
        p(k,is,nu)=sm;
      end
    end
    Bnum=0;
    Bden=0;
    for is=1:s+1
      for js=1:s+1
        if(same&is==js | ~same)
          for mu=1:nd(js)
```

```
                xsm=xw(mu,js)*xp(k-1,is,js,mu);
                if is>1
                  if same
                    xsm=xsm+(is-1)*xp(k-1,is-1,is-1,mu);
                  else
                    xsm=xsm+(is-1)*xp(k-1,is-1,js,mu);
                  end
                end
                for l=1:k-1
                  xsm=xsm-B(l,k-1)*xp(k-l,is,js,mu);
                end
                xp(k,is,js,mu)=xsm;
                if is==js
                  Bnum=Bnum+xw(mu,is+s+1)*xp(k,is,js,mu)*p(k-j+1,is,mu);
                  Bden=Bden+xw(mu,is+s+1)*p(k-j+1,is,mu)^2;
                end
              end
            end
          end
        end
        B(j,k)=Bnum/Bden;
        if k==n, normsq(n-j+1)=Bden; end
      end
    end
```

The routine for evaluating Sobolev Fourier sums and their first two derivatives is clenshaw_sob.m, an extension of Clenshaw's original algorithm (cf. Appendix A.5).

```
function r=clenshaw_sob(n,N,s,x,B,c)
%CLENSHAW_SOB Clenshaw-like algorithm for Sobolev orthogonal
%polynomials

if N<0|n<0|n>N, error('N or n out of range'), end
if s>2, error('s too large'), end
if size(x,2)>1, x=x'; end
lx=length(x);
if max(size(B)<[N N])>0, error('array B too small'), end
if size(c,2)<n+1, error('array c too short'), end
r=zeros(lx,s+1);
u=zeros(lx,n+1); v=zeros(lx,n+1); w=zeros(lx,n+1);
u(:,n+1)=c(n+1); v(:,n+1)=0; w(:,n+1)=0;
for k=n:-1:1
  u(:,k)=(x-B(1,k)).*u(:,k+1)+c(k);
  if k<n
    for l=1:n-k
      u(:,k)=u(:,k)-B(l+1,k+l).*u(:,k+l+1);
    end
  end
end
```

```
r(:,1)=u(:,1);
if s>0
  for k=n:-1:1
    v(:,k)=(x-B(1,k)).*v(:,k+1)+u(:,k+1);
    if k<n
      for l=1:n-k
        v(:,k)=v(:,k)-B(l+1,k+l).*v(:,k+l+1);
      end
    end
  end
  r(:,2)=v(:,1);
  if s>1
    for k=n:-1:1
      w(:,k)=(x-B(1,k)).*w(:,k+1);
      if k>1, w(:,k)=w(:,k)+u(:,k+1); end
      if k<n
        for l=1:n-k
          w(:,k)=w(:,k)-B(l+1,k+l).*w(:,k+l+1);
        end
      end
    end
    for k=n:-1:1
      v(:,k)=(x-B(1,k)).*v(:,k+1)+w(:,k+1);
      if k<n
        for l=1:n-k
          v(:,k)=v(:,k)-B(l+1,k+l).*v(:,k+l+1);
        end
      end
    end
    r(:,3)=2*v(:,1);
  end
end
```

Finally, there is the driver script `Demo5_8.m` generating the results of Demo#8.

```
%DEMO5_8

fprintf('\n')
f0='%1.0f %2.0f %12.3e %12.3e %12.3e %12.3e\n';
f1='%4.0f %12.3e %12.3e %12.3e %12.3e\n';
disp('s  n      E2           Einf0         Einf1         Einf2')
N=5; s=2; same=1;
%
% Sobolev least squares
%
s=2; Ns=N*(s+1);
t=linspace(0,2,N)'; nd=N*ones(1,s+1);
xw=ones(N,2*s+2)/N;
for sig=1:s+1
```

```
  xw(:,sig)=t;
end
xw0=xw(:,s+1:2*s+2);
a0=sum(t)/N;
[B,normsq]=stieltjes_sob(3*N-1,s,nd,xw,a0,same);
f=[exp(t.^2).*erfc(t) 2*t.*exp(t.^2).*erfc(t)-2/sqrt(pi)...
   2*(1+2*t.^2).*exp(t.^2).*erfc(t)-4*t/sqrt(pi)];
for n=0:3*N-1
  [phat,c]=least_squares_sob(n,f,xw0,B);
end
x=linspace(0,2)';
fx=[exp(x.^2).*erfc(x) 2*x.*exp(x.^2).*erfc(x)-2/sqrt(pi)...
    2*(1+2*x.^2).*exp(x.^2).*erfc(x)-4*x/sqrt(pi)];
E2=zeros(3*N,1);
Einf0=zeros(3*N,1); Einf1=zeros(3*N,1); Einf2=zeros(3*N,1);
for n=[1 3 5 10 15]
  for sig=1:s+1
    sigN=sig:s+1:Ns;
    E2(n)=E2(n)+sum(xw0(:,sig+1)'.*...
      (phat(n,sigN)-f(:,sig)').^2);
  end
  E2(n)=sqrt(E2(n));
  p=clenshaw_sob(n-1,3*N-1,s,x,B,c');
  Einf0(n)=max(abs(p(:,1)-fx(:,1)));
  Einf1(n)=max(abs(p(:,2)-fx(:,2)));
  Einf2(n)=max(abs(p(:,3)-fx(:,3)));
  if n==1
    fprintf(f0,s,n-1,E2(n),Einf0(n),Einf1(n),Einf2(n))
  else
    fprintf(f1,n-1,E2(n),Einf0(n),Einf1(n),Einf2(n))
  end
end
%
% Ordinary least squares
%
fprintf('\n')
s=0; Ns=N*(s+1);
t=linspace(0,2,N)'; nd=N*ones(1,s+1);
xw=ones(N,2*s+2)/N;
for sig=1:s+1
  xw(:,sig)=t;
end
xw0=xw;
a0=sum(t)/N;
[B,normsq]=stieltjes_sob(N-1,s,nd,xw,a0,same);
f=[exp(t.^2).*erfc(t) 2*t.*exp(t.^2).*erfc(t)-2/sqrt(pi)...
   2*(1+2*t.^2).*exp(t.^2).*erfc(t)-4*t/sqrt(pi)];
for n=0:N-1
  [phat,c]=least_squares_sob(n,f,xw0,B);
end
```

```
x=linspace(0,2)';
fx=[exp(x.^2).*erfc(x) 2*x.*exp(x.^2).*erfc(x)-2/sqrt(pi)...
    2*(1+2*x.^2).*exp(x.^2).*erfc(x)-4*x/sqrt(pi)];
E2=zeros(N,1);
Einf0=zeros(N,1); Einf1=zeros(N,1); Einf2=zeros(N,1);
for n=[1 3 5]
  for sig=1:s+1
    sigN=sig:s+1:Ns;
    E2(n)=E2(n)+sum(xw0(:,sig+1)'.*...
      (phat(n,sigN)-f(:,sig)').^2);
  end
  E2(n)=sqrt(E2(n));
  p=clenshaw_sob(n-1,N-1,s+2,x,B,c');
  Einf0(n)=max(abs(p(:,1)-fx(:,1)));
  Einf1(n)=max(abs(p(:,2)-fx(:,2)));
  Einf2(n)=max(abs(p(:,3)-fx(:,3)));
  if n==1
    fprintf(f0,s,n-1,E2(n),Einf0(n),Einf1(n),Einf2(n))
  else
    fprintf(f1,n-1,E2(n),Einf0(n),Einf1(n),Einf2(n))
  end
end
```

The output of `Demo5_8.m` (whose format was slightly edited by hand) is displayed below, showing in the last four columns the least squares error (E2) and the maximum errors in the function and the first two derivatives (`Einf0`–`Einf2`), in the upper part for Sobolev least squares approximation and in the lower part for ordinary least squares approximation.

```
>> Demo5_8

 s  n      E2          Einf0        Einf1        Einf2
 2  0   1.153e+000   4.759e-001   1.128e+000   2.000e+000
    2   7.356e-001   8.812e-002   2.860e-001   1.411e+000
    4   1.196e-001   1.810e-002   5.434e-002   1.960e-001
    9   2.178e-005   4.710e-006   3.011e-005   3.159e-004
   14   2.532e-016   1.130e-009   1.111e-008   1.966e-007

 0  0   2.674e-001   4.759e-001   1.128e+000   2.000e+000
    2   2.245e-002   3.865e-002   3.612e-001   1.590e+000
    4   4.300e-017   3.516e-003   5.160e-002   4.956e-001
>>
```

Note that the Sobolev least squares error E2 is zero in theory, when $n = 14$, since the Hermite interpolation polynomial of degree $3N-1$ ($= 14$ when $N = 5$) interpolates the data exactly. Likewise, the ordinary least squares error is zero when $n = 4$. This is clearly visible in the numerical results obtained. On the other hand, the Sobolev least squares polynomial is capable of achieving much smaller maximum errors, when n is taken large enough, than the ordinary least squares polynomial, for which $n \leq 4$ are the only values of n that are reasonable.

5.3 ▪ Moment-preserving approximation

Exercise 5.15. (a) *Determine the coefficients* a_ν *and abscissae* t_ν *in the* n*-term linear combination of Dirac delta functions*

$$d_n(t)=\sum_{\nu=1}^{n} a_\nu \delta(t-t_\nu), \quad 0<t_1<t_2<\cdots<t_n,\ t\in\mathbb{R}_+,$$

such that as many as possible of its moments $\int_0^\infty t^j d_n(t)\mathrm{d}t$, $j=0,1,2,\ldots$, *agree with the moments* $\mu_j=\int_0^\infty t^j f(t)\mathrm{d}t$ *of a given function* $f\geq 0$ *on* $\mathbb{R}_+$.

(b) *Apply the result of* (a) *to the half-range Hermite distribution* $f(t)\mathrm{d}t=e^{-t^2}\mathrm{d}t$ *on* $\mathbb{R}_+$.

Answer

(a) The requirement of moment preservation is

$$\int_0^\infty t^j \sum_{\nu=1}^{n} a_\nu \delta(t-t_\nu)\mathrm{d}t=\int_0^\infty t^j f(t)\mathrm{d}t, \quad j=0,1,2,\ldots,$$

assuming that the moments on the right exist. Thus, by the defining property of Dirac delta functions,

$$\sum_{\nu=1}^{n} a_\nu t_\nu^j=\int_0^\infty t^j f(t)\mathrm{d}t, \quad j=0,1,2,\ldots.$$

The maximum value possible of j is $2n-1$, in which case the $2n$ equations identify a_ν and t_ν as the weights $a_\nu=a_\nu^G$ and nodes $t_\nu=t_\nu^G$ of the n-point Gaussian quadrature rule for the measure $\mathrm{d}\lambda(t)=f(t)\mathrm{d}t$. Since f is nonnegative (and, of course, not identically zero), this rule exists uniquely.

(b) In the case of the half-range Hermite measure $\mathrm{d}\lambda(t)=e^{-t^2}\mathrm{d}t$ on $\mathbb{R}_+$, the first 200 relevant recurrence coefficients are available to 32 decimal digits in the file `ab_hrhermite.txt` (cf. Appendix A.5); see also the answer to Ex. 2.17(d). The required Gaussian quadrature rules are then easily obtained by the `OPQ` routine `gauss.m` or the `SOPQ` routine `sgauss.m` (cf. Appendix A.2, resp., Appendix B.2).

Exercise 5.16.* (Ex. 6* of [27, §5].) *Consider the problem of determining a spline function*

$$s_{n,m}(t)=\sum_{\nu=1}^{n} a_\nu (t_\nu-t)_+^m, \quad t_\nu>0, \quad t\in\mathbb{R}_+, \tag{5.23}$$

such that

$$\int_0^\infty t^j s_{n,m}(t)\mathrm{d}t=\mu_j, \quad \mu_j=\int_0^\infty t^j f(t)\mathrm{d}t, \quad j=0,1,2,\ldots,2n-1. \tag{5.24}$$

Assume that $f\in C^{m+1}(\mathbb{R}_+)$, $\int_0^\infty t^j f(t)\mathrm{d}t$ *exists for* $0\leq j\leq 2n-1$, *and that*

$$f^{(\mu)}(t)=o(t^{-2n-\mu}), \quad \mu=0,1,\ldots,m, \quad \text{as } t\to\infty. \tag{5.25}$$

Explore the connection of this problem with Gauss quadrature and find a way to determine the constants a_ν, t_ν *in* (5.23). {Hint: Apply repeated integrations by parts to both integrals in (5.24).}

Answer. Substituting (5.23) into (5.24), one obtains, since $t_\nu > 0$,

$$\sum_{\nu=1}^{n} a_\nu \int_0^{t_\nu} t^j (t_\nu - t)^m \mathrm{d}t = \int_0^\infty t^j f(t) \mathrm{d}t, \quad j = 0, 1, 2, \ldots, 2n-1. \tag{5.26}$$

To the integrals on the left we apply m integrations by parts to get

$$\begin{aligned}\int_0^{t_\nu} t^j (t_\nu - t)^m \mathrm{d}t &= \frac{m}{j+1} \int_0^{t_\nu} t^{j+1} (t_\nu - t)^{m-1} \mathrm{d}t \\ &= \frac{m(m-1)}{(j+1)(j+2)} \int_0^{t_\nu} t^{j+2} (t_\nu - t)^{m-2} \mathrm{d}t = \cdots \\ &= \frac{m(m-1)\cdots 2}{(j+1)(j+2)\cdots(j+m-1)} \int_0^{t_\nu} t^{j+m-1} (t_\nu - t) \mathrm{d}t \\ &= \frac{m!}{(j+1)(j+2)\cdots(j+m)} \int_0^{t_\nu} t^{j+m} \mathrm{d}t \\ &= \frac{m!}{(j+1)(j+2)\cdots(j+m)(j+m+1)} t_\nu^{j+m+1},\end{aligned}$$

so that the left-hand side of (5.26) becomes

$$\frac{m!}{(j+1)(j+2)\cdots(j+m+1)} \sum_{\nu=1}^{n} a_\nu t_\nu^{j+m+1}. \tag{5.27}$$

To the integral on the right, we similarly apply $m+1$ integrations by parts. To exhibit the arguments involved, we carry out in detail the first integration by parts. For any $b > 0$, there holds

$$\int_0^b t^j f(t) \mathrm{d}t = \frac{t^{j+1}}{j+1} f(t) \bigg|_0^b - \frac{1}{j+1} \int_0^b t^{j+1} f'(t) \mathrm{d}t.$$

The integrated term clearly vanishes at $t = 0$, and also as $t \to \infty$ by (5.25) for $\mu = 0$. Since $j \leq 2n-1$, the integral on the left, by assumption, converges as $b \to \infty$. Therefore, also the integral on the right converges, and we have

$$\int_0^\infty t^j f(t) \mathrm{d}t = -\frac{1}{j+1} \int_0^\infty t^{j+1} f'(t) \mathrm{d}t.$$

Continuing in this manner, using the assumption (5.25) to show convergence to zero of the integrated term at the upper limit $t = \infty$ (its value at the lower limit $t = 0$ always being zero), and using the existence of $\int_0^\infty t^{j+\mu} f^{(\mu)}(t) \mathrm{d}t$ already established to infer the existence of $\int_0^\infty t^{j+\mu+1} f^{(\mu+1)}(t) \mathrm{d}t$, $\mu = 1, 2, \ldots, m$, we arrive at

$$\int_0^\infty t^j f(t) \mathrm{d}t = \frac{(-1)^{m+1}}{(j+1)(j+2)\cdots(j+m+1)} \int_0^\infty t^{j+m+1} f^{(m+1)}(t) \mathrm{d}t.$$

This is required to be equal to the left-hand side of (5.26), that is, to (5.27), for $j = 0, 1, \ldots, 2n-1$, giving

$$\sum_{\nu=1}^{n} (a_\nu t_\nu^{m+1}) t_\nu^j = \int_0^\infty \frac{(-1)^{m+1}}{m!} t^{m+1} f^{(m+1)}(t) \cdot t^j \mathrm{d}t, \quad j = 0, 1, 2, \ldots, 2n-1.$$

These are precisely the conditions for t_ν to be the nodes τ_ν^G, and for $a_\nu t_\nu^{m+1}$ to be the weights λ_ν^G, of the n-point Gaussian quadrature rule with measure

$$\mathrm{d}\lambda^{[m]}(t) = \frac{(-1)^{m+1}}{m!}\, t^{m+1} f^{(m+1)}(t)\mathrm{d}t \quad \text{on } [0,\infty]. \tag{5.28}$$

Thus, our problem has a unique solution, given by

$$t_\nu = \tau_\nu^G, \;\; a_\nu = \frac{\lambda_\nu^G}{[\tau_\nu^G]^{m+1}}, \quad \nu = 1,2,\dots,n, \tag{5.29}$$

precisely if the n-point Gaussian quadrature rule for $\mathrm{d}\lambda^{[m]}$ exists uniquely. (This is not necessarily the case, since $f^{(m+1)}$ may not be of constant sign.)

Exercise 5.17. (Ex. 7 of [27, §5].) *Derive the measure* $\mathrm{d}\lambda^{[m]}$ *for the half-range Hermite distribution* $f(t)\mathrm{d}t = e^{-t^2}\mathrm{d}t$ *on* $\mathbb{R}_+$ (cf. [27, Example 17]).

Answer. We have [1, Eq. 7.1.19]

$$f^{(m+1)}(t) = \frac{\sqrt{\pi}}{2}[\mathrm{erf}(t)]^{(m+2)} = (-1)^{m+1} H_{m+1}(t) e^{-t^2},$$

where H_{m+1} is the Hermite polynomial of degree $m+1$. Therefore,

$$\mathrm{d}\lambda^{[m]}(t) = \frac{(-1)^{m+1}}{m!}\, t^{m+1} f^{(m+1)}(t)\mathrm{d}t = \frac{1}{m!} t^{m+1} H_{m+1}(t) e^{-t^2}\mathrm{d}t \quad \text{on } \mathbb{R}_+.$$

When $m = 0$, the right-hand side, $2t^2 e^{-t^2}\mathrm{d}t$, is twice the so-called Maxwell velocity distribution.

Exercise 5.18. *Specialize the result* (5.28), (5.29) *for the spline function* $s_{n,m}$ *in* (5.23) *to* $m = 0$, *expressing* $s_{n,0}$ *in terms of the Heaviside step function*

$$H(x) = \begin{cases} 1 & \text{if } x \ge 0, \\ 0 & \text{if } x < 0. \end{cases}$$

Apply this for $n = 5$ *to the function* $f(t) = e^{-t^2}$ *on* $[0,\infty]$ *and plot both* f *and* $s_{n,0}$ (the latter in red).

Answer. Clearly

$$s_{n,0}(t) = \sum_{\nu=1}^{n} a_\nu H(t - t_\nu), \tag{5.30}$$

and $\mathrm{d}\lambda^{[0]}(t) = 2te^{-t^2}\mathrm{d}t$ is twice the Maxwell velocity distribution. The recurrence coefficients of the Maxwell distribution are generated from its moments $\mu_k = \Gamma((k+3)/2)/2$, $k = 0,1,2,\dots$, by the routines `r_maxwell.m` and the symbolic analogue `sr_maxwell.m`, both using the classical Chebyshev algorithm. When $n = 5$, and for plotting accuracy, the former routine (in MATLAB double precision) is more than enough. It is combined in the script `sn0_herm.m` below with the routine `stepfunc.m` evaluating a linear combination of Heaviside step functions as on the right of (5.30).

```
%SNO_HERM Step function approximation of the Hermite
%   weight function on [0,Inf].

n=5;
ab=r_maxwell(n); ab(1,2)=2*ab(1,2);
xw=gauss(n,ab);
xw(:,2)=xw(:,2)./xw(:,1);
nx=500;
x=linspace(0,max(xw(:,1)),nx);
y=zeros(1,nx);
y=stepfunc(n,x,xw);
hold on
ymaxwell=exp(-x.^2);
plot(x',ymaxwell','LineWidth',1.5);set(gca,'FontSize',14)
plot(x',y','r-','LineWidth',1.5);set(gca,'FontSize',14)
plot([xw(1,1) xw(1,1)],[y(75) 0],':')
plot([xw(2,1) xw(2,1)],[y(150) 0],':')
plot([xw(3,1) xw(3,1)],[y(250) 0],':')
axis([0 xw(n,1)+.1 -.1 1.1])
text(xw(1,1),-.05,'t_1')
text(xw(2,1),-.05,'t_2')
text(xw(3,1),-.05,'t_3')
text(xw(4,1),-.05,'t_4')
text(xw(5,1),-.05,'t_5')
hold off
```

The resulting plot is shown in Fig. 5.3.

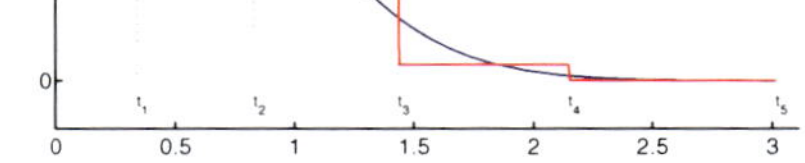

Figure 5.3. *Moment-preserving step function approximation of the half-range Hermite weight function*

5.4 ▪ Summation by integration

Exercise 5.19. Let $(\mathscr{L}f)(s)=\int_0^\infty e^{-st}f(t)\mathrm{d}t$ be the Laplace transform of f.

(a) *Consider the series*

$$s=\sum_{k=1}^{\infty}a_k, \quad a_k=(\mathscr{L}f)(k).$$

Show that

$$s=\int_0^\infty \frac{f(t)}{t}\,\frac{t}{e^t-1}\,\mathrm{d}t. \tag{5.31}$$

(b) (Ex. 12 of [27, §5].) *Consider the series*

$$s=\sum_{k=1}^\infty a_k, \quad a_k=-\frac{\mathrm{d}}{\mathrm{d}s}(\mathscr{L}f)(s)\Big|_{s=k}.$$

Show that

$$s=\int_0^\infty f(t)\frac{t}{e^t-1}\,\mathrm{d}t. \tag{5.32}$$

(c) (Ex. 10 of [27, §5].) *Consider the alternating series*

$$s=\sum_{k=1}^\infty a_k, \quad a_k=(-1)^{k-1}(\mathscr{L}f)(k).$$

Show that

$$s=\int_0^\infty f(t)\frac{1}{e^t+1}\,\mathrm{d}t. \tag{5.33}$$

Answer

(a) We have

$$\begin{aligned} s&=\sum_{k=1}^\infty\int_0^\infty e^{-kt}f(t)\mathrm{d}t=\int_0^\infty\sum_{k=1}^\infty e^{-(k-1)t}\cdot e^{-t}f(t)\mathrm{d}t\\ &=\int_0^\infty\frac{1}{1-e^{-t}}e^{-t}f(t)\mathrm{d}t=\int_0^\infty\frac{f(t)}{t}\,\frac{t}{e^t-1}\,\mathrm{d}t. \end{aligned}$$

(b) We have

$$\begin{aligned} s&=\sum_{k=1}^\infty\int_0^\infty te^{-kt}f(t)\mathrm{d}t=\int_0^\infty t\left(\frac{1}{1-e^{-t}}-1\right)f(t)\mathrm{d}t\\ &=\int_0^\infty t\,\frac{e^{-t}}{1-e^{-t}}f(t)\mathrm{d}t=\int_0^\infty f(t)\frac{t}{e^t-1}\,\mathrm{d}t. \end{aligned}$$

(c) We have

$$\begin{aligned} s&=\sum_{k=1}^\infty(-1)^{k-1}\int_0^\infty e^{-kt}f(t)\mathrm{d}t=\int_0^\infty\sum_{k=1}^\infty(-1)^{k-1}e^{-(k-1)t}\cdot e^{-t}f(t)\mathrm{d}t\\ &=\int_0^\infty\frac{1}{1+e^{-t}}e^{-t}f(t)\mathrm{d}t=\int_0^\infty f(t)\frac{1}{e^t+1}\,\mathrm{d}t. \end{aligned}$$

Exercise 5.20. (Ex. 8 of [27, §5], expanded.) (a) *For the Theodorus constant* (cf. Example 18 of [27, §5])

$$\sum_{k=1}^\infty a_k, \quad a_k=\frac{1}{k^{1/2}+k^{3/2}},$$

determine the function f *in* Ex. 5.19(a). {Hint: Use the convolution theorem for Laplace transforms.}

(b) *Use the result of* (a) *to compute the Theodorus constant to* 32 *decimal digits.* {Hint: For the integral $\int_0^\infty [f(t)/t][t/(e^t-1)]\mathrm{d}t$ find a measure of integration such that the integrand becomes an entire function, and then use Gaussian quadrature.}

Answer

(a) Since

$$a_k = k^{-1/2}\cdot(1+k)^{-1}$$

and

$$k^{-1/2} = \left(\mathcal{L}\,\frac{t^{-1/2}}{\sqrt{\pi}}\right)(k),\quad (1+k)^{-1} = \left(\mathcal{L}e^{-t}\right)(k),$$

we have by the convolution theorem for Laplace transforms,

$$\mathcal{L}g\cdot\mathcal{L}h = \mathcal{L}g*h,\quad (g*h)(t) := \int_0^t g(\tau)h(t-\tau)\mathrm{d}\tau,$$

that

$$a_k = \left(\mathcal{L}\,\frac{t^{-1/2}}{\sqrt{\pi}}\right)(k)\cdot\left(\mathcal{L}e^{-t}\right)(k) = \left(\mathcal{L}\int_0^t \frac{\tau^{-1/2}}{\sqrt{\pi}}e^{-(t-\tau)}\mathrm{d}\tau\right)(k),$$

and hence

$$f(t) = \frac{1}{\sqrt{\pi}}\int_0^t \tau^{-1/2}e^{-(t-\tau)}\mathrm{d}\tau = \frac{e^{-t}}{\sqrt{\pi}}\int_0^t \tau^{-1/2}e^{\tau}\mathrm{d}\tau.$$

In the last integral we change variable $\tau = u^2$, to get

$$f(t) = \frac{e^{-t}}{\sqrt{\pi}}\int_0^{\sqrt{t}} u^{-1}e^{u^2}2u\,\mathrm{d}u = \frac{2}{\sqrt{\pi}}e^{-t}\int_0^{\sqrt{t}} e^{u^2}\mathrm{d}u = \frac{2}{\sqrt{\pi}}F(\sqrt{t}),$$

where

$$F(x) = e^{-x^2}\int_0^x e^{u^2}\mathrm{d}u$$

is Dawson's integral.

(b) According to (a) and (5.31) of Ex. 5.19, we need to compute the integral

$$s = \frac{2}{\sqrt{\pi}}\int_0^\infty \frac{F(\sqrt{t})}{t}\,\frac{t}{e^t-1}\mathrm{d}t,\quad F = \text{Dawson's integral}.$$

Now $F(x)$ is an entire odd function of x, and hence $F(\sqrt{t}) = \sqrt{t}\,\phi(t)$, where $\phi(t)$ is an entire function of t. We therefore write

$$s = \frac{2}{\sqrt{\pi}}\int_0^\infty \frac{F(\sqrt{t})}{\sqrt{t}}\,\frac{t^{1/2}}{e^t-1}\mathrm{d}t = \frac{2}{\sqrt{\pi}}\int_0^\infty \phi(t)\frac{t^{1/2}}{e^t-1}\mathrm{d}t. \tag{5.34}$$

Thus, the appropriate measure of integration is the square root Bose–Einstein measure (cf. Ex. 2.28(c)), for which the first 80 recurrence coefficients are available in the file `ab_BEsqrtm1.txt` to 32-digit accuracy (cf. Appendix A.5). Hence we can use Gaussian quadrature to evaluate (5.34) to an accuracy of 32 digits. This is done in the script `theodorus_const.m`,

```
%THEODORUS_CONST

digits(32); dig=32;
a=loadvpa('ab_BEsqrtm1.txt',3,dig,80,2);
ab=a;
c=2/sqrt(vpa(pi));
thc0=vpa(0); thc=vpa(1); n=0;
while abs(double((thc-thc0)/thc))>.5*10^(-dig)
  n=n+1; thc0=thc;
  abn=ab(1:n,:);
  xw=sgauss(dig,n,abn); x=xw(:,1); w=xw(:,2);
  sqrtx=sqrt(vpa(x));
  thc=c*sum(w.*sdawson(dig,sqrtx)./sqrtx);
end
[n thc]
```

The output

```
>> theodorus_const
ans =
[ 36, 1.8600250792211903071806959157172]
```

shows that 36-point quadrature yields an answer correct to 32 digits except for the last digit, which should be 1 instead of 2 (compare with the 64-digit answer given in [29, §1]).

Exercise 5.21. (Ex. 9 of [27, §5], expanded.) (a) *For the series*

$$H(x)=\sum_{k=1}^{\infty} a_k(x), \quad a_k(x)=\frac{1}{k}\sin\frac{x}{k}$$

(the Hardy–Littlewood function; cf. [27, Example 19]), *determine the function* $f(\cdot;x)$ *for which* $a_k(x)=(\mathscr{L}f(\cdot;x))(k)$.

(b) *Plot the function* $H(x)$ *for* $0\le x\le 50$. {Hint: Evaluate the integral (5.31) with f as determined in (a) in variable-precision arithmetic by Gauss quadrature relative to the Bose–Einstein measure $d\lambda(t)=t\,dt/(e^t-1)$ on $\mathbb{R}_+$, using the file `ab_BE.txt` of 32-digit recurrence coefficients (cf. answer to Ex. 2.28(d) and Appendix A.5) to generate the required quadrature rules to the same 32-digit accuracy.}

Answer

(a) We note from [1, Eq. 29.3.81] that

$$\frac{1}{k}e^{x/k}=(\mathscr{L}\,I_0(2\sqrt{xt}))(k),$$

where I_0 is the modified Bessel function of order zero. There follows, by Euler's formula,

$$\frac{1}{k}\sin\frac{x}{k}=\frac{1}{k}\frac{1}{2\mathrm{i}}\left(e^{\mathrm{i}x/k}-e^{-\mathrm{i}x/k}\right)=\frac{1}{2\mathrm{i}}\left(\mathscr{L}[I_0(2\sqrt{\mathrm{i}xt})-I_0(2\sqrt{-\mathrm{i}xt})]\right)(k),$$

that is,

$$f(t)=f(t;x)=\frac{1}{2\mathrm{i}}[I_0(2\sqrt{\mathrm{i}xt})-I_0(2\sqrt{-\mathrm{i}xt})]=\operatorname{Im} I_0(2\sqrt{\mathrm{i}xt}).$$

(b) As x becomes larger, the quadrature sums in the Gaussian rules are progressively subject to internal cancellation of terms. For this reason, the script HLfun.m below, evaluating and plotting the Hardy–Littlewood function, is run in full 32-digit arithmetic (the precision of the input file ab_BE.txt). The variable canmax measures the maximum degree of cancellation.

```
%HLFUN Hardy-Littlewood function.

f0='%8.1f %4.0f %19.14f\n';
digits(32); dig=32;
ab=loadvpa('ab_BE.txt',3,dig,80,2)
nmax=0; eps0=.5e-6;
y=zeros(501,1);
k=0;
for x=0:.1:50
  x=vpa(x); k=k+1; canmax=0;
  hl0=vpa(0); hl=vpa(1); n=30; canm=0;
  while abs(double((hl-hl0)/hl))>eps0
    hl0=hl; n=n+5;
    if n>80
      fprintf('n exceeds 80 for x=%3.2f\n',double(x))
      error('halt')
    end
    abn=ab(1:n,:);
    xw=sgauss(dig,n,abn); xg=xw(:,1); wg=xw(:,2);
    tm=max(abs(double(wg.*imag(besseli(0,2*sqrt(i*x*xg)))./xg)));
    hl=sum(wg.*imag(besseli(0,2*sqrt(i*x*xg)))./xg);
    can=abs(tm/double(hl));
    if can>canm, canm=can; end
  end
  y(k)=double(hl);
  if canm>canmax, canmax=canm; end
end
canmax
x=(0:.1:50)';
plot(x,y);set(gca,'FontSize',14)
axis([0 50 -.5 3.5])
xlabel('x'), ylabel('H')
```

The output canmax $=9.87\times10^9$ suggests cancellation errors as large as 10^{10}. The plot produced (in about 40 minutes of run time) is shown in Fig. 5.4.

Exercise 5.22. (a) (Ex. 11 of [27, §5].) *Determine* f *in* Ex. 5.19(c) *for the series* (cf. [27, Example 20])

$$\sum_{k=1}^{\infty} a_k, \quad a_k = \frac{(-1)^{k-1}}{k}\, e^{-1/k}.$$

(b) *Compute the series in* (a) *using Gaussian quadrature on the integral* (5.33). {Hint: Use the file ab_Fermi.txt (cf. answer to Ex. 2.29(d) and Appendix A.5) to generate the required Gaussian rules.}

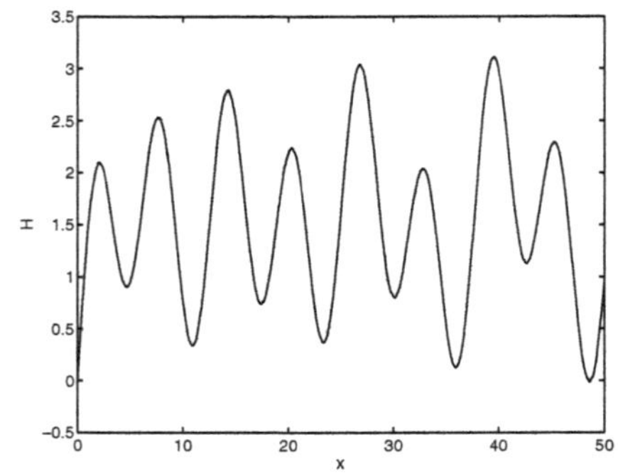

Figure 5.4. *The Hardy-Littlewood function on the interval* [0,50]

Answer

(a) From [1, Eq. 29.3.80] we have

$$\frac{1}{k}e^{-1/k} = (\mathscr{L}J_0(2\sqrt{t}))(k),$$

where J_0 is the Bessel function of order zero. Therefore, the function f in question is

$$f(t) = J_0(2\sqrt{t}).$$

(b) The script `S3.m` computes the series in the manner outlined in the hint.

```
%S3

f0='     n=%2.0f   s3=%17.15f\n';
ab=loadvpa('ab_Fermi.txt',3,32,80,2);
ab=double(ab);
s0=0; s=1; n=0;
while abs((s-s0)/s)>.5e-15
  s0=s; n=n+1;
  xw=gauss(n,ab); x=xw(:,1); w=xw(:,2);
  s=sum(w.*besselj(0,2*sqrt(x)));
end
xw=gauss(n,ab); x=xw(:,1); w=xw(:,2);
s3=sum(w.*besselj(0,2*sqrt(x)));
fprintf(f0,n,s3)
```

The output

```
>> S3
   n=11   s3=0.197107936397951
>>
```

suggests that the 11-point Gauss rule already achieves an accuracy of 15 decimal digits. In fact, all digits of s3 are correct (compare with the 24-digit result shown at the bottom of p. 183 of [35]).

Exercise 5.23. (a) (Ex. 13 of [27, §5].) *For the series* (cf. [27, Example 21])

$$\sum_{k=1}^{\infty} a_k, \quad a_k = (\tfrac{3}{2}k+1)k^{-2}(k+1)^{-1/2},$$

determine the function f *in* Ex. 5.19(b).

(b) *Compute the series in* (a) *using Gaussian quadrature on* (5.32), *choosing the measure of integration in such a way that the integrand becomes an entire function.* {Hint: Use the file `ab_BEsqrtm1.txt` (cf. answer to Ex. 2.28(d) and Appendix A.5) to generate the Gaussian rules.}

Answer

(a) From [1, Eq. (29.3.44)] with $a=0$, $b=1$ we have

$$s^{-1}(s+1)^{-1/2} = \left(\mathcal{L}\,\mathrm{erf}\sqrt{t}\right)(s).$$

Evaluating the negative derivative of this at $s=k$ gives

$$\left(\tfrac{3}{2}k+1\right)k^{-2}(k+1)^{-1/2} = -\frac{\mathrm{d}}{\mathrm{d}s}\left(\mathcal{L}\,\mathrm{erf}\sqrt{t}\right)(s)\Big|_{s=k},$$

that is,

$$f(t) = \mathrm{erf}\sqrt{t}.$$

(b) By (5.32) and the result of (a), the series s in (a) equals

$$s = \int_0^\infty \mathrm{erf}\sqrt{t}\,\frac{t}{e^t-1}\,\mathrm{d}t.$$

Since $\sqrt{t}\,\mathrm{erf}\sqrt{t}$ is an entire function, we write

$$s = \int_0^\infty \sqrt{t}\,\mathrm{erf}\sqrt{t}\,\mathrm{d}\lambda(t), \quad \mathrm{d}\lambda(t) = \frac{t^{1/2}}{e^t-1}\,\mathrm{d}t,$$

involving the square root Bose–Einstein measure $\mathrm{d}\lambda$. The script for evaluating the integral s therefore is as follows.

```
%S4

f0='    n=%2.0f   s4=%17.15f\n';
ab=loadvpa('ab_BEsqrtm1.txt',3,32,80,2);
ab=double(ab);
s0=0; s=1; n=0;
while abs((s-s0)/s)>.5e-15
  s0=s; n=n+1;
  xw=gauss(n,ab); x=xw(:,1); w=xw(:,2);
  s=sum(w.*sqrt(x).*erf(sqrt(x)));
end
xw=gauss(n,ab); x=xw(:,1); w=xw(:,2);
s4=sum(w.*sqrt(x).*erf(sqrt(x)));
fprintf(f0,n,s4)
```

The output

```
>> S4
    n=18   s4=1.285974816155241
>>
```

shows that 18-point Gaussian quadrature yields a result accurate to all digits but the last, which should be 0 instead of 1 (compare with the 25-digit result on p. 185 of [35] divided by 2).

Exercise 5.24. (Ex. 14 of [27, §5], expanded.) *Let*

$$T_p(x;b)=\sum_{k=0}^{\infty}\frac{1}{(2k+1)^p}\frac{\cosh(2k+1)x}{\cosh(2k+1)b},\quad 0\le x\le b,\ b>0,\ p=2,3. \tag{5.35}$$

(a) *Prove*

$$T_p(x,b)=\frac{1}{2^p(p-1)!}\sum_{n=0}^{\infty}(-1)^n \mathrm{e}^{(2n+1)b}[\varphi_n(-x)+\varphi_n(x)], \tag{5.36}$$

where

$$\varphi_n(s)=\mathrm{e}^s\int_0^1\frac{\mathrm{d}\lambda^{[p]}(t)}{\mathrm{e}^{2[(2n+1)b+s]}-t},\quad -b\le s\le b,\quad \mathrm{d}\lambda^{[p]}(t)=\frac{[\ln(1/t)]^{p-1}}{\sqrt{t}}\,\mathrm{d}t. \tag{5.37}$$

{Hint: Expand the ratio of hyperbolic cosines into a series in exponentials and express the factor $(k+1/2)^p$ as a Laplace transform.}

(b) *For $p=2$, write a MATLAB script that produces a table of $T_p(x;b)$ to an accuracy of 12 decimal digits, for selected values of b, say, $b=.1,.3,.5,.75,1,1.5,2.25,3$, and for each b, for $x=(i/N)b$, $i=0,1,2,\dots,N$ (say, with $N=10$). Print the number n of terms required in the series for $T_p(x;b)$ to achieve the given accuracy and also the maximum value* `numax` *of the starting index ν in the backward recursion to evaluate φ_n in* (5.37). *Prepare respective plots of the function $T_p(x;b)$.* {Hint: To evaluate the Cauchy integral in (5.37), use the `OPQ` routine `cauchy.m` (cf. Appendix A.1). When $x=b$, use the fact that $T_p(x;b)=\sum_{k=0}^{\infty}(2k+1)^{-p}=(1-2^{-p})\zeta(p)$.}

(c) *Do the same as in* (b), *but for $p=3$.*

Answer

(a) Using the definition of the hyperbolic cosine, we have

$$\begin{aligned}\frac{\cosh(2k+1)x}{\cosh(2k+1)b}&=\frac{e^{(2k+1)x}+e^{-(2k+1)x}}{e^{(2k+1)b}+e^{-(2k+1)b}}=\frac{e^{(2k+1)x}\left[1+e^{-2(2k+1)x}\right]}{e^{(2k+1)b}\left[1+e^{-2(2k+1)b}\right]}\\&=e^{(2k+1)(x-b)}\left[1+e^{-2(2k+1)x}\right]\sum_{n=0}^{\infty}(-1)^n e^{-2(2k+1)nb}\\&=\sum_{n=0}^{\infty}(-1)^n\left\{e^{-(2k+1)[(2n+1)b-x]}+e^{-(2k+1)[(2n+1)b+x]}\right\}.\end{aligned}$$

Inserting this in (5.35) and using the fact that

$$\frac{1}{(k+1/2)^p}=(\mathcal{L}f)(k),$$

where

$$f(t)=\frac{1}{(p-1)!}t^{p-1}e^{-t/2},$$

yields

$$T_p(x;b) = \frac{1}{2^p(p-1)!}\sum_{k=0}^{\infty}\int_0^{\infty} e^{-kt}t^{p-1}e^{-t/2}\mathrm{d}t$$
$$\times\sum_{n=0}^{\infty}(-1)^n\left\{e^{-(2k+1)[(2n+1)b-x]}+e^{-(2k+1)[(2n+1)b+x]}\right\}$$
$$= \frac{1}{2^p(p-1)!}\sum_{n=0}^{\infty}(-1)^n\sum_{k=0}^{\infty}\int_0^{\infty} e^{-kt}t^{p-1}e^{-t/2}\mathrm{d}t$$
$$\times\left\{e^{-(2k+1)[(2n+1)b-x]}+e^{-(2k+1)[(2n+1)b+x]}\right\}$$
$$= \frac{1}{2^p(p-1)!}\sum_{n=0}^{\infty}(-1)^n\sum_{k=0}^{\infty}\int_0^{\infty}\Big(e^{-kt-2k[(2n+1)b-x]}e^{-[(2n+1)b-x]}$$
$$+e^{-kt-2k[(2n+1)b+x]}e^{-[(2n+1)b+x]}\Big)t^{p-1}e^{-t/2}\mathrm{d}t$$
$$= \frac{1}{2^p(p-1)!}\sum_{n=0}^{\infty}(-1)^n\int_0^{\infty}\left(\sum_{k=0}^{\infty}e^{-k[t+2((2n+1)b-x)]}e^{-[(2n+1)b-x]}\right.$$
$$\left.+\sum_{k=0}^{\infty}e^{-k[t+2((2n+1)b+x)]}e^{-[(2n+1)b+x]}\right)t^{p-1}e^{-t/2}\mathrm{d}t$$
$$= \frac{1}{2^p(p-1)!}\sum_{n=0}^{\infty}(-1)^n\int_0^{\infty}\left(\frac{e^{-[(2n+1)b-x]}}{1-e^{-[t+2(2n+1)b-2x]}}\right.$$
$$\left.+\frac{e^{-[(2n+1)b+x]}}{1-e^{-[t+2(2n+1)b+2x]}}\right)t^{p-1}e^{-t/2}\mathrm{d}t.$$

By a change of variable, $e^{-t}=\tau$, and defining $\mathrm{d}\lambda^{[p]}$ as in (5.37), we obtain

$$T_p(x;b) = \frac{1}{2^p(p-1)!}\sum_{n=0}^{\infty}(-1)^n e^{(2n+1)b}\int_0^1\left(\frac{e^{-x}}{e^{2[(2n+1)b-x]}-\tau}\right.$$
$$\left.+\frac{e^{x}}{e^{2[(2n+1)b+x]}-\tau}\right)\mathrm{d}\lambda^{[p]}(\tau).$$

This is the same as (5.36), (5.37).

(b) The MATLAB script for $T_2(x;b)$ is as follows (cf. also Ex. 2.27*(c)):

```
%T2XB The series T_2(x,b)

f0='%12.2f %7.3f %17.12f %5.0f %6.0f\n';
dig=25; ab=loadvpa('ab_jaclog.txt',3,dig,100,2);
ab=double(ab);
eps0=.5e-12;
disp('           b         x            T_2(x,b)          n     numax')
for b=[.1 .3 .5 .75 1 1.5 2.25 3]
  for i=1:11
    x=(i-1)*b/10;
%  for i=1:51
%    x=(i-1)*b/50;
    X(i)=x;
    if x==b
```

```
        s1=pi^2/8; n=0; numax=0;
        T(i)=s1;
      else
        s1=0; s0=1; n=-1;
        numax=0;
        while abs(s1-s0)>eps0
          s0=s1; n=n+1;
          zm=exp(2*((2*n+1)*b-x)); zp=exp(2*((2*n+1)*b+x));
          [rhom,r,num]=cauchy(0,ab,zm,eps0,10,100);
          [rhop,r,nup]=cauchy(0,ab,zp,eps0,10,100);
          nu=max(num,nup);
          if nu>numax, numax=nu; end
          s1=s0+(-1)^n*exp((2*n+1)*b)*(exp(-x)*rhom ...
            +exp(x)*rhop);
        end
        s1=s1/4;
        T(i)=s1;
      end
      fprintf(f0,b,x,s1,n,numax)
    end
  % plot(X,T)
  % hold on
  % plot([0 3],[pi^2/8 pi^2/8],'--')
    fprintf('\n')
  end
```

The results produced by the script are shown below.

```
>> T2xb
          b       x          T_2(x,b)       n     numax
        0.10    0.000     1.175388369330   152      20
        0.10    0.010     1.175781879205   152      25
        0.10    0.020     1.176972302139   152      25
        0.10    0.030     1.178990609694   152      25
        0.10    0.040     1.181893164680   152      25
        0.10    0.050     1.185770672808   152      30
        0.10    0.060     1.190765862567   152      30
        0.10    0.070     1.197110745763   152      35
        0.10    0.080     1.205218037002   152      40
        0.10    0.090     1.215983232204   152      50
        0.10    0.100     1.233700550136     0       0

        0.30    0.000     1.058764035144    51      15
        0.30    0.030     1.059944564433    51      15
        0.30    0.060     1.063515832228    51      15
        0.30    0.090     1.069570753246    51      15
        0.30    0.120     1.078278415956    51      20
        0.30    0.150     1.089910937546    51      20
        0.30    0.180     1.104896503551    51      20
        0.30    0.210     1.123931149468    51      25
        0.30    0.240     1.148253019210    51      25
```

0.30	0.270	1.180548600630	51	35
0.30	0.300	1.233700550136	0	0
0.50	0.000	0.942172572405	30	15
0.50	0.050	0.944139716404	30	15
0.50	0.100	0.950090624921	30	15
0.50	0.150	0.960180185472	30	15
0.50	0.200	0.974690260791	30	15
0.50	0.250	0.994074445906	30	15
0.50	0.300	1.019046465886	30	20
0.50	0.350	1.050766476498	30	20
0.50	0.400	1.091298159253	30	20
0.50	0.450	1.145119111283	31	30
0.50	0.500	1.233700550136	0	0
0.75	0.000	0.797683004096	20	15
0.75	0.075	0.800618029890	20	15
0.75	0.150	0.809497708392	20	15
0.75	0.225	0.824555520378	20	15
0.75	0.300	0.846216144323	20	15
0.75	0.375	0.875162545610	20	15
0.75	0.450	0.912468509806	20	15
0.75	0.525	0.959878014914	20	15
0.75	0.600	1.020490781739	20	20
0.75	0.675	1.101027755328	20	25
0.75	0.750	1.233700550136	0	0
1.00	0.000	0.659670293981	15	15
1.00	0.100	0.663493465731	15	15
1.00	0.200	0.675064668962	15	15
1.00	0.300	0.694701816430	15	15
1.00	0.400	0.722981974405	15	15
1.00	0.500	0.760830549282	15	15
1.00	0.600	0.809697635561	15	15
1.00	0.700	0.871930060468	15	15
1.00	0.800	0.951684921679	15	20
1.00	0.900	1.057949634997	15	20
1.00	1.000	1.233700550136	0	0
1.50	0.000	0.427609801913	10	15
1.50	0.150	0.432669028734	10	15
1.50	0.300	0.448015646228	10	15
1.50	0.450	0.474175384301	10	15
1.50	0.600	0.512094093728	10	15
1.50	0.750	0.563269746852	10	15
1.50	0.900	0.630014467623	10	15
1.50	1.050	0.716009408935	10	15
1.50	1.200	0.827671210343	10	15
1.50	1.350	0.978677631030	10	20
1.50	1.500	1.233700550136	0	0

```
          2.25   0.000   0.208743660293    7     15
          2.25   0.225   0.214105473153    7     15
          2.25   0.450   0.230490517943    7     15
          2.25   0.675   0.258826623497    7     15
          2.25   0.900   0.300766391060    7     15
          2.25   1.125   0.358886709291    7     15
          2.25   1.350   0.437081810270    7     15
          2.25   1.575   0.541392397913    7     15
          2.25   1.800   0.682047378120    7     15
          2.25   2.025   0.880238562757    7     15
          2.25   2.250   1.233700550136    0      0

          3.00   0.000   0.099355376323    5     15
          3.00   0.300   0.103870667292    5     15
          3.00   0.600   0.117835271160    5     15
          3.00   0.900   0.142551565974    5     15
          3.00   1.200   0.180355888158    5     15
          3.00   1.500   0.234917159643    5     15
          3.00   1.800   0.311798908586    5     15
          3.00   2.100   0.419599395660    5     15
          3.00   2.400   0.572694258663    5     15
          3.00   2.700   0.800264659676    5     15
          3.00   3.000   1.233700550136    0      0
>>
```

(c) The script `T3xb.m` differs from `T2xb.m` in (b), apart from loading a different file, only in the definition of `s1` when $x = b$ and in the final adjustment of `s1` near the end of the script. When run, it produces the following results (cf. also Ex. 2.27*(c)):

```
>> T3xb
           b       x        T_2(x,b)        n    numax
          0.10   0.000   1.041821677493   159     20
          0.10   0.010   1.041916167907   159     20
          0.10   0.020   1.042200102573   159     20
          0.10   0.030   1.042674903000   159     20
          0.10   0.040   1.043343049458   159     25
          0.10   0.050   1.044208275374   159     25
          0.10   0.060   1.045275910948   159     25
          0.10   0.070   1.046553502726   159     30
          0.10   0.080   1.048052038260   159     35
          0.10   0.090   1.049788881926   159     40
          0.10   0.100   1.051799790265     0      0

          0.30   0.000   0.986844451277    53     15
          0.30   0.030   0.987446131883    53     15
          0.30   0.060   0.989255375648    53     15
          0.30   0.090   0.992285069586    53     15
          0.30   0.120   0.996557691750    53     15
          0.30   0.150   1.002107060719    53     20
          0.30   0.180   1.008981427920    53     20
          0.30   0.210   1.017249053629    53     20
```

0.30	0.240	1.027009228632	53	25
0.30	0.270	1.038418696294	53	30
0.30	0.300	1.051799790265	0	0
0.50	0.000	0.904082092756	32	15
0.50	0.050	0.905424703205	32	15
0.50	0.100	0.909464400788	32	15
0.50	0.150	0.916237564656	32	15
0.50	0.200	0.925807601723	32	15
0.50	0.250	0.938269803451	32	15
0.50	0.300	0.953759931602	32	15
0.50	0.350	0.972469702011	32	20
0.50	0.400	0.994677394091	32	20
0.50	0.450	1.020821238266	32	25
0.50	0.500	1.051799790265	0	0
0.75	0.000	0.780521584827	21	15
0.75	0.075	0.782923231488	21	15
0.75	0.150	0.790155904137	21	15
0.75	0.225	0.802304550393	21	15
0.75	0.300	0.819516972108	21	15
0.75	0.375	0.842014721446	21	15
0.75	0.450	0.870112374670	21	15
0.75	0.525	0.904252314415	21	15
0.75	0.600	0.945073533085	21	20
0.75	0.675	0.993576675326	21	20
0.75	0.750	1.051799790265	0	0
1.00	0.000	0.651846571067	16	15
1.00	0.100	0.655271616248	16	15
1.00	0.200	0.665598691116	16	15
1.00	0.300	0.682986709521	16	15
1.00	0.400	0.707711541240	16	15
1.00	0.500	0.740185308470	16	15
1.00	0.600	0.780990544406	16	15
1.00	0.700	0.830941889062	16	15
1.00	0.800	0.891208231884	16	15
1.00	0.900	0.963605908993	16	20
1.00	1.000	1.051799790265	0	0
1.50	0.000	0.425927835811	11	15
1.50	0.150	0.430806571192	11	15
1.50	0.300	0.445569894802	11	15
1.50	0.450	0.470606254229	11	15
1.50	0.600	0.506588258072	11	15
1.50	0.750	0.554516663477	11	15
1.50	0.900	0.615797928693	11	15
1.50	1.050	0.692383756263	11	15
1.50	1.200	0.787046565805	11	15
1.50	1.350	0.904039650971	11	20

```
        1.50   1.500   1.051799790265   0   0

        2.25   0.000   0.208569359325   7  15
        2.25   0.225   0.213889539128   7  15
        2.25   0.450   0.230129419391   7  15
        2.25   0.675   0.258145615199   7  15
        2.25   0.900   0.299431107599   7  15
        2.25   1.125   0.356226534487   7  15
        2.25   1.350   0.431710749739   7  15
        2.25   1.575   0.530333399068   7  15
        2.25   1.800   0.658454340777   7  15
        2.25   2.025   0.825847850786   7  15
        2.25   2.250   1.051799790265   0   0

        3.00   0.000   0.099337073785   5  15
        3.00   0.300   0.103844420104   5  15
        3.00   0.600   0.117778259442   5  15
        3.00   0.900   0.142414040476   5  15
        3.00   1.200   0.180017067679   5  15
        3.00   1.500   0.234075973501   5  15
        3.00   1.800   0.309691395526   5  15
        3.00   2.100   0.414229823152   6  15
        3.00   2.400   0.558531695052   6  15
        3.00   2.700   0.759658820037   6  15
        3.00   3.000   1.051799790265   0   0
>>
```

Plots of the two tables (with higher resolution) are shown in Fig. 5.5.

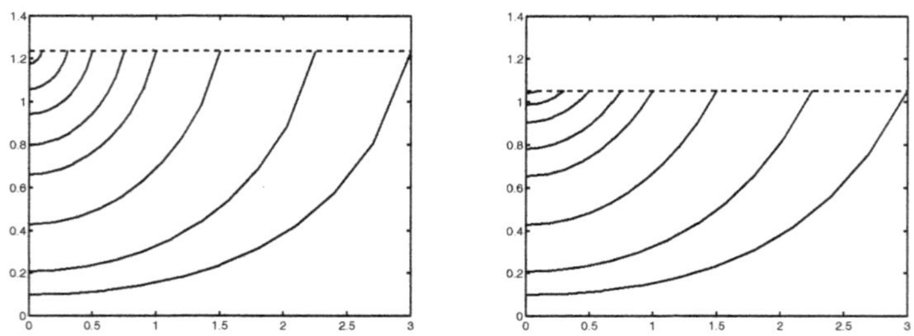

Figure 5.5. *The functions* $T_2(\cdot, b)$ *(left) and* $T_3(\cdot, b)$ *(right) on* $[0, b]$ *for* $b = .1, .3, .5, .75, 1, 1.5, 2.25, 3$

5.5 ▪ The spiral of Theodorus

Exercise 5.25. The spiral of Theodorus (cf. [29]), as smoothed by P. J. Davis [7, Part I, Lecture I], can be represented parametrically in the complex plane as

$$T(\alpha) = \sqrt{\alpha}\exp(\mathrm{i}\theta(\alpha)), \quad \alpha > 0, \tag{5.38}$$

where

$$\theta(\alpha)=\frac{\alpha-1}{\sqrt{\pi}}\int_0^\infty \frac{1-e^{-(\alpha-1)t}}{(\alpha-1)t}\,\frac{F(\sqrt{t})}{\sqrt{t}}\,w(t)\mathrm{d}t. \tag{5.39}$$

Here, F is Dawson's integral [1, p. 298] and $w(t)=t^{1/2}/(e^t-1)$ the square root Bose-Einstein weight function (cf. Ch. 2, Ex. 2.28). A second branch of the spiral—the analytic continuation of (5.38) into the second sheet of the Riemann surface for the square root function—is given by

$$S(\alpha)=\frac{1+\text{ß}/\sqrt{\alpha}}{1-\text{ß}/\sqrt{\alpha}}\,T(\alpha)=\frac{T(\alpha+1)}{1-\text{ß}/\sqrt{\alpha}},\quad \alpha>0. \tag{5.40}$$

Both spirals together constitute what may be called the "twin-spiral of Theodorus."

(a) *Write a variable-precision MATLAB routine computing Dawson's integral from* $F(x)=\frac{1}{2}\sqrt{\pi}\,\mathrm{i}e^{-x^2}\mathrm{erf}(-\mathrm{i}x)$, *where* erf *is the error function.* (The MATLAB double-precision routine for erf does not allow complex arguments.)

(b) *Wishing to plot both spirals until they cross the negative real axis for the second time, determine* $\alpha=\alpha_0$ *such that* $\theta(\alpha_0)=3\pi$. {Hint: Use the method of bisection.}

(c) *Plot both spirals for* $0\le\alpha\le\alpha_0$, *using Gauss quadrature to evaluate the integral in* (5.39). {Hint: Generate the data for $T(\alpha)$ first for $1\le\alpha\le 2$ by using (5.39), then the rest of the data by using the recurrence relation $T(\alpha+1)=(1+\mathrm{i}/\sqrt{\alpha})T(\alpha)$ for $1<\alpha\le\alpha_0-1$, resp., for $0<\alpha<1$, when solved for $T(\alpha)$. Use the file `ab_BEsqrtm1.txt` (cf. answer to Ex. 2.28(d) and Appendix A.5) to generate the necessary Gauss quadrature rule.}

Answer

(a) There are two routines, the symbolic variable-precision routine `sdawson.m`, and the routine `dawson.m` using the former to produce results in MATLAB double-precision accuracy.

```
function y=sdawson(dig,x)
%SDAWSON Symbolic Dawson's integral.

digits(dig);
y=vpa(zeros(size(x))); x=vpa(x);
y=(sqrt(vpa(pi))/2)*i*exp(vpa(-x.^2)).*erf(vpa(-i*x));
```

```
function y=dawson(x)
%DAWSON  Dawson's integral.

dig=16; sx=size(x);
sy=sdawson(dig,x); y=zeros(sx);
y=double(sy);
```

(b) We need a routine `theta.m` for evaluating the function $\theta(\alpha)$ in (5.39) to plotting accuracy.

```
function th=theta(a)
%THETA  Polar angle of the Theodorus spiral.
%   TH=THETA(A) computes the angular part TH in the polar
%   representation of the Theodorus spiral.
```

```
ab=loadvpa('ab_BEsqrtm1.txt',3,16,80,2);
ab=double(ab); eps0=.5e-3;
c=(a-1)/sqrt(pi);
n=20; th0=1; th1=0;
while abs(th1-th0)>eps0
  th0=th1; n=n+10;
  xw=gauss(n,ab); x=xw(:,1); w=xw(:,2);
  th1=0;
  for k=1:n
    xa=(a-1)*x(k); sqx=sqrt(x(k));
    th1=th1+w(k)*exp0(xa)*dawson(sqx)/sqx;
  end
  th=c*th1;
end
```

Here, exp0.m evaluates the function $(1-e^{-x})/x$.

The bisection routine we use is similar to the symbolic one in [30, p. 263], but transcribed into MATLAB double precision.

```
function x=bisection(fun,a,b,tol)
%BISECTION Bisection method.
%   BISECTION(FUN,A,B,TOL) computes to an absolute accuracy
%   TOL a solution of FUN(X)=0 by the method of bisection
%   assuming A < B and FUN(A)*FUN(B) < 0. In the calling
%   routine the name of the function FUN has to be prefixed
%   by the symbol @.

ntol=floor(1+log((b-a)/tol)/log(2))
s=1;
if fun(a)>0, s=-s; end
for n=1:ntol
  x=(a+b)/2;
  if s*fun(x)<0
    a=x;
  else
    b=x;
  end
end
```

For fun we use theta.m, in which the last statement is replaced by th=c*th1-3*pi and a=33, b=34, tol=.5e-3. The result is $\alpha_0 = 33.385$.

(c) The script plotting the spiral and twin spiral of Theodorus is plot_Theo.m. It calls on the routine Theo.m.

```
%PLOT_THEO  Plot the spiral of Theodorus and its
%   twin spiral.
%
% spiral
%
for j=1:100
  a=1+j/100; T(100+j)=Theo(a);
end
```

```
for j=1:3139
  a=2+j/100; T(200+j)=(1+i/sqrt(a-1))*T(100+j);
end
for j=1:100
  a=j/100; T(j)=T(100+j)/(1+i/sqrt(a));
end
T=[0 T];
%plot(real(T),imag(T),'k-','LineWidth',1.5);set(gca,'FontSize',14)
plot(real(T),imag(T),'r-','LineWidth',1.5);set(gca,'FontSize',14)
axis equal
axis([-7 6 -5 6])
hold on
plot([-7 6],[0 0],'k-')
plot([0 0],[-5 6],'k-')
%error('quit')
%
% twin-spiral
%
for j=1:3139
  a=j/100; S(j)=T(100+j)/(1-i/sqrt(a));
end
S=[0 S];
plot(real(S),imag(S),'b-','Linewidth',1.5);set(gca,'FontSize',14)

function T=Theo(a)
%THEO Spiral of Theodorus.

T=sqrt(a)*exp(i*theta(a));
```

The two plots produced are shown in Fig. 5.6.

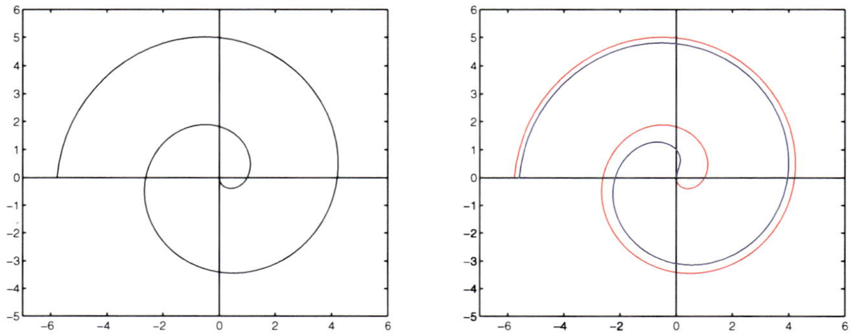

Figure 5.6. *The spiral of Theodorus and its twin spiral*

Appendix A

The Software Package OPQ (Orthogonal Polynomials and Quadrature)

A.1 ▪ Orthogonal polynomials

`HTjac.m`: Hilbert transform of the Jacobi measure.
This script uses the routine `SPTH.m` to compute the Hilbert transform of the Chebyshev measures of the first and second kind for selected values of the argument.

`SPHT.m`: Stieltjes–Perron inversion and Hilbert transform.
`[E,w,nu]=SPHT(n,x,iopt,ab,nu0,numax,eps0)`, if `iopt`=1, applies the Stieltjes–Perron inversion formula to compute to an accuracy of `eps0` the weight function at a point `x` from its first `numax` recurrence coefficients input by the `numax`×2 array `ab`. The Cauchy integral of the weight function is generated for a sequence of `n` complex arguments approaching `x` from above. To the negative real parts of these (divided by π) the epsilon algorithm is applied to accelerate convergence. The accelerated values are returned in the even-numbered columns of the `n`×(`n`+1) array `E`. If `iopt` is different from 1, the routine computes the Hilbert transform of the weight function at the point `x` using the same procedure as above, but applying the epsilon algorithm to the negative real parts of the Cauchy integrals. The input parameter `nu0` is an estimate of the backward recurrence index to be used in the routine `cauchy.m` for computing the Cauchy integrals. The output variable `w` holds the `n`×1 array of the vector to which the epsilon algorithm is applied, while the output variable `nu` is inherited from the routine `cauchy.m`.

`SPjac.m`: Stieltjes–Perron inversion for Jacobi polynomials.
This script tests the Stieltjes–Perron inversion formula, as implemented by the routine `SPHT.m`, on Jacobi measures with parameters α and β in $(-1,1]$, recovering the Jacobi weight function for selected values of x in $(-1,1)$ from its recurrence coefficients.

`althammer.m`: Althammer polynomials generated by Stieltjes's algorithm.
This script uses Stieltjes's algorithm to generate the recurrence matrix B of order 21 for the Sobolev orthogonal polynomials with the ground measure being Lebesgue's measure on $[-1,1]$ and the first derivative measure 100 times the ground measure.

`cauchy.m`: Cauchy integral of orthogonal polynomials.

`[rho,r,nu]=cauchy(N,ab,z,eps0,nu0,numax)` generates to a relative accuracy `eps0` the Cauchy integrals of the first `N+1` orthogonal polynomials associated with the recurrence coefficients in the `numax`×2 array `ab`, i.e., the `N+1` integrals

$$\rho_k(z;w) = \int \pi_k(t)w(t)\mathrm{d}t/(z-t), \quad k=0,1,\ldots,N,$$

and their ratios

$$r_k = \rho_k(z;w)/\rho_{k-1}(z;w), \quad k=0,1,\ldots,N,$$

where $r_0 = \rho_0(z,w)$ and π_k is the monic polynomial of degree k orthogonal relative to the weight function w. The procedure uses a continued fraction algorithm, which involves backward recursion from ν down to 0. The starting index ν is increased by 5 until convergence of the continued fraction algorithm is obtained. The initial value of ν to be used is `nu0`, and ν is not allowed to exceed `numax`. The output variable `nu` is either `numax`, if there is no convergence, or the final value of ν achieving convergence. The Cauchy integrals are returned in the (`N+1`)×1 output array `rho` and their ratios in the (`N+1`)×1 output array `r`.

`chebyshev.m`: Modified Chebyshev algorithm.

`[ab,normsq]=chebyshev(N,mom,abm)` uses the modified Chebyshev algorithm to generate the `N`×2 array `ab` of recurrence coefficients for a given weight function w from its first `2N` modified moments. The latter, input by the 1×(`2N`) array `mom`, are relative to monic orthogonal polynomials whose first `2N`−1 recurrence coefficients are given by the (`2N`−1)×2 input array `abm`. The call `[ab,normsq]=chebyshev(N,mom)` does the same, but using the classical Chebyshev algorithm and ordinary moments. The `N`×1 output array `normsq` in either case contains the squared norm of the first `N` (monic) orthogonal polynomials relative to the weight function w.

If `N` is larger than the sizes of `mom` and `abm` warrant, `N` is reduced accordingly.

`chebyshev_sob.m`: Modified Chebyshev algorithm for Sobolev orthogonal polynomials.

`[B,normsq]=chebyshev_sob(N,mom,abm)` uses the modified Chebyshev algorithm to generate the `N`×`N` array `B` of recurrence coefficients for monic Sobolev orthogonal polynomials involving in their inner product a ground measure $\mathrm{d}\lambda_0$ and a measure $\mathrm{d}\lambda_1$ associated with the first derivative ($s=1$). The 2×(`2N`) input array `mom` furnishes the first `2N` modified moments of $\mathrm{d}\lambda_0$ and $\mathrm{d}\lambda_1$ in its first and second rows, resp. Both modified moments are relative to monic orthogonal polynomials whose first `2N`−1 recurrence coefficients are given by the (`2N`−1)×2 input array `abm`. The recurrence coefficient β_j^k, $0 \le j \le k$, $k = 0,1,\ldots,\mathtt{N}-1$, occupies the position `B(j+1,k+1)` of the array `B`, all remaining elements being zero. The routine also returns the `N` × 1 array `normsq` containing the squared norm of the first `N` Sobolev orthogonal polynomials. The call

[B,normsq]=chebyshev_sob(N,mom) does the same, but using the classical Chebyshev algorithm and ordinary moments [36, §2].

chri1.m: Modification by a linear factor.

ab=chri1(N,ab0,z) generates the N×2 array ab of the first N recurrence coefficients for the modified weight function $(t-\mathtt{z})w(t)$, z real or complex outside the support interval of w, where the weight function w is specified by the (N+1)×2 array ab0 of recurrence coefficients.

chri2.m: Modification by a quadratic factor.

ab=chri2(N,ab0,x,y) generates the N×2 array ab of the first N recurrence coefficients for the modified weight function $((t-\mathtt{x})^2+\mathtt{y}^2)w(t)$, x real and y > 0, where the weight function w is specified by the (N+2)×2 array ab0 of recurrence coefficients.

chri3.m: Modification by a simplified quadratic factor.

ab=chri3(N,ab0,y) generates the N×2 array ab of the first N recurrence coefficients for the modified weight function $(t^2+\mathtt{y}^2)w(t)$, y > 0, where the symmetric weight function w is specified by the (N+2)×2 array ab0 of recurrence coefficients.

chri4.m: Modification by a linear divisor.

[ab,nu]=chri4(N,ab0,z,eps0,nu0,numax,rho0,iopt) generates the N×2 array ab of the first N recurrence coefficients for the modified weight function $w(t)/(t-\mathtt{z})$, z real or complex outside the support interval of w, where the weight function w is specified by the array ab0 of recurrence coefficients. If iopt=1, the routine computes the desired recurrence coefficients by means of ratios of appropriate Cauchy integrals. These are computed to a relative accuracy eps0 by the continued fraction algorithm implemented in the routine cauchy.m. This calls for input parameters nu0, numax, and for an input array ab0 of size numax×2. The output parameter nu is inherited from the routine cauchy.m. If iopt is not equal to 1, the routine uses forward recursion to compute the required ratios of Cauchy integrals, which, however, may be unstable. In that case, the routine needs the value rho0 of the first Cauchy integral and an input array ab0 of size N×2 only and outputs nu=0.

chri5.m: Modification by a quadratic divisor.

[ab,nu]=chri5(N,ab0,z,eps0,nu0,numax,rho0,iopt) generates the N×2 array ab of the first N recurrence coefficients for the modified weight function $w(t)/((t-x)^2+y^2)$, $\mathtt{z}=x+iy$ outside the support interval of w and $y>0$, where the weight function w is specified by the input array ab0 of recurrence coefficients. If iopt=1, the routine computes the desired recurrence coefficients by means of ratios of appropriate Cauchy integrals. These are computed to a relative accuracy eps0 by the continued fraction algorithm implemented in the routine cauchy.m. This calls for input parameters nu0, numax, and for an input array ab0 of size numax×2. The output parameter nu is inherited from the routine cauchy.m. If iopt is not equal to 1, the routine uses forward recursion to compute the required ratios of Cauchy integrals, which, however, may be unstable. In that case, the routine needs the value rho0 of the first

Cauchy integral and an input array `ab0` of size `N`$\times$2 only and outputs `nu=0`.

`chri6.m`: Modification by a simplified quadratic divisor.

`[ab,nu]=chri6(N,ab0,y,eps0,nu0,numax,rho0,iopt)` generates the `N`$\times$2 array `ab` of the first `N` recurrence coefficients for the modified weight function $w(t)/(t^2+\mathtt{y}^2)$, $\mathtt{y} > 0$, where the symmetric weight function w is specified by the array `ab0` of recurrence coefficients. If `iopt=1`, the routine computes the desired recurrence coefficients by means of ratios of appropriate Cauchy integrals. These are computed to a relative accuracy `eps0` by the continued fraction algorithm implemented in the routine `cauchy.m`. This calls for input parameters `nu0`, `numax`, and for an input array `ab0` of size `numax`$\times$2. The output parameter `nu` is inherited from the routine `cauchy.m`. If `iopt` is not equal to 1, the routine uses forward recursion to compute the required ratios of Cauchy integrals, which, however, may be unstable. In that case, the routine needs the value `rho0` of the first Cauchy integral and an input array `ab0` of size `N`$\times$2 only and outputs `nu=0`.

`chri7.m`: Modification by a special quadratic factor.

`ab=chri7(N,ab0,x)` generates the `N`$\times$2 array `ab` of the first `N` (>1) recurrence coefficients for the modified weight function $(t-\mathtt{x})^2 w(t)$, where the weight function w is specified by the (N+1)$\times$2 array `ab0` of recurrence coefficients.

`chri8.m`: Modification by a special quadratic divisor.

`[ab,nu]=chri8(N,ab0,x,eps0,nu0,numax,rho0p,rho0m,iopt)` generates the `N`$\times$2 array `ab` of the first `N` recurrence coefficients for the modified weight function $w(t)/(t^2-\mathtt{x}^2)$, `x` real outside the support interval of w, where the weight function w is specified by the array `ab0` of recurrence coefficients. If `iopt=1`, the routine computes the desired recurrence coefficients by means of ratios of appropriate Cauchy integrals. These are computed to a relative accuracy `eps0` by the continued fraction algorithm implemented in the routine `cauchy.m`. This calls for input parameters `nu0`, `numax`, and for an input array `ab0` of size `numax`$\times$2. The output parameter `nu` is inherited from the routine `cauchy.m`. If `iopt` is not equal to 1, the routine uses forward recursion to compute the required ratios of Cauchy integrals, which, however, may be unstable. In that case, the routine needs the values `rho0p` $= \rho_0(\mathtt{x};w)$ and `rho0m` $= \rho_0(-\mathtt{x};w)$ of the first Cauchy integral and an input array `ab0` of size `N`$\times$2 only and outputs `nu=0`.

`fejer.m`: The Fejér quadrature rule.

`uv=fejer(N)` generates the `N`$\times$2 array `uv` of nodes and weights of the `N`-point Fejér quadrature rule.

`gchri1.m`: Alternative modification by a linear divisor.

`[ab,rho,nu]=gchri1(N,ab0,x,eps0,nu0,numax)` generates the `N`$\times$2 array `ab` of the first `N` recurrence coefficients for the modified weight function $w(t)/(t-\mathtt{x})$, `x` outside the support interval of w, where the weight function w is specified by the array `ab0` of recurrence coefficients.

The routine computes the desired recurrence coefficients by means of the modified Chebyshev algorithm based on modified moments relative to the polynomials π_k orthogonal with respect to the weight function w. The latter are, except for sign, the Cauchy integrals of π_k and can be computed to a relative accuracy `eps0` by a continued fraction algorithm implemented in the routine `cauchy.m`. This calls for input parameters `nu0`, `numax`, and for an input array `ab0` of size `numax`×2. The Cauchy integrals are returned in the (2N)×1 output array `rho`. The output parameter `nu` is inherited from the routine `cauchy.m`.

`gchri2.m`: Alternative modification by a quadratic divisor.

`[ab,rho,nu]=gchri2(N,ab0,x,y,eps0,nu0,numax)` generates the N×2 array `ab` of the first N recurrence coefficients for the modified weight function $w(t)/((t-\mathrm{x})^2+\mathrm{y}^2)$, $z=\mathrm{x}+i\mathrm{y}$ outside the support interval of w, where the weight function w is specified by the input array `ab0` of recurrence coefficients. The routine computes the desired recurrence coefficients by means of the modified Chebyshev algorithm based on modified moments relative to the polynomials π_k orthogonal with respect to the weight function w. The latter are, except for sign, the Cauchy integrals of π_k and can be computed to a relative accuracy `eps0` by a continued fraction algorithm implemented in the routine `cauchy.m`. This calls for input parameters `nu0`, `numax`, and for an input array `ab0` of size `numax`×2. The Cauchy integrals are returned in the (2N)×1 output array `rho`. The output parameter `nu` is inherited from the routine `cauchy.m`.

`indop.m`: Induced orthogonal polynomials.

`ab=indop(N,m,ab0)` generates the N×2 array `ab` of the first N recurrence coefficients for the mth induced orthogonal polynomial relative to a weight function w, i.e., the polynomial orthogonal with respect to the weight function $[\pi_{\mathrm{m}}(t)]^2 w(t)$, where π_{m} is the mth-degree monic orthogonal polynomial relative to the weight function w. This requires the (N+m)×2 input array `ab0` of recurrence coefficients for the weight function w.

`lanczos.m`: Lanczos procedure.

`ab=lanczos(N,xw)` generates the N×2 array `ab` of recurrence coefficients for the discrete measure $d\lambda_{\mathrm{N}}$ identified by the N×2 input array `xw` containing the abscissae and jumps of $d\lambda_{\mathrm{N}}$.

The script is adapted from the routine RKPW in [41].

`mcchebyshev.m`: Multicomponent discretized Chebyshev algorithm.

`[ab,Mcap,kount]=mcchebyshev(N,eps0,quad,Mmax)` generates to a relative accuracy `eps0` (in the β-coefficients) the first N recurrence coefficients for a measure using the modified Chebyshev algorithm with the modified moments approximated by a multicomponent discretization procedure, and places them into the N×2 output array `ab`. The successive M-point discretizations of the modified moments are effected by the quadrature routine `quad`. If convergence to within the accuracy `eps0` is achieved for a value $M \le$ `Mmax` of M, this value is output as `Mcap` along with the number `kount` of iterations. If not, the routine displays the message "`Mcap exceeds Mmax in mccheb`" prior to exiting.

The details of the discretization are specified by a number of global parameters as follows:

`mc`	the number of component intervals
`mp`	the number of points in the discrete part of the measure (`mp`=0 if there is none)
`iq`	a parameter to be set equal to 1 if the user provides his or her own quadrature routine, and different from 1 otherwise
`idelta`	a parameter whose default value is 1, but is preferably set equal to 2 if `iq`=1 and the user provides Gauss-type quadrature routines

The component intervals have to be specified (in the order left to right) by a global `mc` × 2 array `AB=[[a1 b1];[a2 b2];...;[amc bmc]]`, where for infinite extreme intervals `a1=-Inf`, resp., `bmc=Inf`. The discrete spectrum (if `mp`>0) is similarly specified by a global `mp`×2 array `DM=[[x1 y1];[x2 y2];...;[xmp ymp]]` containing the abscissae and jumps of the discrete measure.

If the user provides his or her own quadrature routine "`quadown`", the routine `mcchebyshev.m` must be called with the input parameter "`quad`" replaced by "`@quadown`", otherwise with "`quad`" replaced by "`@quadgp`", a general-purpose routine provided by `OPQ`. The quadrature routine must have the form

$$\texttt{function xw} = \texttt{quad(M,i)}$$

where `M` is the number of nodes and `i` identifies the interval to which the routine is to be applied.

The routine `mcchebyshec.m` also applies to measures given originally in multicomponent form.

`mcdis.m`: Multicomponent discretization procedure.

`[ab,Mcap,kount]=mcdis(N,eps0,quad,Mmax)` generates to a relative accuracy `eps0` (in the β-coefficients) the first `N` recurrence coefficients for a measure, using a multicomponent discretization procedure, and places them into the `N`×2 output array `ab`. The successive M-point discretizations are accomplished by the quadrature routine `quad`. If convergence to within the accuracy `eps0` is achieved for a value $M \le$ `Mmax` of M, this value is output as `Mcap` along with the number `kount` of iterations. If not, the routine displays the message "`Mcap exceeds Mmax in mcdis`" prior to exiting. The recurrence coefficients in `ab` are produced from the discretized measure either by Stieltjes's or by Lanczos's method, the choice being made by setting the global parameter `irout` equal to 1, resp., different from 1.

The details of the discretization are specified by a number of global parameters (other than irout) as follows:

`mc`	the number of component intervals
`mp`	the number of points in the discrete part of the measure (`mp=0` if there is none)
`iq`	a parameter to be set equal to 1 if the user provides his or her own quadrature routine, and different from 1 otherwise
`idelta`	a parameter whose default value is 1, but is preferably set equal to 2 if `iq=1` and the user provides Gauss-type quadrature routines

The component intervals have to be specified (in the order left to right) by a global `mc` × 2 array `AB=[[a1 b1];[a2 b2];...;[amc bmc]]`, where for infinite extreme intervals `a1=-Inf`, resp., `bmc=Inf`. The discrete spectrum (if `mp>0`) is similarly specified by a global `mp`×2 array `DM=[[x1 y1];[x2 y2];...;[xmp ymp]]` containing the abscissae and jumps of the discrete measure.

If the user provides his or her own quadrature routine "`quadown`", the routine `mcdis.m` must be called with the input parameter "`quad`" replaced by "`@quadown`", otherwise with "`quad`" replaced by "`@quadgp`", a general-purpose routine provided by `OPQ`. The quadrature routine must have the form

$$\texttt{function xw} = \texttt{quad}(\texttt{M},\texttt{i})$$

where `M` is the number of nodes and `i` identifies the interval to which the routine is to be applied.

The routine `mcdis.m` also applies to measures given originally in multi-component form.

`nu0her.m`: Initial backward recurrence index for the Hermite weight.

`nu0=nu0her(N,z,eps0)` computes the initial index `nu0` to be used in the continued fraction algorithm of the routine `cauchy.m` for generating to a relative accuracy `eps0` the Cauchy integral $\int \pi_{\texttt{N}}(t)w(t)\mathrm{d}t/(\texttt{z}-t)$ for large `N` and complex `z` outside of $[-\infty,\infty]$. Here, $\pi_{\texttt{N}}$ is the `N`th-degree Hermite polynomial and w the Hermite weight function.

`nu0jac.m`: Initial backward recurrence index for the Jacobi weight.

`nu0=nu0jac(N,z,esp0)` computes the initial index `nu0` to be used in the continued fraction algorithm of the routine `cauchy.m` for generating to a relative accuracy `eps0` the Cauchy integral $\int \pi_{\texttt{N}}(t)w(t)\mathrm{d}t/(\texttt{z}-t)$ for large `N` and complex `z` outside of $[-1,1]$. Here, $\pi_{\texttt{N}}$ is the `N`th-degree Jacobi polynomial and w the Jacobi weight function.

`nu0lag.m`: Initial backward recurrence index for the Laguerre weight.

`nu0=nu0lag(N,z,a,eps0)` computes the initial index `nu0` to be used in the continued fraction algorithm of the routine `cauchy.m` for generating to a relative accuracy `eps0` the Cauchy integral $\int \pi_{\texttt{N}}(t)w(t)\mathrm{d}t$ $/(\texttt{z}-t)$ for large `N` and complex `z` outside of $[0,\infty]$. Here, $\pi_{\texttt{N}}$ is the `N`th-degree generalized Laguerre polynomial with parameter `a`, and w the generalized Laguerre weight function.

`quadgj.m`: Quadrature routine used in `r_gjacobi.m`.

`quadgp.m`: General-purpose quadrature routine.

`xw=quadgp(N,i)` accomplishes the discretization in `mcdis.m` and `mccheb.m` by providing the necessary Fejér quadrature nodes `xw(:,1)` and weights `xw(:,2)` in the `i`th component interval. A routine for the weight function on this interval must be provided at the bottom of the routine.

`quad_jacplusdis.m`: Quadrature routine used in `r_jacplusdis.m`.

`xw=quad_jacpludis(N,i)` accomplishes the discretization in `mcdis.m` by providing the necessary quadrature nodes `xw(:,1)` and weights `xw(:,2)` for obtaining the recurrence coefficients of a normalized Jacobi weight function plus a discrete measure.

`r_ellcheb.m`: Recurrence coefficients for elliptic Chebyshev polynomials.

`ab=r_ellcheb(N,om2)` generates the N×2 array `ab` of the first N recurrence coefficients for polynomials orthogonal with respect to the weight function $w(t)=[(1-\mathrm{om2}\,t^2)(1-t^2)]^{-1/2}$ on $[-1,1]$, $0<\mathrm{om2}<1$.

`r_gjacobi.m`: Recurrence coefficients for generalized Jacobi polynomials.

`ab=r_gjacobi(N,a,b)` generates to a relative accuracy `eps0`, using multicomponent discretization, the N×2 array `ab` of the first N recurrence coefficients for polynomials orthogonal on $[-1,1]$ with respect to the generalized Jacobi weight function

$$w(t)=(t-1)^{\mathrm{a}}\,|t-t_{\mathrm{mc}}|^{\gamma_{\mathrm{mc}}}\cdots|t-t_2|^{\gamma_2}(t+1)^{\mathrm{b}}.$$

In the calling program the user has to provide the global (mc+1)×2 array `tc` of abscissae, $-1<t_2<\cdots<t_{\mathrm{mc}}<1$ (in the first column), and the corresponding exponents, $\mathrm{b},\gamma_2,\ldots,\gamma_{\mathrm{mc}},\mathrm{a}$ (in the second column), as well as the global parameters `eps0` and `irout`, where `irout`=1 selects Stieltjes's algorithm, and `irout`≠1 Lanczos's algorithm for obtaining the desired recurrence coefficients from the discretized generalized Jacobi measure. See also `gjactest.m`.

`r_hahn.m`: Recurrence coefficients for Hahn polynomials.

`ab=r_hahn(N,a,b)` generates the $(\mathrm{N}+1)\times 2$ array `ab` of the first N+1 recurrence coefficients for the Hahn polynomials with parameters `a` and `b`, orthogonal on the discrete set of N+1 points $0,1,2,\ldots,\mathrm{N}$ with weights $\binom{\mathrm{a}+k}{k}\binom{\mathrm{b}+\mathrm{N}-k}{\mathrm{N}-k}$, $k=0,1,2,\ldots,\mathrm{N}$. The call `ab=r_hahn(N,a)` is the same as `ab= r_hahn(N,a,a)`, and `ab=r_hahn(N)` the same as `ab=r_hahn(N, 0,0)`, which produces the recurrence coefficients of the discrete Chebyshev polynomials for the points $0,1,2,\ldots,\mathrm{N}$.

`r_hermite.m`: Recurrence coefficients for generalized Hermite polynomials.

`ab=r_hermite(N,mu)` generates the N×2 array `ab` of the first N recurrence coefficients for polynomials orthogonal on $[-\infty,\infty]$ with respect to the weight function $w(t)=|t|^{2\mathrm{mu}}e^{-t^2}$, $\mathrm{mu}>-1/2$. The call `ab=r_hermite(N)` is the same as `ab=r_hermite(N,0)`.

Supplied by Dirk Laurie, 6-22-1998; edited by Walter Gautschi, 4-4-2002.

`r_hrhermite.m`: Recurrence coefficients for half-range Hermite polynomials.

`ab=r_hrhermite(N,Mmax,eps0)` generates the `N`×2 array `ab` of the first `N` recurrence coefficients for polynomials orthogonal on $[0,\infty]$ with respect to the weight function $w(t)=e^{-t^2}$. The routine uses the multi-component discretization routine `mcdis.m`, which requires the input parameter `Mmax` to control the discretization, and the input parameter `eps0` to control the accuracy. At the bottom of the quadrature routine `quadgp.m`, the user has to provide the weight function `y=exp(-x.^2)`.

`r_jaclog.m`: Recurrence coefficients for an algebraic/logarithmic weight function.

`ab=r_jaclog(N,a)` generates the `N`×2 array `ab` of the first `N` recurrence coefficients for polynomials orthogonal on $[0,1]$ with respect to the weight function $w(t)=t^{\mathrm{a}}\log(1/t)$, a> −1. The call `ab=r_jaclog(N)` is the same as `ab=r_jaclog(N,0)`.

`r_jacobi.m`: Recurrence coefficients for Jacobi polynomials.

`ab=r_jacobi(N,a,b)` generates the `N` × 2 array `ab` of the first `N` recurrence coefficients for polynomials orthogonal on $[-1,1]$ with respect to the weight function $w(t)=(1-t)^{\mathrm{a}}(1+t)^{\mathrm{b}}$. The call `ab=r_jacobi(N,a)` is the same as `ab=r_jacobi(N,a,a)`, and `ab=r_jacobi(N)` the same as `ab=r_jacobi(N,0,0)`.

Supplied by Dirk Laurie, 6-22-1998; edited by Walter Gautschi, 4-4-2002.

`r_jacobi01.m`: Recurrence coefficients for shifted Jacobi polynomials.

`ab=r_jacobi01(N,a,b)` generates the `N`×2 array `ab` of the first `N` recurrence coefficients for polynomials orthogonal on $[0,1]$ with respect to the weight function $w(t)=(1-t)^{\mathrm{a}}t^{\mathrm{b}}$. The call `ab=r_jacobi01(N,a)` is the same as `ab=r_jacobi01(N,a,a)`, and `ab=r_jacobi01(N)` the same as `ab=r_jacobi01(N,0,0)`.

`r_jacplusdis.m`: Recurrence coefficients for normalized Jacobi weight functions plus a discrete measure.

`ab=r_jacplusdis(N,alpha,beta,ty)` generates the `N`×2 array `ab` of the first `N` recurrence coefficients for polynomials orthogonal on $[-1,1]$ with respect to the normalized Jacobi weight function with parameters `alpha`, `beta` and a discrete measure added on to it. The latter is given by the two-column input array `ty` containing in the first column the abscissae, and in the second column the jumps of the discrete measure. The routine uses a Gauss–Jacobi quadrature discretization, calling on the routines `mcdis.m` and `quad_jacplusdis.m`. The user has a choice on whether to use Stieltjes's or Lanczos's algorithm to obtain the desired recurrence coefficients from the discretized measure. In the first case, the global parameter `irout` must be set equal to 1, in the other case different from 1.

`r_laguerre.m`: Recurrence coefficients for generalized Laguerre polynomials.

`ab=r_laguerre(N,a)` generates the `N`×2 array `ab` of the first `N` recurrence coefficients for polynomials orthogonal on $[0,\infty]$ with respect to the weight function $w(t)=t^{\mathrm{a}}e^{-t}$, a> −1. The call `ab=r_laguerre(N)` is the same as `ab=r_laguerre(N,0)`.

Supplied by Dirk Laurie, 6-22-1998; edited by Walter Gautschi, 4-4-2002.

`r_logistic.m`: Recurrence coefficients for the logistic weight function.

`ab=r_logistic(N)` generates the `N`×2 array `ab` of the first `N` recurrence coefficients for polynomials orthogonal on $[-\infty,\infty]$ with respect to the weight function $w(t)=e^{-t}/(1+e^{-t})^2$.

`r_meixner_pollaczek.m`: Recurrence coefficients for Meixner–Pollaczek polynomials.

`ab=r_meixner_pollaczek(N,lambda,phi)` generates the `N`×2 array `ab` of the first `N` recurrence coefficients for polynomials orthogonal on $[-\infty,\infty]$ with respect to the weight function $w(t)=(2\pi)^{-1}\exp((2\phi-\pi)t)|\Gamma(\lambda+it)|^2$. The call `ab=r_meixner_pollaczek(N,lambda)` is the same as `ab=r_meixner_pollaczek(N,lambda,pi/2)`.

`r_modbess.m`: Recurrence coefficients for a weight function involving a modified Bessel function.

`ab=r_modbess(N,a,Mmax,eps0)` generates to a relative accuracy `eps0` the `N` × 2 array `ab` of the first `N` recurrence coefficients for polynomials orthogonal on $[0,\infty]$ with respect to the weight function $w(t)=t^{\mathtt{a}}K_0(t)$, `a`$>-1$, where K_0 is the modified Bessel function of order 0. The routine uses a multicomponent discretization procedure, calling on the routines `mcdis.m` and `quadbess.m`. The input parameter `Mmax` is the maximum allowed value of M in `mcdis.m`.

`sobzeros.m`: Zeros of Sobolev orthogonal polynomials.

`z=sobzeros(n,N,B)` generates the $n\times 1$ array `z` of zeros of the nth-degree Sobolev orthogonal polynomial specified by the `N`×`N` array `B`, `N`$\geq n$, of recurrence coefficients. If all real, the zeros are ordered increasingly, otherwise with increasing moduli.

`stieltjes.m`: Discretized Stieltjes procedure.

`ab=stieltjes(N,xw)` generates the `N`×2 array `ab` of the first `N` recurrence coefficients for polynomials orthogonal with respect to a discrete measure specified by the `N`×2 input array `xw` containing in the first column the abscissae, and in the second column the jumps of the discrete measure.

`stieltjes_sob.m`: Stieltjes algorithm for Sobolev orthogonal polynomials.

`[B,normsq]=stieltjes_sob(n,s,nd,xw,a,same)` generates the $n\times n$ matrix B of recurrence coefficients β_j^k for monic Sobolev orthogonal polynomials, with β_j^k, $0\leq j\leq k$, $k=0,1,\dots,n-1$, occupying the position $(j+1,k+1)$ of the matrix B, all other elements of B being zero except for those in the first lower side diagonal, which are 1. Also returned is the $n\times 1$ array `normsq` of the squared norms of the Sobolev orthogonal polynomials. The routine uses an inner product representation of the coefficients β_j^k and evaluates the inner products exactly by Gaussian quadrature. The number n_σ of quadrature points may be different for the `s`+1 individual measures making up the Sobolev inner product. The input variable `nd` collects the numbers n_σ in a vector of dimension `s`+1. The input variable `xw` is an array of dimension $N\times(2\mathtt{s}+2)$, where $N=\max n_\sigma$. It holds the nodes of the quadrature rules in the first half, and the weights in the second half of the array. The input parameter `a0` is the first α-coefficient of the "ground measure" of the Sobolev inner product, and the variable `same` has to be set equal to 1 if all quadrature rules have the same set of nodes, and equal to `0` otherwise.

The procedure is an extended version of the Stieltjes algorithm described in Section 4 of [36].

A.2 ▪ Quadrature

`NewtCotes.m`: Weighted Newton–Cotes formula.

> `w=NewtCotes(n,x,ab,eps0)` computes the `n`×1 array `w` of weights for the `n`-point Newton–Cotes quadrature formula having prescribed distinct nodes specified in the `n`×1 input array `x` and involving a weight function w. The [floor((n+1)/2)]×2 array `ab` is to hold the recurrence coefficients for the polynomials orthogonal with respect to the weight function w. The input parameter `eps0`, assumed close to machine precision, is used to detect any coincidence between Gauss quadrature nodes generated and the given Newton–Cotes nodes.

`bose_einstein.m`: Rational Gauss quadrature rule for evaluating the generalized Bose–Einstein integral.

`bose_einstein_diffpole.m`: Rational Gauss quadrature rule for evaluating the generalized Bose–Einstein integral with a difficult pole.

`cauchyPVI.m`: Cauchy principal value integral.

> `cpvi=cauchyPVI(N,x,f,ddf,iopt,ab,rho0)`, if `iopt`=1, computes an `N`-point approximation, otherwise an (`N`+1)-point approximation (containing `x` as a node), to the Cauchy principal value integral $\int f(t)d\lambda(t)/(\mathtt{x}-t)$. The integration measure is input via the (`N`+1)×2 array `ab` of recurrence coefficients. The Hilbert transform $\int d\lambda(t)/(\mathtt{x}-t)$ of the measure has to be input through the parameter `rho0`. The routine assumes the function f to be provided by the M-file `f.m`, and it also relies on the M-file `ddf.m` for providing a carefully programmed routine for the divided difference $(f(x)-f(y))/(x-y)$ of f. Both routines must be able to accept vector-valued arguments. See also `Example3_44.m`.

`fermi_dirac.m`: Rational Gauss quadrature rule for evaluating the generalized Fermi–Dirac integral.

`fex31.m`: Function used in `Example3_31.m`.

`fex32.m`: Function used in `Example3_32.m`.

`fourier_gauss.m`: Fourier coefficients based on Gauss quadrature points and weights.

`fourier_lobatto.m`: Fourier coefficients based on Gauss–Lobatto quadrature points and weights.

`gauss.m`: Gauss quadrature rule.

`xw=gauss(N,ab)` generates the N×2 array `xw` of Gauss quadrature nodes and weights for a given weight function w. The nodes, in increasing order, are placed into the first column of `xw`, and the corresponding weights into the second column. The weight function w is specified by the N×2 input array `ab` of recurrence coefficients for the polynomials orthogonal with respect to the weight function w.

`gauss_rational.m`: Rational Gauss quadrature rule.

`xw=gauss_rational(N,abmod)` generates from the first N modified recurrence coefficients, provided in the N×2 input array `abmod`, the N×2 array `xw` of the rational Gauss-type quadrature rule according to Theorem 3.25 of [24].

`globatto.m`: Generalized Gauss–Lobatto quadrature formula.

`xw=globatto(N,ab,r,endl,endr)` generates the generalized Gauss–Lobatto quadrature rule with left and right endpoints `endl`, `endr` of multiplicity `r`, an integration measure $d\lambda$ given by the (N+2r)×2 array `ab` of its recurrence coefficients, and N interior points. The N interior nodes and weights are placed into the last N rows of the (N+2r)×2 output array `xw`, and the fixed nodes (each repeated `r` times) and the 2r weights associated with the 2r endpoint derivatives into the first 2r rows of `xw`. It is assumed that N>2 and r>0.

`gradau.m`: Generalized Gauss–Radau quadrature formula.

`xw=gradau(N,ab,r,endl)` generates the generalized Gauss–Radau quadrature rule with a left endpoint `endl` of multiplicity `r`, an integration measure $d\lambda$ given by the (N+r)×2 array `ab` of its recurrence coefficients, and N interior points. The N interior nodes and weights are placed into the last N rows of the (N+r)×2 output array `xw`, and the fixed node `endl` (repeated `r` times) and the `r` weights associated with the `r` endpoint-derivatives into the first `r` rows of `xw`.

`kernel.m`: Gauss remainder kernels.

`[ker,nu]=kernel(N,ab,z,eps0,nu0,numax)` generates to a relative accuracy `eps0` the (N+1)×1 array `ker` of the first N+1 Gauss remainder kernels for a weight function w, i.e., the ratios $\rho_k(z)/\pi_k(z)$, $k = 0, 1, \ldots, N$, where π_k is the monic polynomial of degree k orthogonal with respect to the weight function w, and ρ_k the Cauchy integral of π_k. The latter can be computed by the routine `cauchy.m`, which requires the input parameters `z`, `eps0`, `nu0`, `numax`, and the (numax)×2 array `ab` of recurrence coefficients for the weight function w. The output parameter `nu` is inherited from the routine `cauchy.m`.

`kronrod.m`: Gauss–Kronrod quadrature formula.

`xw=kronrod(N,ab)` generates the (2N+1)-point Gauss–Kronrod quadrature rule for a weight function w specified by the [ceil(3N/2)+1]×2 array `ab` of its recurrence coefficients. The 2N+1 nodes, in increasing order, are placed into the first column, the corresponding weights into the second column of the (2N+1)×2 output array `xw`.

Supplied by Dirk Laurie, 6-22-1998.

`kronrod_rational.m`: Rational Gauss–Kronrod quadrature rule.

`xw=kronrod_rational(N,abmod)` generates from the first 2N+1 modified recurrence coefficients, provided in the (2N+1)×2 input array `abmod`, the (2N+1)×2 array `xw` of the rational Gauss–Kronrod formula according to Theorem 3.41 of [24].

`lanczos_algorithm.m`: Lanczos algorithm.

`[H,a,g]=lanczos_algorithm(N,A,h0)` generates the N×N matrix `H` of the successive Lanczos vectors $h_0, h_1, \dots, h_{N-1}$ of the matrix A, as well as the N×1 array `a` of coefficients $\alpha_0, \alpha_1, \dots, \alpha_{N-1}$ and the (N+1)×1 array `g` of coefficients $\gamma_0, \gamma_1, \dots, \gamma_N$ generated by the Lanczos algorithm with starting vector `h0`. (In theory, $\gamma_N = 0$.)

`least_squares.m`: Polynomial least squares approximation.

`[phat,c]=least_squares(n,f,xw,ab,d)` generates the N×(n+1) array `phat` of least squares approximants of degrees $0, 1, \dots, n$ evaluated at the N abscissae of the N-point discrete inner product given by the N×2 array `xw` of abscissae and weights, and the (n+1)×1 array `c` of Fourier coefficients. The input array `ab` is to contain the first n+1 recurrence coefficients for the monic discrete orthogonal polynomials. Those actually used are scaled to have leading coefficients supplied by the 1×(n+1) array `d`. The data are input by the N×1 array `f`.

`least_squares_sob.m`: Polynomial least squares approximation in Sobolev space.

`[phat,c]=least_squares_sob(n,f,xw,B)` generates the (n+1) ×(N(s+1)) array `phat` of Sobolev least squares approximants of degrees $0, 1, \dots, n$ and their derivatives of orders $0, 1, \dots, s$ evaluated at the N abscissae t_k of the discrete Sobolev inner product given by the N×($s+2$) array `xw` of abscissae and weights, and the (n+1)×1 array `c` of Sobolev-Fourier coefficients. Here, s is the highest-order derivative appearing in the Sobolev inner product and is determined by the routine automatically from the size of the input array `xw`. The N×N input array `B` is the upper Hessenberg matrix of recurrence coefficients for the Sobolev orthogonal polynomials, which for $s = 1$ can be generated by the routine `chebyshev_sob.m`, and for arbitrary s by the routine `stieltjes_sob.m`. The N values of the derivative of order `sig` of the νth-degree approximant, $\nu \le n$, are output in positions (ν+1,`sig`$+1{:}s+1{:}N(s+1)$) of the array `phat`. The N×(s+1) input array `f` contains the N values of the given function and its first s derivatives at the N support points t_k.

`lobatto.m`: Gauss–Lobatto quadrature rule.

`xw=lobatto(N,ab,endl,endr)` generates the (N+2)-point Gauss–Lobatto quadrature rule for a weight function w having two prescribed nodes `endl`, `endr` (typically the left and right endpoint of the support interval of w, or points to the left, resp., to the right, thereof). The N+2 nodes, in increasing order, are placed into the first column, the corresponding N+2 weights into the second column of the (N+2)×2 output array `xw`. The weight function w is specified by the (N+2)×2 array `ab` of its recurrence coefficients.

For Jacobi weight functions, see also `lobatto_jacobi.m`.

`lobatto_jacobi.m`: Gauss–Lobatto quadrature rule for Jacobi weight function.

`xw=lobatto_jacobi(N,a,b)` generates the (N+2)-point Gauss–Lobatto quadrature rule for the weight function $w(t)=(1-t)^{a}(1+t)^{b}$, $a>-1$, $b>-1$, on $[-1,1]$ (having fixed nodes at +1 and −1). The N+2 nodes, in increasing order, are placed into the first column, the corresponding N+2 weights into the second column of the (N+2)×2 output array `xw`. The call `xw=lobatto_jacobi(N,a)` is the same as `xw=lobatto_jacobi(N,a,a)`, and `xw=lobatto_jacobi(N)` the same as `xw=lobatto_jacobi(N,0,0)` [21].

`lobatto_rational.m`: Rational Gauss–Lobatto quadrature rule.

`xw=lobatto_rational(N,abmod,endl,endr)` generates from the first N+2 modified recurrence coefficients, supplied by the (N+2)×2 input array `abmod`, the (N+2)×2 array `xw` of the rational Gauss–Lobatto quadrature rule according to Theorem 3.40 of [24].

`quadrat.m`: Quadrature routine used in `r_mod.m`.

`r_kronrod.m`: Jacobi–Kronrod matrix.

`ab=r_kronrod(N,ab0)` generates, for a weight function w, the (2N+1) ×2 array

$$\mathtt{ab} = [\,[\alpha_0^K\ \beta_0^k];[\alpha_1^K\ \beta_1^K];\ldots;[\alpha_{2N}^K\ \beta_{2N}^K]\,]$$

of α-Kronrod and β-Kronrod elements, so that the Jacobi–Kronrod matrix of order 2N+1 is the symmetric tridiagonal matrix with $\alpha_0^K,\alpha_1^K,\ldots,\alpha_{2N}^K$ on the diagonal, and $\sqrt{\beta_1^K},\sqrt{\beta_2^K},\ldots,\sqrt{\beta_{2N}^K}$ on the two side diagonals (β_0^K is the integral $\int w(t)dt$ of w). The weight function w is identified by the [ceil(3N/2)+1]×2 input array `ab0`.

Supplied by Dirk Laurie, 6-22-1998.

`r_mod.m`: Modified recurrence coefficients.

`[abmod,Ncap,kount]=r_mod(N,ab)` generates the N×2 array `abmod` of the first N recurrence coefficients of the modified weight function w/p, where p is the polynomial of degree m defined by [24, Eq. (3.1.75)] incorporating the M poles selected. Their negative reciprocals (the quantities ζ_μ in [24, Eq. (3.1.71)]) must be provided by the user in the first column of the global $M\times 2$ array `Z`, and their multiplicities in the second column. It is assumed that complex poles occur in conjugate complex pairs and that both poles of each pair are included in the array `Z`. Other global parameters the user must provide are `mc`, `mp`, `iq`, `idelta`, `irout`, `AB`, `eps0`, `Nmax`, used in the routine `mcdis.m` with their meaning explained there, and `M`. Be sure that the variables `ab`, `M`, and `Z` are declared global variables in the quadrature routine `quadrat.m`. The output variables `Ncap` and `kount` are inherited from the routine `mcdis.m` with meanings explained there. The measure $d\lambda$ in [24, Eq. (3.1.59)] is identified by the (Nmax)×2 input array `ab` of recurrence coefficients.

`r_multidomain_cheb.m`: Multidomain modified Chebyshev algorithm.

`ab=r_multidomain_cheb(N,abmd,abmm)` generates the N×2 array `ab` of recurrence coefficients for polynomials orthogonal on the union of a finite number d of intervals (disjoint or not) with measures identified by the N$\times(2d)$ input array abmd, where columns $2j-1,2j$ hold the N$\times$2 matrix of recurrence coefficients for the jth interval, $j=1,2,\ldots,d$. The routine uses a matrix formulation of the modified Chebyshev algorithm in combination with Gauss quadratures. The 2N-1 recurrence coefficients for the polynomials defining the modified moments are input through the (2N-1)$\times$2 input array abmm.

r_multidomain_sti.m: Multidomain Stieltjes algorithm.

ab=r_multidomain_sti(N,abmd) generates the N$\times$2 array ab of recurrence coefficients for polynomials orthogonal on the union of a finite number d of intervals (disjoint or not) with measures identified by the N$\times(2d)$ input array abmd, where columns $2j-1,2j$ hold the N$\times$2 matrix of recurrence coefficients for the jth interval, $j=1,2,\ldots,d$. The routine uses a matrix formulation of Stieltjes's algorithm in combination with Gauss quadratures.

radau.m: Gauss–Radau quadrature rule.

xw=radau(N,ab,end0) generates the (N+1)-point Gauss–Radau quadrature rule for a weight function w having a prescribed node end0 (typically at one of the endpoints of the support interval of w, or outside thereof). The N+1 nodes, in increasing order, are placed into the first column, the corresponding N+1 weights into the second column of the (N+1)$\times$2 output array xw. The weight function w is specified by the (N+1)$\times$2 input array ab of its recurrence coefficients.

For Jacobi weight functions, see also radau_jacobi.m.

radau_jacobi.m: Gauss–Radau quadrature rule for Jacobi weight function.

xw=radau_jacobi(N,iopt,a,b) generates the (N+1)-point Gauss–Radau quadrature rule for the weight function $w(t)=(1-t)^{\mathrm{a}}(1+t)^{\mathrm{b}}$ on $[-1,1]$, a>-1, b>-1. If iopt=1, the fixed node is at the left endpoint, otherwise at the right endpoint of the interval $[-1,1]$. The N+1 nodes, in increasing order, are placed into the first column, the corresponding N+1 weights into the second column of the (N + 1) $\times$ 2 output array xw. The call xw=radau_jacobi(N,iopt,a) is the same as xw=radau_jacobi(N,iopt,a,a), and xw=radau_jacobi(N,iopt) the same as xw=radau_jacobi(N,iopt,0,0) [22].

radau_laguerre.m: Gauss–Radau quadrature rule for Laguerre weight function.

xw=radau_laguerre(N,a) generates the (N+1)-point Gauss–Radau quadrature rule for the weight function $w(t)=t^{\mathrm{a}}\exp(-t)$ on $[0,\infty]$, a>-1, with fixed node 0. The N+1 nodes, in increasing order, are placed into the first column, the corresponding N+1 weights into the second column of the (N+1)$\times$2 output array xw. The call xw=radau_laguerre(N) is the same as xw=radau_laguerre(N,0) [22].

radau_rational.m: Rational Gauss–Radau quadrature rule.

xw=radau_rational(N,abmod,end0) generates from the first N+1 modified recurrence coefficients, supplied in the (N+1)$\times$2 input array abmod, the (N+1)$\times$2 array xw of the rational Gauss–Radau rule with fixed node end0 according to Theorem 3.39 of [24].

turan.m: Gauss–Turán quadrature rule.

xw=turan(n,s,eps0,ab0,hom) generates to a relative accuracy eps0 the n×(2s+2) array xw of the n-point Gauss–Turán quadrature rule involving successive derivatives of orders up to, and including, 2s and a weight function w. The latter is specified by the ((s+1)n)×2 array ab0 of its recurrence coefficients. If hom=1, a discrete homotopy in the variable s is used to facilitate convergence of Newton's method for computing the quadrature nodes. Otherwise, no homotopy is used. The n nodes of the quadrature rule are placed into the first column of the output array xw, the successive weights into the remaining 2s+1 columns.

A.3 ▪ Examples and tests

(All examples and tests are from [24].)

Example2_29.m: Script used in Example 2.29.

Example2_33.m: Script used in Example 2.33.

At the bottom of the routine quadgp.m the weight function wf has to be specified as follows:

```
y=x.^x.*(1-x).^(1-x);
```

Example2_34.m: Script used in Example 2.34.

At the bottom of the routine quadgp.m the weight function wf has to be specified as follows:

```
y=exp(-x.*x);
```

Example2_37.m: Recomputation of Table 2.9 with the Lanczos algorithm.

Example2_38.m: Script used in Example 2.38.

Example2_42.m: Script used in Example 2.42.

Example2_43.m: Script used in Example 2.43.

Example2_46.m: Script used in Example 2.46.

Example2_63.m: Script used in Example 2.63.

Example2_64.m: Script used in Example 2.64.

Example3_19.m: Script used in Example 3.19.

Example3_23.m: Script used in Example 3.23.

Example3_24.m: Script used in Example 3.24.

Example3_31.m: Script used in Example 3.31.

Example3_32.m: Script used in Example 3.32.

Example3_33.m: Script used in Example 3.33.

`Example3_34.m`: Script used in Example 3.34.

`Example3_35.m`: Script used in Example 3.35.

`Example3_36.m`: Script used in Example 3.36.

`Example3_37.m`: Script used in Example 3.37.

`Example3_42.m`: Script used in Example 3.42.

`Example3_44.m`: Script used in Example 3.44.

`Example3_45.m`: Script used in Example 3.45.

Example 3.45 is computed using the multidomain Stieltjes algorithm if iopt=1, and the multidomain modified Chebyshev algorithm otherwise. Timings are provided for both algorithms.

`Example3_48.m`: Script used in Example 3.48.

`Example3_49.m`: Script used in Example 3.49.

`Example3_51.m`: Script used in Example 3.51.

`Example3_52.m`: Script used in Example 3.52.

`Example3_53.m`: Script used in Example 3.53.

`Example3_55.m`: Script used in Example 3.55.

`Example3_59.m`: Script used in Example 3.59.

`Example3_60.m`: Script used in Example 3.60.

`Example3_63.m`: Script used in Example 3.63.

`Example3_64.m`: Script used in Example 3.64.

`Example3_65.m`: Script used in Example 3.65.

`Example3_66.m`: Script used in Example 3.66.

`Example3_67.m`: Script used in Example 3.67.

`Example3_68.m`: Script used in Example 3.68.

`Example3_69.m`: Script used in Example 3.69.

`Example3_70.m`: Script used in Example 3.70.

`Example3_71.m`: Script used in Example 3.71.

`ddfex44.m`: Function used in Example 3.44.

`fHL.m`: Function used in Example 3.64.

`fex31.m`: Function used in Example 3.31.

`fex33.m`: Function used in Example 3.33.

`fex34.m`: Function used in Example 3.34.

`fex35.m`: Function used in Example 3.35.

`fex37.m`: Function used in Examples 3.36 and 3.37.

`fex44.m`: Function used in Example 3.44.

`gjactest.m`: Test for the generalized Jacobi measure.

`qherm.m`: Quadrature routine used in Example 3.60.

`quadch.m`: Quadrature routine used in Example 2.46.

`quadeinstein.m`: Quadrature routine used in Example 2.41.

`quadfermi.m`: Quadrature routine used in Example 2.41.

`quadsqeinstein.m`: Quadrature routine used in Example 2.42.

`quadsqfermi.m`: Quadrature routine used in Example 2.42.

`test_globatto_lobatto.m`: Test of generalized Gauss–Lobatto for $r = 1$ against Lobatto.

`test_globatto_req2.m`: Test of generalized Gauss–Lobatto against Lobatto with double endpoints for the Legendre and Chebyshev measures.

`test_globatto_rgt0.m`: Moment test for generalized Gauss–Lobatto.

`test_gradau_radau.m`: Test of generalized Gauss–Radau for $r = 1$ against Radau.

`test_gradau_req2.m`: Test of generalized Gauss–Radau against Radau with double endpoint for the Legendre and Chebyshev measures.

`test_gradau_rgt0.m`: Moment test for generalized Gauss–Radau.

A.4 ▪ Tables and figures

(All tables and figures are from [24].)

`Table2_1.m`: Script for computing Table 2.1.

`Table2_2.m`: Script for computing Table 2.2.

`Table2_3.m`: Script for computing Table 2.3.

`Table2_4.m`: Script for computing Table 2.4.

`Table2_5.m`: Script for computing Table 2.5.

`Table2_6.m`: Script for computing Table 2.6.

`Table2_7.m`: Script for computing Table 2.7.

`Table2_8.m`: Script for computing Table 2.8.

`Table2_9.m`: Script for computing Table 2.9.

Table2_10.m: Script for computing Table 2.10.

Table2_11.m: Script for computing Table 2.11.

Table2_12.m: Script for computing Table 2.12.

Table2_14.m: Script for computing Table 2.14.

Table2_15.m: Script for computing Table 2.15.

Table2_18.m: Script for computing Table 2.18.

Table2_19.m: Script for computing Table 2.19.

Table2_20.m: Script for computing Table 2.20.

Table2_22.m: Script for computing Table 2.22.

Table2_23.m: Script for computing Table 2.23.

Table2_24.m: Script for computing Table 2.24.

Table2_25.m: Script for computing Table 2.25.

Table2_26.m: Script for computing Table 2.26.

Table2_27.m: Script for computing the first example in Example 2.61.

Table2_28.m: Script for computing the second example in Example 2.61.

Table2_29.m: Script for computing Table 2.29.

befej160: 18-digit recurrence coefficients for the discrete N-point measure, with $N = 160$, whose abscissae and jumps are the nodes and weights of the N-point Fejér quadrature rule.

befej320: 18-digit recurrence coefficients for the discrete N-point measure, with $N = 320$, whose abscissae and jumps are the nodes and weights of the N-point Fejér quadrature rule.

befej40: 18-digit recurrence coefficients for the discrete N-point measure, with $N = 40$, whose abscissae and jumps are the nodes and weights of the N-point Fejér quadrature rule.

befej80: 18-digit recurrence coefficients for the discrete N-point measure, with $N = 80$, whose abscissae and jumps are the nodes and weights of the N-point Fejér quadrature rule.

Figure2_2.m: Script for generating Figure 2.2.

Figure2_3.m: Script for generating Figure 2.3.

Figure3_3.m: Script for generating Figure 3.3.

Figure3_4.m: Script for generating Figure 3.4.

quadbess.m: Quadrature routine used in Table2_19.m.

quadcl.m: Quadrature routine used in Figure2_2m.

quadlag.m: Quadrature routine used in Table2_18.m.

A.5 ▪ Utility

`Tinv.m`: Inverse of a confluent Vandermonde matrix.

`ab_Airy.txt`: The first 200 recurrence coefficients to 32 digits for the weight function $w(t)=c\,t^{-1/2}e^{-t}K_{-1/3}(t)$ on $[0,\infty]$, where $c=2^{-1/2}\pi^{-3/2}$ and $K_{-1/3}$ is the modified Bessel function of order $-1/3$.

`ab_BE.txt`: The first 80 recurrence coefficients to 32 digits for the weight function $w(t)=t/(e^t-1)$ on $[0,\infty]$.

`ab_BEsq.txt`: The first 80 recurrence coefficients to 32 digits for the weight function $w(t)=[t/(e^t-1)]^2$ on $[0,\infty]$.

`ab_BEsqrtm1.txt`: The first 80 recurrence coefficients to 32 digits for the weight function $w(t)=t^{-1/2}[t/(e^t-1)]$ on $[0,\infty]$.

`ab_BEsqrtm1_sq.txt`: The first 80 recurrence coefficients to 25 digits for the weight function $w(t)=t^{-1/2}[t/(e^t-1)]^2$ on $[0,\infty]$.

`ab_BEsqrtp1.txt`: The first 80 recurrence coefficients to 25 digits for the weight function $w(t)=t^{1/2}\times[t/(e^t-1)]$ on $[0,\infty]$.

`ab_BEsqrtp1_sq.txt`: The first 80 recurrence coefficients to 25 digits for the weight function $w(t)=t^{1/2}\times[t/(e^t-1)]^2$ on $[0,\infty]$.

`ab_Fermi.txt`: The first 80 recurrence coefficients to 32 digits for the weight function $w(t)=1/(e^t+1)$ on $[0,\infty]$.

`ab_Fermisq.txt`: The first 80 recurrence coefficients to 32 digits for the weight function $w(t)=[1/(e^t+1)]^2$ on $[0,\infty]$.

`ab_Fermisqrtm1.txt`: The first 80 recurrence coefficients to 25 digits for the weight function $w(t)=t^{-1/2}/(e^t+1)$ on $[0,\infty]$.

`ab_Fermisqrtm1_sq.txt`: The first 80 recurrence coefficients to 25 digits for the weight function $w(t)=t^{-1/2}\times[1/(e^t+1)]^2$ on $[0,\infty]$.

`ab_Fermisqrtp1.txt`: The first 80 recurrence coefficients to 25 digits for the weight function $w(t)=t^{1/2}/(e^t+1)$ on $[0,\infty]$.

`ab_Fermisqrtp1_sq.txt`: The first 80 recurrence coefficients to 25 digits for the weight function $w(t)=t^{1/2}\times[1/(e^t+1)]^2$ on $[0,\infty]$.

`ab_hrf3mod.txt`: The first 200 recurrence coefficients to 32 digits for the weight function $w(t)=\exp(-t^3/3)$ on $[0,\infty]$.

`ab_hrhermite.txt`: The first 200 recurrence coefficients to 32 digits for the weight function $w(t)=\exp(-t^2)$ on $[0,\infty]$.

`ab_jaclog.txt`: The first 100 recurrence coefficients to 25 digits for the weight function $w(t)=t^{-1/2}\log(1/t)$ on $[0,1]$.

`ab_jaclog_sq.txt`: The first 100 recurrence coefficients to 25 digits for the weight function $w(t)=t^{-1/2}[\log(1/t)]^2$ on $[0,1]$.

`acondG.m`: Absolute condition number of the map G_n.

`ac=acondG(n,ab,ab0)` computes the absolute condition number `ac` of the map from the first 2`n` normalized modified moments of a measure $d\lambda$ to the `n`-point Gauss quadrature formula for the same measure [24, §2.1.3]. The measure is identified by the `n`×2 array `ab` of its first `n` recurrence coefficients. The orthogonal polynomials defining the modified moments are specified by the (2`n`)×2 array `ab0` of their recurrence coefficients. The routine calls on the function `g_n.m`.

`acondGlow.m`: Lower bound for the absolute condition number of the map G_n.

`acl=acondGlow(n,ab)` computes a lower bound `acl` for the absolute condition number of the map from the first 2`n` ordinary moments of a measure $d\lambda$ to the `n`-point Gauss quadrature formula for the same measure [24, §2.1.3]. The measure is specified by the `n`×2 array `ab` of its first `n` recurrence coefficients.

`clenshaw.m`: Clenshaw algorithm.

`s=clenshaw(N,x,y0,ym1,ab,c)` evaluates the sum

$$\mathtt{s} = \sum_{j=0}^{\mathtt{N}} c_j y_j(\mathtt{x}),$$

where $y_j(\mathtt{x})$, $j=-1,0,1,\ldots,\mathtt{N}$, satisfy a three-term recurrence relation

$$y_{k+1}(\mathtt{x}) = (\mathtt{x}-a_k)y_k(\mathtt{x}) - b_k y_{k-1}(\mathtt{x}), \quad k=0,1,\ldots,\mathtt{N}-1,$$

with initial values given by $y_0 = \mathtt{y0}$, $y_{-1} = \mathtt{ym1}$. The recurrence coefficients (if $\mathtt{N}>0$) are input by the $\mathtt{N}\times 2$ array `ab` and the coefficients in the series by the $1\times(\mathtt{N}+1)$ array `c`. The argument `x` is allowed to be (row-) vector-valued.

`clenshaw_cheb.m`: Clenshaw algorithm for Chebyshev series.

`s=clenshaw_cheb(N,x,c)` evaluates the sum

$$\mathtt{s} = (1/2)c_0 + \sum_{j=1}^{\mathtt{N}} c_j T_j(\mathtt{x})$$

in Chebyshev polynomials T_j of the first kind, where the coefficients c_j are specified by the $1\times(\mathtt{N}+1)$ input array `c`. The argument `x` is allowed to be (row-) vector-valued.

`clenshaw_sob.m`: Clenshaw-like algorithm for Sobolev orthogonal polynomials.

`r=clenshaw_sob(n,N,s,x,B,c)` evaluates a linear combination of the first `N`+1 monic Sobolev orthogonal polynomials $\pi_k(\mathtt{x})$, with coefficients given by the $1\times(\mathtt{N}+1)$ array `c`, and its first `s` derivatives, $\mathtt{s}\le 2$, placing the results into the $(\mathtt{s}+1)\times 1$ array `r`. The Sobolev orthogonal polynomials are specified by the $\mathtt{N}\times\mathtt{N}$ ($\mathtt{N}\ge\mathtt{n}$) input array `B` of the matrix of recurrence coefficients. The input variable `x` is allowed to be (row) vector-valued.

`convert.m`: Conversion algorithm for series of orthogonal polynomials.

Given two sets of orthogonal polynomials, π_k^0 and π_k, $k=0,1,2,\ldots,\mathtt{N}$, specified, resp., by the `N`×2 arrays `ab0` and `ab` of recurrence coefficients, and given a polynomial p of degree `N` as a linear combination of the polynomials π_k^0, with coefficients specified by the $(\mathtt{N}+1)\times 1$ array `c0`, the call `c=convert(N,c0,ab0,ab)` determines and stores in the $(\mathtt{N}+1)\times 1$ output array `c` the coefficients of p expanded in the polynomials π_k. The choice `ab0=zeros(N,2)` is permitted and corresponds to p being given as a linear combination of the first `N`+1 powers.

`epsalg.m`: Epsilon algorithm.

`E=epsalg(n,s)` applies to a sequence of `n` numbers given in the `n`×1 array `s` the epsilon algorithm, returning the results in the `n`×(`n` +1) array `E`. The values of interest are located in the even-numbered columns of `E`. Preferably, `n` is to be chosen an odd integer.

`g_n.m`: Function used in the routine `acondG.m`.

`g=g_n(n,t,xw)` is a function needed to compute the absolute condition number for the map $G_{\mathtt{n}}$ from the first 2`n` normalized modified moments of a measure $d\lambda$ to the `n`-point Gauss quadrature formula for the same measure. The nodes and weights of the Gauss formula are input by the `n`×2 array `xw`. The input variable `t` must be a single variable; it cannot be an array.

`gcondG.m`: Global condition number of the map G_n.

`gc=gcondG(n,ab,mom)` computes the global condition number `gc` of the map from the first 2`n` ordinary moments of a measure $d\lambda$ to the Gauss quadrature formula for the same measure [24, §2.1.4]. The measure is specified by the `n`×2 input array `ab` of its first `n` recurrence coefficients, and the moments by the (2`n`)×1 input array `mom`.

`heaviside.m`: The Heaviside function.

`mm_ellcheb.m`: Modified moments for the elliptic Chebyshev weight functions.

`mm=mm_ellcheb(N,om2,eps0)` computes to a relative accuracy `eps0` the first 2`N` Chebyshev moments of the weight function $w(t)=((1-\omega^2t^2)\times(1-t^2))^{-1/2}$ on $[-1,1]$, $0<\omega<1$, and places them into the 1×(2`N`) array `mm`. The input parameter `om2` has the meaning of ω^2 [15].

`mm_log.m`: Modified moments for an algebraic/logarithmic weight function.

`mm=mm_log(N,a)` generates the 1×2`N` array `mm` of the first 2`N` modified moments relative to monic shifted Legendre polynomials of the weight function $w(t)=t^{\mathrm{a}}\log(1/t)$ on $[0,1]$, $\mathtt{a}>-1$ [14].

`mm_logsq.m`: Modified moments for an algebraic/square-logarithmic weight function.

`mm=mm_logsq(N,a)` generates the 1×2`N` array `mm` of the first 2`N` modified moments relative to monic shifted Legendre polynomials of the weight function $w(t)=t^{\mathrm{a}}[\log(1/t)]^2$ on $[0,1]$, $\mathtt{a}>-1$ [18].

`moll.m`: Mollifier function.

`m=moll(t)` replaces the values of `t` that are larger than, or equal to 1 in absolute value by |`t`|, and those less than 1 in absolute value by 1.

`rcondG.m`: Relative condition number of the map G_n.

`rc=rcondG(n,iopt,ab,ab0,mom)` computes the relative condition number `rc` of the map from the first `2n` moments of a measure $d\lambda$ to the `n`-point Gauss quadrature formula for the same measure [24, §2.1]. The measure is specified by the `n`×2 input array `ab` of its first `n` recurrence coefficients. If `iopt`=1, the moments, input by the 1×(2`n`) array `mom`, are ordinary moments, otherwise modified moments relative to orthogonal polynomials specified by the (2`n`)×2 array `ab0` of their recurrence coefficients. In the case of ordinary moments, use the array `zeros(2*n,2)` for `ab0`.

`splinefunc.m`: Spline function.

`y=splinefunc(N,M,x,ta)` evaluates the spline function

$$\mathtt{y} = s_{\mathtt{N},\mathtt{M}}(\mathtt{x}) = \sum_{\nu=1}^{\mathtt{N}} a_\nu (t_\nu - \mathtt{x})_+^{\mathtt{M}}$$

of degree `M` > `0`. The positive knots t_ν, ordered increasingly, and the corresponding coefficients a_ν are specified by the `N`×2 array `ta`. The argument `x` is permitted to be vector-valued with all components positive.

`stepfunc.m`: Step function.

`y=stepfunc(N,x,ta)` evaluates the step function

$$\mathtt{y} = s_{\mathtt{N}}(\mathtt{x}) = \sum_{\nu=1}^{\mathtt{N}} a_\nu H(t_\nu - \mathtt{x}),$$

where H is the Heaviside function. The positive knots t_ν, ordered increasingly, and the corresponding coefficients a_ν are specified by the `N`×2 array `ta`. The argument `x` is permitted to be vector-valued with all components positive.

Appendix B

The Software Package SOPQ (Symbolic Orthogonal Polynomials and Quadrature)

B.1 ▪ Orthogonal polynomials

`dig_hrh.m`: Number of digits needed for half-range Hermite polynomials.

> `[ab,dig]=dig_hrh(N,dig0,dd,nofdig)` determines the number `dig` of digits needed to obtain the first `N` recurrence coefficients for the weight function $w(t)=e^{-t^2}$ on $[0,\infty]$ to an accuracy of `nofdig` decimal digits using the classical Chebyshev algorithm. It successively increments an initial estimate `dig0` by `dd` units until the required accuracy is achieved. The recurrence coefficients to `nofdig` digits are placed into the $N\times 2$ array `ab`.

`schebyshev.m`: Symbolic version of the `OPQ` function `chebyshev.m`.

`slanczos.m`: Symbolic version of the `OPQ` function `lanczos.m`.

`smcdis.m`: Symbolic version of the `OPQ` function `mcdis.m`.

`squad_logist.m`: Quadrature routine used by `sr_logist.m` in `smcdis.m`.

`sr_Airy.m`: Recurrence coefficients for an Airy weight function.

> `ab=sr_Airy(dig,nofdig,N)` generates in `dig`-digit arithmetic the first `N` recurrence coefficients for the weight function $w(t) = c\,t^{-5/6}e^{-t}$ $\mathrm{Ai}((3t/2)^{2/3})$ on $[0,\infty]$, where $c = 2^{-1/6}3^{1/6}/\sqrt{\pi}$, and places them to `nofdig` digits into the $N\times 2$ array `ab`.

`sr_BE.m`: Recurrence coefficients for Bose–Einstein weight functions.

> `ab=sr_BE(i)` generates the first 80 recurrence coefficients in 99-digit arithmetic for polynomials orthogonal with respect to the weight function $w(t) = t/(e^t-1)$ on $[0,\infty]$ if i=1, in 135-digit arithmetic for polynomials orthogonal with respect to the weight function $w(t) = [t/(e^t-1)]^2$ on $[0,\infty]$ if i=2, and in 98-digit arithmetic for polynomials orthogonal with respect to the weight function $w(t)=t^{1/2}/(e^t-1)$ on $[0,\infty]$ if i=3, and places them, rounded to 32 digits, into the 80×2 array `ab`.

`sr_Enu.m`: Recurrence coefficients for the exponential integral weight function.

`ab=sr_Enu(dig,N,nu)` generates in `dig`-digit arithmetic the $\mathtt{N}\times 2$ array `ab` of the first `N` recurrence coefficients for the weight function $w(t)=\int_1^\infty e^{-tu}\mathrm{d}u/u^{\mathrm{nu}}$ on $[0,\infty]$, $\mathrm{nu}>0$.

`sr_Enufin.m`: Recurrence coefficients for the finite exponential integral weight function.

`ab=sr_Enufin(dig,N,nu,c)` generates in `dig`-digit arithmetic the $\mathtt{N}\times 2$ array `ab` of the first `N` recurrence coefficients for the weight function $w(t)=\int_1^\infty e^{-tu}\mathrm{d}u/u^{\mathrm{nu}}$ on $[0,\mathtt{c}]$, $\mathrm{nu}>0$, $\mathtt{c}>0$.

`sr_Knu.m`: Recurrence coefficients for the modified Bessel weight function.

`ab=sr_Knu(dig,nofdig,N,nu)` generates in `dig`-digit arithmetic the first `N` recurrence coefficients for the weight function $w(t)=c_\nu K_\nu(t)$ on $[0,\infty]$, where $c_\nu=2\cos(\nu\pi/2)$, and places them to `nofdig` digits into the $\mathtt{N}\times 2$ array `ab`.

`sr_boseeinstein.m`: Recurrence coefficients for Bose–Einstein-type weight functions generated by the Chebyshev algorithm.

`[ab,dig]=sr_boseeinstein(N,r,nofdig)` generates to an accuracy of `nofdig` digits the first `N` recurrence coefficients for polynomials orthogonal with respect to the weight function $w(t)=[t/(e^t-1)]^{\mathtt{r}}$ on $[0,\infty]$, $\mathtt{r}>0$ an integer, and places them into the $\mathtt{N}\times 2$ array `ab`. The output variable `dig` is the number of digits required in Chebyshev's algorithm to achieve the desired accuracy.

`sr_exp0.m`: Variable-precision recurrence coefficients for the weight function $w(t)=\exp(1/t)$ on $[0,1]$.

`sr_exp0inf.m`: Variable-precision recurrence coefficients for the weight function $w(t)=\exp(-1/t-t)$ on $[0,\infty]$.

`sr_exp0infab.m`: Variable-precision recurrence coefficients for the weight function $w(t)=\exp(-a/t-bt)$ on $[0,\infty]$, $a>0$, $b>0$.

`sr_fermi.m`: Recurrence coefficients for Fermi-type weight functions.

`ab=sr_fermi(i)` generates to 32 decimal digits the first eighty recurrence coefficients for polynomials orthogonal with respect to the weight function $w(t)=1/(e^t+1)$ on $[0,\infty]$ if `i`=1, for polynomials orthogonal with respect to the weight function $w(t)=[1/(e^t+1)]^2$ on $[0,\infty]$ if `i`=2, and for polynomials orthogonal with respect to the weight function $w(t)=t^{-1/2}/(e^t+1)$ on $[0,\infty]$ if `i`=3, and places them into the 80×2 array `ab`.

`sr_fermidirac.m`: Recurrence coefficients for Fermi–Dirac-type weight functions generated by the Chebyshev algorithm.

`[ab,dig]=sr_fermidirac(N,r,nofdig)` generates to an accuracy of `nofdig` digits the first `N` recurrence coefficients for polynomials orthogonal with respect to the weight function $w(t)=[1/(e^t+1)]^{\mathtt{r}}$ on $[0,\infty]$, $\mathtt{r}>0$ an integer, and places them into the $\mathtt{N}\times 2$ array `ab`. The output variable `dig` is the number of digits required in Chebyshev's algorithm to achieve the desired accuracy.

`sr_freud.m`: Recurrence coefficients for Freud weight functions.

> `ab=sr_freud(dig,nofdig,N,mu,nu)` generates in `dig`-digit arithmetic the first `N` recurrence coefficients for the weight function $w(t) = |t|^{\mathrm{mu}} \exp(-|t|^{\mathrm{nu}})$ on $[-\infty, \infty]$, $\mathtt{mu} > -1$, $\mathtt{nu} > 0$, and places them to `nofdig` digits into the `N`×2 array `ab`.

`sr_hermite.m`: Recurrence coefficients for the Hermite weight function.

> `ab=sr_hermite(dig,N,mu)` generates in `dig`-digit arithmetic the `N`×2 array `ab` of the first `N` recurrence coefficients for generalized Hermite polynomials orthogonal on $[-\infty, \infty]$ with respect to the weight function $w(t) = |t|^{2\mathrm{mu}} \exp(-t^2)$, $\mathtt{mu} > -1/2$.

`sr_hrf.m`: Recurrence coefficients for the half-range Freud weight function.

> `ab=sr_hrf(dig,nofdig,N,mu,nu)` generates in `dig`-digit arithmetic the first `N` recurrence coefficients for the weight function $w(t) = t^{\mathrm{mu}} \exp(-t^{\mathrm{nu}})$ on $[0, \infty]$, $\mathtt{mu} > -1$, $\mathtt{nu} > 0$, and places them to `nofdig` digits into the `N`×2 array `ab`.

`sr_hrh.m`: Recurrence coefficients for the half-range Hermite weight function.

> `ab=sr_hrh(dig,nofdig,N)` generates in `dig`-digit arithmetic the first `N` recurrence coefficients for the weight function $w(t) = \exp(-t^2)$ on $[0, \infty]$ and places them to `nofdig` digits into the `N`×2 array `ab`.

`sr_jaclog.m`: Recurrence coefficients for an algebraic/logarithmic weight function.

> `ab=sr_jaclog(dig,N,a)` generates in `dig`-digit arithmetic the `N`×2 array `ab` of the first `N` recurrence coefficients for polynomials orthogonal with respect to the weight function $w(t) = t^{\mathrm{a}} \log(1/t)$ on $[0, 1]$, $\mathtt{a} > -1$. The call `ab=sr_jaclog(dig,N)` is the same as `ab=sr_jaclog(dig,N,0)`.

`sr_jaclogsq.m`: Recurrence coefficients for an algebraic/square-logarithmic weight function.

> `ab=sr_jaclogsq(dig,N,a)` generates in `dig`-digit arithmetic the first `N` recurrence coefficients for polynomials orthogonal with respect to the weight function $w(t) = t^{\mathrm{a}} \log^2(1/t)$ on $[0, 1]$, $\mathtt{a} > -1$, and places them to 25 digits into the `N`×2 array `ab`. The call `ab=sr_jaclogsq(dig,N)` is the same as `ab=sr_jaclogsq(dig,N,0)`.

`sr_jacobi.m`: Recurrence coefficients for Jacobi polynomials.

> `ab=sr_jacobi(dig,N,a,b)` generates in `dig`-digit arithmetic the `N`×2 array `ab` of the first `N` recurrence coefficients for Jacobi polynomials orthogonal on $[-1, 1]$ with respect to the weight function $w(t) = (1-t)^{\mathrm{a}}(1+t)^{\mathrm{b}}$, $\mathtt{a} > -1$, $\mathtt{b} > -1$.

`sr_jacobi01.m`: Recurrence coefficients for shifted Jacobi polynomials.

> `ab=sr_jacobi01(dig,N,a,b)` generates in `dig`-digit arithmetic the `N`×2 array `ab` of the first `N` recurrence coefficients for Jacobi polynomials orthogonal on $[0, 1]$ with respect to the weight function $w(t) = (1-t)^{\mathrm{a}} t^{\mathrm{b}}$, $\mathtt{a} > -1$, $\mathtt{b} > -1$.

`sr_l1sf.m`: Recurrence coefficients for lower 1-sided Freud weight functions.

> `ab=sr_l1sf(dig,nofdig,N,c,mu,nu)` generates in `dig`-digit arithmetic the first `N` recurrence coefficients for the weight function $w(t) = t^{\mathrm{mu}}$ $\exp(-t^{\mathrm{nu}})$ on $[0, c]$, mu> -1, nu> 0, $c > 0$, and places them to `nofdig` digits into the N×2 array `ab`.

`sr_laguerre.m`: Recurrence coefficients for Laguerre polynomials.

> `ab=sr_laguerre(dig,N,a)` generates in `dig`-digit arithmetic the N×2 array `ab` of the first `N` recurrence coefficients for generalized Laguerre polynomials orthogonal on $[0, \infty]$ with respect to the weight function $w(t) = t^{\mathrm{a}} \exp(-t)$, a$> -1$.

`sr_logist.m`: Recurrence coefficients for the logistic weight function.

> `sr_logist(dig,N)` generates in `dig`-digit arithmetic the first `N` recurrent coefficients for the weight function $w(t) = e^{-t}/((1 + e^{-t})^2)$ on $[-\infty, \infty]$ to an accuracy of 12 decimal digits.

`sr_lsf.m`: Recurrence coefficients for lower symmetric Freud weight functions.

> `ab=sr_lsf(dig,nofdig,N,c,mu,nu)` generates in `dig`-digit arithmetic the first `N` recurrence coefficients for the weight function $w(t) = |t|^{\mathrm{mu}}$ $\exp(-|t|^{\mathrm{nu}})$ on $[-c, c]$, mu> -1, nu> 0, $c > 0$.

`sr_sin0.m`: Recurrence coefficients for a special trigonometric weight function.

> `ab=sr_sin0(dig,nofdig,N)` generates in `dig`-digit arithmetic the first `N` recurrence coefficients for the weight function $w(t) = 1 + \sin(1/t)$ on $[0, 1]$ and places them to `nofdig` digits into the N×2 array `ab`.

`sr_u1sf.m`: Recurrence coefficients for upper 1-sided Freud weight functions.

> `ab=sr_u1sf(dig,nofdig,N,c,mu,nu)` generates in `dig`-digit arithmetic the first `N` recurrence coefficients for the weight function $w(t) = t^{\mathrm{mu}}$ $\exp(-t^{\mathrm{nu}})$ on $[0, \infty]$, mu> -1, nu> 0, $c > 0$, and places them in `nofdig` digits into the N×2 array `ab`.

`sr_usf.m`: Recurrence coefficients for upper symmetric Freud weight functions.

> `ab=sr_usf(dig,nofdig,N,c,mu,nu)` generates in `dig`-digit arithmetic the first `N` recurrence coefficients for the weight function $w(t) = |t|^{\mathrm{mu}}$ $\exp(-|t|^{\mathrm{nu}})$ on $[-\infty, -c] \cup [c, \infty]$, mu$> -1$, nu$> 0$, $c > 0$, and places them to `nofdig` digits into the N×2 array `ab`.

`sstieltjes.m`: Symbolic version of the `OPQ` function `stieltjes.m`.

B.2 ▪ Quadrature

`sgauss.m`: Symbolic version of the `OPQ` function `gauss.m`.

`slanczos_algorithm.m`: Symbolic version of the `OPQ` function `lanczos_algorithm.m`.

B.3 ▪ Utility

`loadvpa.m`: Loading a multiprecision source file.

`a=loadvpm(filename,nheader_lines,dig,n,m)` loads into the MATLAB working window a source text file named `filename`, which contains `nheader_lines` comment lines followed by an `n×m` array of `dig`-digit numbers, ignoring the comment lines. The input argument `filename` has to be placed between single quotes when the function is called.

`momboseeinstein.m`: Moments of Bose–Einstein-type weight functions.

`mom=momboseeinstein(dig,N,r)` generates in `dig`-digit arithmetic the 1×2`N` array `mom` of the first 2`N` moments of the weight function $w(t) = [t/(e^t-1)]^{\mathtt{r}}$ on $[0,\infty]$, `r` > 0 an integer.

`momfermidirac.m`: Moments of Fermi–Dirac-type weight functions.

`mom=momfermidirac(dig,N,r)` generates in `dig`-digit arithmetic the 1×2`N` array `mom` of the first 2`N` moments of the weight function $w(t) = [1/(e^t+1)]^{\mathtt{r}}$ on $[0,\infty]$, `r` > 0 an integer.

`smom_Airy.m`: Moments of an Airy weight function.

`mom=smom_Airy(dig,N)` generates in `dig`-digit arithmetic the 2`N`$\times 1$ array `mom` of the first 2`N` moments of the weight function $w(t) = c\,t^{-5/6}e^{-t}\mathrm{Ai}((3t/2)^{2/3})$ on $[0,\infty]$, $c = 2^{-1/6}3^{1/6}/\sqrt{\pi}$.

`smm_log.m`: Symbolic version of the `OPQ` function `mm_log.m`.

`smm_logsq.m`: Symbolic version of the `OPQ` function `mm_logsq.m`.

`smom_Enu.m`: Moments of exponential integral weight functions.

`mom=smom_Enu(dig,N,nu)` generates in `dig`-digit arithmetic the 2`N`$\times 1$ array `mom` of the first 2`N` moments of the weight function $w(t) = \int_1^\infty e^{-tu}du/u^{\mathrm{nu}}$ on $[0,\infty]$, nu > 0.

`smom_Enufin.m`: Moments of finite exponential integral weight functions.

`mom=smom_Enufin(dig,N,nu,c)` generates in `dig`-digit arithmetic the 2`N`$\times 1$ array `mom` of the first 2`N` moments of the weight function $w(t) = \int_1^\infty e^{-tu}du/u^{\mathrm{nu}}$ on $[0,\mathrm{c}]$, nu > 0, c > 0.

`smom_freud.m`: Moments of Freud weight functions.

`mom=smom_freud(dig,N,mu,nu)` generates in `dig`-digit arithmetic the 2`N`$\times 1$ array `mom` of the first 2`N` moments of the weight function $w(t) = |t|^{\mathrm{mu}}\exp(-|t|^{\mathrm{nu}})$ on $[-\infty,\infty]$, mu > -1, nu > 0.

`smom_hrf.m`: Moments of half-range Freud weight functions.

`mom=smom_hrf(dig,N,mu,nu)` generates in `dig`-digit arithmetic the 2`N`$\times 1$ array `mom` of the first 2`N` moments of the weight function $w(t) = t^{\mathrm{mu}}\exp(-t^{\mathrm{nu}})$ on $[0,\infty]$, mu > -1, nu > 0.

`smom_hrh.m`: Moments of the half-range Hermite weight function.

`mom=smom_hrh(dig,N)` generates in `dig`-digit arithmetic the 2`N`$\times 1$ array `mom` of the first 2`N` moments of the weight function $w(t) = e^{-t^2}$ on $[0,\infty]$.

`smom_Knu.m`: Moments of the modified Bessel function.

`mom=smom_Knu(dig,N,nu)` generates in `dig`-digit arithmetic the 2N×1 array `mom` of the first 2N moments of the weight function $w(t) = K_\nu(t)$ on $[0, \infty]$, $0 < \nu < 1$.

`smom_l1sf.m`: Moments of lower 1-sided Freud weight functions.

`mom=smom_l1sf(dig,N,c,mu,nu)` generates in `dig`-digit arithmetic the 2N×1 array `mom` of the first 2N moments of the weight function $w(t) = t^{\mathrm{mu}} \exp(-t^{\mathrm{nu}})$ on $[0, \mathrm{c}]$, $\mathrm{mu} > -1$, $\mathrm{nu} > 0$, $\mathrm{c} > 0$.

`smom_lsf.m`: Moments of lower symmetric Freud weight functions.

`mom=smom_lsf(dig,N,c,mu,nu)` generates in `dig`-digit arithmetic the 2N×1 array `mom` of the first 2N moments of the weight function $w(t) = |t|^{\mathrm{mu}} \exp(-|t|^{\mathrm{nu}})$ on $[-c, c]$, $\mathrm{mu} > -1$, $\mathrm{nu} > 0$, $c > 0$.

`smom_sin0.m`: Moments of a special trigonometric weight function.

`mom=smom_sin0(dig,N)` generates in `dig`-digit arithmetic the 2N×1 array `mom` of the first 2N moments of the weight function $w(t) = 1 + \sin(1/t)$ on $[0, 1]$.

`smom_u1sf.m`: Moments of upper 1-sided Freud weight functions.

`mom=smom_u1sf(dig,N,c,mu,nu)` generates in `dig`-digit arithmetic the 2N×1 array `mom` of the first 2N moments of the weight function $w(t) = t^{\mathrm{mu}} \exp(-t^{\mathrm{nu}})$ on $[0, \infty]$, $\mathrm{mu} > -1$, $\mathrm{nu} > 0$, $c > 0$.

`smom_usf.m`: Moments of upper symmetric Freud weight functions.

`mom=smom_usf(dig,N,c,mu,nu)` generates in `dig`-digit arithmetic the 2N×1 array `mom` of the first 2N moments of the weight function $w(t) = |t|^{\mathrm{mu}} \exp(-|t|^{\mathrm{nu}})$ on $[-\infty, -c] \cup [c, \infty]$, $\mathrm{mu} > -1$, $\mathrm{nu} > 0$, $c > 0$.

Bibliography

[1] ABRAMOWITZ, MILTON AND IRENE A. STEGUN (eds.) *Handbook of Mathematical Functions*, Dover Publications, New York, 1992. (Cited on pp. 94, 111, 150, 236, 238, 274, 278, 280, 281, 289)

[2] ALFARO, M., F. MARCELLÁN, M. L. REZOLA, AND A. RONVEAUX. *On orthogonal polynomials of Sobolev type: Algebraic properties and zeros*, SIAM J. Math. Anal. 23 (1992), 737–757. (Cited on pp. 174, 177)

[3] CHANDRASEKHAR, S. *Radiative Transfer*, Oxford University Press, Oxford, 1950. (Cited on p. 88)

[4] CHIHARA, T. S. *An Introduction to Orthogonal Polynomials*, Mathematics and Its Applications 13, Gordon and Breach, New York, 1978. (Cited on pp. vii, 16)

[5] DANLOY, BERNARD. *Numerical construction of Gaussian quadrature formulas for* $\int_0^1(-\mathrm{Log}\, x)\cdot x^\alpha\cdot f(x)\cdot dx$ *and* $\int_0^\infty E_m(x)f(x)dx$, Math. Comp. 27 (1973), 861–869. (Cited on pp. 88, 91)

[6] DAVIS, PHILIP J. *Interpolation and Approximation*, republication, with minor corrections, of the 1963 original, with a new preface and bibliography, Dover Publications, New York, 1975. (Cited on p. 214)

[7] DAVIS, PHILIP J. *Spirals: from Theodorus to Chaos*, A K Peters, Wellesley, MA, 1993. (Cited on p. 288)

[8] DAVIS, P. J. AND P. RABINOWITZ. *Some geometrical theorems for abscissas and weights of Gauss type*, J. Math. Anal. Appl. 2 (1961), 428–437. (Cited on p. 185)

[9] ERDÉLYI, A. (ed.) *Tables of Integral Transforms*, v. II, McGraw-Hill, New York, 1954. (Cited on pp. 97, 244)

[10] ERDŐS, P. *On the distribution of the roots of orthogonal polynomials*, in Proceedings of the Conference on the Constructive Theory of Functions (Approximation Theory) (Budapest, 1969) Akadémiai Kiadó, Budapest, 1972, 145–150. (Cited on p. 124)

[11] FREUD, GÉZA. *Orthogonale Polynome*, Lehrbücher und Monographien aus dem Gebiete der Exakten Wissenschaften, Mathematische Reihe 33, Birkhäuser, Basel, 1969. (Cited on p. vii)

[12] FREUD, GÉZA. *On the coefficients in the recursion formulae of orthogonal polynomials*, Proc. R. Irish Acad. Sect. A 76, 1976, no. 1, 1–6. (Cited on p. 81)

[13] GAUTSCHI, WALTER. *Efficient computation of the complex error function*, SIAM J. Numer. Anal. 7 (1970), 187–198. [Also in *Selected works*, v. 1, 267–278.] (Cited on p. 239)

[14] GAUTSCHI, WALTER. *On the preceding paper "A Legendre polynomial integral" by James L. Blue*, Math. Comp. 33 (1979), 742–743. [Also in *Selected works*, v. 2, 55–56.] (Cited on pp. 97, 99, 314)

[15] GAUTSCHI, WALTER. *On generating orthogonal polynomials*, SIAM J. Sci. Stat. Comput. 3 (1982), 289–317. [Also in *Selected works*, v. 2, pp. 71–99.] (Cited on pp. 5, 314)

[16] GAUTSCHI, WALTER. *How and how not to check Gaussian quadrature formulae*, BIT 23 (1983), 209–216. [Also in *Selected works*, v. 2, 674–681.] (Cited on pp. 87, 88)

[17] GAUTSCHI, WALTER. *On some orthogonal polynomials of interest in theoretical chemistry*, BIT 24 (1984), 473–483. [Also in *Selected works*, v. 2, 101–111.] (Cited on pp. 49, 54, 56)

[18] GAUTSCHI, WALTER. *On certain slowly convergent series occurring in plate contact problems*, Math. Comp. 57 (1991), 325–338. [Also in *Selected works*, v. 3, 652–665.] (Cited on pp. 97, 99, 102, 103)

[19] GAUTSCHI, WALTER. *Algorithm 726—ORTHPOL: A package of routines for generating orthogonal polynomials and Gauss-type quadrature rules*, ACM Trans. Math. Software 20 (1994), 21–62. (Cited on pp. 12, 13)

[20] GAUTSCHI, WALTER. *On the computation of special Sobolev-type orthogonal polynomials*, in The Heritage of P. L. Chebyshev: A Festschrift in Honor of the 70th Birthday of T. J. Rivlin (C. A. Micchelli, ed.), Ann. Numer. Math. 4 (1997), 329–342. [Also in *Selected works*, v. 2, 216–229.] (Cited on pp. 166, 167)

[21] GAUTSCHI, WALTER. *High-order Gauss–Lobatto formulae*, Numer. Algorithms 25 (2000), 213–222. [Also in *Selected works*, v. 2, pp. 834–843.] (Cited on p. 306)

[22] GAUTSCHI, WALTER. *Gauss–Radau formulae for Jacobi and Laguerre weight functions*, Math. Comput. Simulation 54 (2000), 403–412. [Also in *Selected works*, v. 2, pp. 845–854.] (Cited on p. 307)

[23] GAUTSCHI, WALTER. *Computation of Bessel and Airy functions and of related Gaussian quadrature formulae*, BIT 42 (2002), 110–118. [Also in *Selected works*, v. 1, 481–489.] (Cited on pp. 117, 118)

[24] GAUTSCHI, WALTER. *Orthogonal Polynomials: Computation and Approximation*, Numerical Mathematics and Scientific Computation, Oxford University Press, Oxford, 2004. (Cited on pp. vii, 4, 7, 19, 23, 31, 33, 37, 44, 61, 82, 86, 97, 127, 138, 149, 153, 174, 181, 182, 186, 190, 195, 196, 199, 203, 208, 214, 260, 261, 262, 264, 304, 305, 306, 307, 308, 310, 313, 314, 315)

[25] GAUTSCHI, WALTER. *Generalized Gauss–Radau and Gauss–Lobatto formulae*, BIT 44 (2004), 711–720. [Also in *Selected works*, v. 2, 873–882.] (Cited on p. 203)

[26] GAUTSCHI, WALTER. *Numerical quadrature computation of the Macdonald function for complex orders*, BIT 45 (2005), 593–603. [Also in *Selected works*, v. 1, 491–501.] (Cited on p. 123)

[27] GAUTSCHI, WALTER. *Orthogonal polynomials, quadrature, and approximation: Computational methods and software (in Matlab)*, in Orthogonal Polynomials and Special Functions: Computation and Applications (F. Marcellán and W. Van Assche, eds.), Lecture Notes in Mathematics 1883, Springer, Berlin, 2006, 1–77. [Also in *Selected works*, v. 3, 232–308.] (Cited on pp. vii, viii, 1, 36, 37, 38, 39, 40, 42, 43, 44, 45, 48, 58, 60, 88, 135, 136, 138, 139, 141, 158, 161, 185, 188, 190, 192, 194, 195, 197, 199, 200, 201, 203, 208, 213, 214, 215, 216, 218, 223, 259, 260, 263, 264, 265, 272, 274, 276, 278, 279, 280, 282)

[28] GAUTSCHI, WALTER. *The circle theorem and related theorems for Gauss-type quadrature rules*, Electron. Trans. Numer. Anal. 25 (2006), 129–137. [Also in *Selected works*, v. 2, 884–892.] (Cited on p. 185)

[29] GAUTSCHI, WALTER. *The spiral of Theodorus, numerical analysis, and special functions*, J. Comput. Appl. Math. 235 (2010), 1042–1052. [Also in *Selected works*, v. 3, 689–699.] (Cited on pp. 278, 288)

[30] GAUTSCHI, WALTER. *Numerical Analysis*, 2d ed., Birkhäuser, New York, 2012. (Cited on p. 290)

[31] GAUTSCHI, WALTER. *Repeated modifications of orthogonal polynomials by linear divisors*, Numer. Algorithms 63 (2013), 369–383. [Also in *Selected works*, v. 2, 287–301.] (Cited on pp. 144, 145, 148, 149)

[32] GAUTSCHI, WALTER. *Polynomials orthogonal with respect to exponential integrals*, Numer. Algorithms 70 (2015), 215–226. (Not cited)

[33] GAUTSCHI, WALTER. *Algorithm: Evaluation of the repeated integrals of the coerror function by half-range Gauss–Hermite quadrature*, ACM Trans. Math. Software, to appear. (Cited on p. 236)

[34] GAUTSCHI, WALTER. *Monotonicity properties of the zeros of Freud and sub-range Freud polynomials: Analytic and empirical results*, submitted for publication. (Cited on pp. 68, 71, 77)

[35] GAUTSCHI, WALTER AND GRADIMIR V. MILOVANOVIĆ. *Gaussian quadrature involving Einstein and Fermi functions with an application to summation of series*, Math. Comp. 44 (1985), 177–190. [Also in *Selected works*, v. 2, 701–725.] (Cited on pp. 107, 110, 280, 282)

[36] GAUTSCHI, WALTER AND MINDA ZHANG. *Computing orthogonal polynomials in Sobolev spaces*, Numer. Math. 71 (1995), 159–183. [Also in *Selected works*, v. 2, 113–137.] (Cited on pp. 157, 158, 162, 295, 303)

[37] GERONIMUS, L. YA. *Orthogonal Polynomials: Estimates, Asymptotic Formulas, and Series of Polynomials Orthogonal on the Unit Circle and on an Interval*, authorized translation from the Russian, Consultants Bureau, New York, 1961. (Cited on p. vii)

[38] GIL, AMPARO, JAVIER SEGURA, AND NICO M. TEMME. *On nonoscillating integrals for computing inhomogeneous Airy functions*, Math. Comp. 70 (2001), 1183–1194. (Not cited)

[39] GORDON, ROY G. *Constructing wavefunctions for nonlocal potentials*, J. Chem. Phys. 52 (1970), 6211–6217. (Cited on p. 117)

[40] GRADSHTEYN, I. S. AND I. M. RYZHIK. *Tables of Integrals, Series, and Products*, 6th ed., Academic Press, San Diego, 2000. (Cited on pp. 49, 117, 119, 147, 149, 169)

[41] GRAGG, W. B. AND W. J. HARROD. *The numerically stable reconstruction of Jacobi matrices from spectral data*, Numer. Math. 44 (1984), 317–335. (Cited on p. 297)

[42] GROENEVELT, W. G. M. *Zeros of Sobolev orthogonal polynomials of Gegenbauer type*, J. Approx. Theory 114 (2002), 115–140. (Cited on p. 168)

[43] ISMAIL, MOURAD E. H. *Classical and Quantum Orthogonal Polynomials in One Variable*, with two chapters by Walter Van Assche. With a foreword by Richard A. Askey. Reprint of the 2005 original. Encyclopedia of Mathematics and Its Applications 98, Cambridge University Press, Cambridge, 2009. (Cited on p. vii)

[44] JOULAK, HÉDI AND BERNHARD BECKERMANN. *On Gautschi's conjecture for generalized Gauss–Radau and Gauss–Lobatto formulae*, J. Comput. Appl. Math. 233 (2009), 768–774. (Cited on p. 208)

[45] KEGEL, WILHELM H. *Zur numerischen Berechnung der Integrale* $\int_0^\tau f(x)K_n(x)dx$, Z. Astrophys. 54 (1962), 10–28. (Cited on pp. 88, 97)

[46] KHRUSHCHEV, SERGEY. *Orthogonal Polynomials and Continued Fractions, from Euler's Point of View*. Encyclopedia of Mathematics and Its Applications 122, Cambridge University Press, Cambridge, 2008. (Cited on p. vii)

[47] LEE, SOO-Y. *The inhomogeneous Airy functions* Gi(z) *and* Hi(z), J. Chem. Phys. 72 (1980), 332–336. (Cited on p. 87)

[48] LUBINSKY, D. S., H. N. MHASKAR, AND E. B. SAFF. *A proof of Freud's conjecture for exponential weights*, Construct. Approx. 4 (1988), 65–83. (Cited on p. 81)

[49] MARCELLÁN, FRANCISCO, TERESA E. PÉREZ, AND MIGUEL A. PIÑAR. *Gegenbauer-Sobolev Orthogonal Polynomials*, Nonlinear Numerical Methods and Rational Approximation II Mathathematics and Its Applications 296, Kluwer Academic Publishers, Dordrecht, 1994, 71–82. (Cited on p. 162)

[50] MÁTÉ, ATTILA, PAUL NEVAI, AND THOMAS ZASLAVSKY, *Asymptotic expansions of ratios of coefficients of orthogonal polynomials with exponential weights*, Trans. Amer. Math. Soc. 287 (1985), 495–505. (Cited on p. 81)

[51] MEIJER. H. G. *Sobolev orthogonal polynomials with a small number of real zeros*, J. Approx. Theory 77 (1994), 305–313. (Cited on p. 30)

[52] NIKIFOROV, A. F., S. K. SUSLOV, AND V. B. UVAROV. *Classical orthogonal polynomials of a discrete variable*, translated from the Russian, Springer Series in Computational Physics, Springer, Berlin, 1991. (Cited on p. vii)

[53] OLVER, FRANK W. J., DANIEL W. LOZIER, RONALD F. BOISVERT, AND CHARLES W. CLARK. *NIST Handbook of Mathematical Functions*, National Institute of Standards and Technology and Cambridge University Press, 2010. (Cited on pp. 82, 104, 107, 108, 119, 242, 244, 247, 253, 255)

[54] PEHERSTORFER, FRANZ AND KNUT PETRAS. *Stieltjes polynomials and Gauss-Kronrod quadrature for Jacobi weight functions*, Numer. Math. 95 (2003), 689–706. (Cited on p. 186)

[55] REIZ, ANDERS. *Quadrature formulae for the numerical calculation of mean intensities and fluxes in a stellar atmosphere*, Ark. Astronomi 1 (1950), 147–153. (Cited on pp. 88, 91)

[56] SCHULTEN, Z., D. G. M. ANDERSON, AND ROY G. GORDON. *An algorithm for the evaluation of the complex Airy function*, J. Comput. Phys. 31 (1979), 60–75. (Cited on p. 119)

[57] SCORER, R.S. *Numerical evaluation of integrals of the form* $I = \int_{x_1}^{x_2} f(x)e^{i\phi(x)}dx$ *and the tabulation of the function* $\mathrm{Gi}(z) = (1/\pi)\int_0^\infty \sin(uz + \frac{1}{3}u^3)du$, Quart. J. Mech. Appl. Math. 3 (1950), 107–112. (Cited on pp. 247, 251, 252)

[58] SHIZGAL, BERNARD. *Spectral Methods in Chemistry and Physics: Applications to Kinetic Theory and Quantum Mechanics*, Springer Science+Business Media, Dordrecht, 2015. (Cited on pp. 5, 129, 131)

[59] STIELTJES, T. J. *Quelques recherches sur la théorie des quadratures dites mécaniques*, Ann. Sci. École Norm. Paris, Sér. 3, 1 (1884), 409–426. [Also in *Œuvres I*, 377–396.] (Cited on p. 5)

[60] SUETIN, P. K. *Classical Orthogonal Polynomials* (Russian), 2nd ed., Nauka, Moscow, 1979. (Cited on p. vii)

[61] SZEGÖ, G.. *Orthogonal Polynomials*, 4th ed., AMS Colloquium Publications 23, American Mathematical Society, Providence, RI, 1975. (Cited on pp. vii, 49, 145, 186)

[62] TROGDON, THOMAS AND SHEEHAN OLVER. *A Riemann–Hilbert approach to Jacobi operators and Gaussian quadrature*, IMA J. Numer. Anal. 36 (2016), 174–196. (Cited on pp. 120, 121, 122)

[63] VIGNERON, J. P. AND PH. LAMBIN. *Gaussian quadrature of integrands involving the error function*, Math. Comp. 35 (1980), 1299–1307. (Cited on p. 115)

Software Index

Subject Index